AF581737

Temperature Reduction Technologies Meet Asphalt Pavement: Green and Sustainability

Temperature Reduction Technologies Meet Asphalt Pavement: Green and Sustainability

Editors

Markus Oeser
Michael Wistuba
Pengfei Liu
Di Wang

MDPI • Basel • Beijing • Wuhan • Barcelona • Belgrade • Manchester • Tokyo • Cluj • Tianjin

Editors

Markus Oeser
Institute of Highway Engineering
RWTH Aachen University
Aachen
Germany

Michael Wistuba
Department of Architecture, Civil Engineering and Environmental Sciences
Technische Universität Braunschweig
Braunschweig
Germany

Pengfei Liu
Institute of Highway Engineering
RWTH Aachen University
Aachen
Germany

Di Wang
Department of Civil Engineering
Aalto University
Espoo
Finland

Editorial Office
MDPI
St. Alban-Anlage 66
4052 Basel, Switzerland

This is a reprint of articles from the Special Issue published online in the open access journal *Materials* (ISSN 1996-1944) (available at: www.mdpi.com/journal/materials/special_issues/Temper_Reduct_Tech_Asphalt_Pave).

For citation purposes, cite each article independently as indicated on the article page online and as indicated below:

LastName, A.A.; LastName, B.B.; LastName, C.C. Article Title. *Journal Name* **Year**, *Volume Number*, Page Range.

ISBN 978-3-0365-6520-0 (Hbk)
ISBN 978-3-0365-6519-4 (PDF)

Contents

About the Editors

Markus Oeser

Univ.-Prof. Dr.-Ing. habil. Markus Oeser studied Civil Engineering at the TU Dresden with a focus on structural engineering. In 2004, he completed his doctorate in Dresden on the topic "Numerical simulation of the non-linear behavior of multi-layered flexible transport pathways". This was followed in 2010 by his habilitation on the topic "Nonlinear numerical simulation models for traffic pathways with regard to mechanical, thermal and hydraulic effects". In 2006, Markus Oeser was a Research Member of the Technical University of Delft University, the Netherlands, and from 2007 to 2011 he was a University Lecturer at the Institute of Geotechnics, Road Construction and Traffic Engineering at the University of New South Wales (UNSW) in Sydney, Australia. In 2011, he became Head of the Chair of Highway Engineering and Director of the Institute of Highway Engineering, and in 2016, Dean at the Faculty of Civil Engineering at the Rheinisch-Westfälische Technische Hochschule Aachen (RWTH). He took up a visiting professorship at the Harbin Institute of Technology (HIT), in Harbin, Heilongjang, China in 2015, which ended as per contract in 2022. In 2021, he was appointed to the professorship of Pavement Engineering and Construction, TU Dresden. He has been President of the Federal Highway Research Institute since November 2021.

Michael Wistuba

Michael P. Wistuba has a Ph.D. in Civil Engineering from Vienna University of Technology, 2002. Since 2008, he has been the head chair of the Braunschweig Pavement Engineering Centre in the Department of Architecture, Civil Engineering and Environmental Sciences at Technische Universität Braunschweig, Germany. He is the author or co-author of more than 55 research reports, 60 peer-reviewed papers in scientific journals and 120 articles published in conference proceedings. His publications address the sustainability of road infrastructures, and cover the fundamental mechanisms that control the behaviour and durability of asphalt materials and pavement systems, the technical testing to address performance properties, the composition of asphalt mixtures considering various additives and re-using reclaimed asphalt, the design of highway and airport pavements, and the development of road management concepts. He has acted in consulting projects in pavement engineering including federal highways in Germany, Austria, and Switzerland. In 2017, he was awarded the Prize of the Construction Industry Association of Lower-Saxony and Bremen for his significant contributions to technology transfer from science to application in the pavement industry sector. He has been a Convenor (2018-2020), and is an Expert (2016-2018 and 2020-2022) of RILEM Cluster F on bituminous materials and polymers. He is steering group member of the European Asphalt Technology Association (EATA), scientific board member of Deutsches Asphaltinstitut (DAI), member of the German Road and Transportation Research Association (FGSV), and scientific board member of the Swiss Expert Association for Road and Transport Engineering (VSS). From the year 2009, he is the Editorial Board Member of the Journal on Road Materials and Pavement Design (RMPD). He has supervised more than 20 PhD students, and 150 MS.

Pengfei Liu

Dr. Pengfei Liu is the Junior Professor (W1) in Modeling and Design of Functionalized Pavement Materials at the Institute of Highway Engineering (ISAC), RWTH Aachen University. He is also the general manager in Sino-European Research Center for Advanced Transportation Infrastructure Technology GmbH. He received his bachelor degree from University of Science and Technology

Be(USTB) in 2008. Afterward, Dr. Liu received his master and doctoral degree from RWTH Aachen University in 2012 and 2017, respectively. His main research areas can be summarized as numerical methods applied in asphalt pavement, multiscale modelling and the characterization of asphalt mixtures. He makes fundamental contributions to the application and realization of national and international scientific research projects in ISAC and has published more than 100 SCI papers as well as 4 books. He also serves as an academic/topic editor for 7 technical journals and as a reviewer for 63 SCI journals. Dr. Liu has been recognized with the 2017 Excellent Self-Funded Student Scholarship of the Ministry of Education P. R. China. In 2017 and 2019, he won the RWTH Aachen –Tsinghua Junior Research Fellowships and RWTH Aachen –Tsinghua Senior Research Fellowships, respectively. Additionally, he received the Outstanding Associate Member Award from the Academy of Pavement Science and Engineering (APSE) in 2022.

Di Wang

Dr. Wang is a Postdoc researcher at the Department of Civil Engineering at Aalto University, Finland. Before he joined Aalto University, he worked as a post-doc researcher at the Technical University of Braunschweig, Germany, and he obtained the doctoral degree in civil engineering at the same university. In 2022, he gained the national scientific qualification to function as Associate Professor at Italian Universities. He has more than eight years of experience in the areas of asphalt pavement, particularly to the characterization and modeling of asphalt binders at low temperatures as well as the incorporation of circular and sustainable materials in asphalt roadways. He has participated in several European, Nordic, German, Finnish, and RILEM projects in the past six years. He also has rich experience in teaching and supervising. He is currently serving in several sector associations, such as the CEN Task Group, RILEM, NRS, COTA and IACIP. As a result, he has published more than 60 contributions in scientific journals, conference proceedings, and technical reports with multiple collaborators in the area of infrastructure and waste management. He is very active in professional activities, including conference organization member, scientific committee, and session chair in more than 10 international conferences world widely in the areas of pavement and waste management. Currently, he is serving as editorial board members and scientific reviewers in more than 50 scientific journals.

Preface to "Temperature Reduction Technologies Meet Asphalt Pavement: Green and Sustainability"

Temperature reduction technologies have been used worldwide over the last two decades in asphalt pavement engineering [1,2]. The aim of these technologies is to use a lower temperature throughout production processes without affecting the performance of asphalt materials. Currently, different temperature reduction technologies, including warm mix asphalt, half-warm mix asphalt and cold mix asphalt, can be used with varieties of organic/chemical additives and foaming/emulsion techniques. The comparable ad even better overall performance properties and durability were achieved in the temperature reduction technologies compared to the conventional hot mix asphalt [3,4,5,6]. Significant economic, social, and environmental benefits can be achieved, including but not restricted to energy, greenhouse gas and fume emissions reduction, and increased field workability [1,3,7]. However, there are still several knowledge gaps to overcome. For example, the incomplete drying of aggregates caused by the lower production temperature may ultimately lead to serious rutting and moisture damage. These limitations hinder the mega-scale application of temperature reduction technologies in asphalt pavement constructions [8,9].

In this respect, the aim of this collection of papers is to report recent innovative studies and practices based on the use of temperature reduction technologies in asphalt industries. It includes twelve original research articles from five different countries and regions: China, Finland, Germany, Taiwan, and U.S.A., with international distribution accepted for this purpose. Various subjects related to advanced temperature reduction technologies in bituminous materials are covered. We believe that this Special Issue could help civil engineers and material scientists to better identify underlying views for sustainable pavement construction.

In the paper by Gao et al. [10], the fracture properties of a cold recycled mixture (CRM) were investigated via the designed rotary test device and Finite Element Method (FEM). It was found that the mixed-mode fracture test method can effectively evaluate the cracking resistance of CRM by the proposed fracture parameters, while the notch length on the initial crack angle and the crack propagation process of the CRM are mainly related to the distribution characteristics of its meso-structure. Jin et al. [11] experimentally evaluated the performance properties of Cold In-Place Recycling (CIR) asphalt mixtures over a wide range of temperatures. The results indicated that the CIR technology could significantly improve the low temperature and fatigue cracking properties. Mechanistic–empirical (M-E) pavement design methods show that the accompanied moisture damage accelerates the rutting and low-temperature fracture when distressed; however, such behaviors are acceptable for low-volume roads. Yan et al. [12] attempted using the Graphite Nanoplatelets (GNP) and/or aggregate packing technology to reduce the asphalt materials'compaction temperatures. The results show that the combined use of these two technologies could significantly reduce the compaction temperatures, while only optimizing the aggregate has the least impact.

For the purpose of mitigating the urban heat island (UHI) effect, Yang et al. [13] used different types of permeable road pavements to improve heat storage and dissipation efficiency. It was found that a fully permeable pavement has a higher efficiency than a semi-permeable pavement and a reasonable design depth of permeable road pavement could be 30 cm. Lai et al. [14] also worked on porous asphalt materials; the noise reduction properties were characterized by the sound pressure level sensors. Different mix designs showed a significant impact on the noise reduction responses;

however, such an effect was mitigated with the increase in the wheel and decrease in the road structure depths. In particular, this was found for the open-graded friction courses. In Wang et al.'s [15] work, water-retentive and thermal-resistant cement (WTC) materials were developed, and their cooling effect was experimentally evaluated. The WTC was prepared with water-retentive material and a high aluminum refractory aggregate (RA) with porous cement concrete (PCC). The experimental results indicated superior cooling effects, and this material has the potential to mitigate the UHI effect under medium-traffic roads. Liu et al. [16] attempted to use numerical methods to study the temperature field effect on the mechanical properties of conventional and cool roads with a reflective coating. Cool pavement shows the potential to improve the rutting resistance, while only leading to limited benefits on the fatigue properties.

The road surface functions were studied by Yu et al. [17] and Guo et al. [18]. Yu et al. [17] focused on the effect of texturing parameters on road skid-resistance performance and driving stability. The width of the groove group can be adjusted to balance these two requirements. The optimal width, depth, spacing, and groove group width were defined in this study and validated by the actual engineering construction. Guo and his co-authors [18] use a Molecular Dynamics (MD) simulation to evaluate the adhesion properties and moisture effect between the interface of binder and aggregates. To balance the dry adhesion and moisture resistance, the energy ratio (ER) is an option to select the optimal SBS additive.

Fan et al. [19] evaluated the environmental benefits of low-emission mixed epoxy asphalt pavement. Overall, better anti-fatigue and rutting properties were found in the epoxy asphalt pavement compared to the conventional HMA, while a 30% reduction in carbon emissions was observed. Both Wei et al. [20] and Wang et al. [21] studied the rheological response of modified asphalt binders. Wei et al. [20] attempted to use polyphosphoric acid (PPA) as an alternative to Styrene-Butadiene-Styrene (SBS) and Styrene-Butadiene Rubber (SBR). It was found that the application of PPA could remarkably reduce the ratio of SBS and/or SBR; meanwhile, better anti-aging properties were found in the PPA-modified binders. Wang et al. [21] used polyphosphates acid and waste cooking oil (WCO) to enhance the conventional and rheological properties of asphalt binders. It was found that different components of WCO lead to a significant improvement in the low-temperature performance of PPA-modified binders. The solid–liquid phase change was observed at low temperatures.

Funding

This research received no external funding.

Acknowledgments

We would like to thank all the authors, reviewers, and staff of the *Materials* Editorial Office for their great support during the preparation of this Special Issue. Special acknowledgments go to the contact editor, Ms. Emma Fang, for her continuous support during the organization of this Special Issue.

Conflicts of Interest

The authors declare no conflicts of interest.

1. Rubio, M. C.; Martínez, G.; Baena, L.; Moreno, F. Warm mix asphalt: an overview. *J. Cleaner Prod.* 2012, 24, 76-84. https://doi.org/10.1016/j.jclepro.2011.11.053

2. Office, J. E.; Chen, J.; Dan, H.; Ding, Y.; Gao, Y.; Guo, M.; ...; Zhu, X. New innovations in pavement materials and engineering: A review on pavement engineering research 2021. *J. Traffic Transp. Eng.* 2021, 8(6), 815-999. https://doi.org/10.1016/j.jtte.2021.10.001

3. Wang, D.; Riccardi, C.; Jafari, B.; Cannone Falchetto, A.; Wistuba, M. P. Investigation on the effect of high amount of Re-recycled RAP with Warm mix asphalt (WMA) technology. *Constr. Build. Mater.* 2021, 312, 125395. https://doi.org/10.1016/j.conbuildmat.2021.125395

4. Jamshidi, A.; Hamzah, M. O.; You, Z. Performance of warm mix asphalt containing Sasobit®: State-of-the-art. *Constr. Build. Mater.* 2013, 38, 530-553. https://doi.org/10.1016/j.conbuildmat.2012.08.015

5. Jain, S.; Singh, B. Cold mix asphalt: An overview. *J. Cleaner Prod.* 2021, 280, 124378. https://doi.org/10.1016/j.jclepro.2020.124378

6. Prowell, B. D.; Hurley, G. C.; Crews, E. Field performance of warm-mix asphalt at national center for asphalt technology test track. *Transp. Res. Rec.* 2007, 1998(1), 96-102. https://doi.org/10.3141/1998-12

7. Vidal, R.; Moliner, E.; Martínez, G.; Rubio, M. C. Life cycle assessment of hot mix asphalt and zeolite-based warm mix asphalt with reclaimed asphalt pavement. *Resour. Conserv. Recycl.* 2013, 74, 101-114. https://doi.org/10.1016/j.resconrec.2013.02.018

8. Diab, A.; Sangiorgi, C.; Ghabchi, R.; Zaman, M.; Wahaballa, A. M. Warm mix asphalt (WMA) technologies: Benefits and drawbacks—A literature review *Functional Pavement Design* 2016, 1145-1154.

9. Cheraghian, G.; Cannone Falchetto, A. C.; You, Z.; Chen, S.; Kim, Y. S.; Westerhoff, J.; ...; Wistuba, M. P. Warm mix asphalt technology: An up to date review. *J. Cleaner Prod.* 2020, 268, 122128. https://doi.org/10.1016/j.jclepro.2020.122128

10. Gao, L.; Deng, X.; Zhang, Y.; Ji, X.; Li, Q. Fracture Parameters and Cracking Propagation of Cold Recycled Mixture Considering Material Heterogeneity Based on Extended Finite Element Method. *Materials* 2021, 14(8), 1993. https://doi.org/10.3390/ma14081993

11. Jin, D.; Ge, D.; Chen, S.; Che, T.; Liu, H.; Malburg, L.; You, Z. Cold in-place recycling asphalt mixtures: laboratory performance and preliminary ME design analysis. *Materials* 2021, 14(8), 2036. https://doi.org/10.3390/ma14082036

12. Yan, T.; Turos, M.; Le, J. L.; Marasteanu, M. Reducing Compaction Temperature of Asphalt Mixtures by GNP Modification and Aggregate Packing Optimization. *Materials* 2022, 15(17), 6060. https://doi.org/10.3390/ma15176060

13. Yang, C. C.; Siao, J. H.; Yeh, W. C.; Wang, Y. M. A Study on Heat Storage and Dissipation Efficiency at Permeable Road Pavements. *Materials* 2021, 14(12), 3431. https://doi.org/10.3390/ma14123431

14. Lai, F.; Huang, Z.; Guo, F. Noise Reduction Characteristics of Macroporous Asphalt Pavement Based on A Weighted Sound Pressure Level Sensor. *Materials* 2021, 14(16), 4356. https://doi.org/10.3390/ma14164356

15. Wang, X.; Hu, X.; Ji, X.; Chen, B.; Chen, H. Development of water retentive and thermal resistant cement concrete and cooling effects evaluation. *Materials* 2021, 14(20), 6141. https://doi.org/10.3390/ma14206141

16. Liu, P.; Kong, X.; Du, C.; Wang, C.; Wang, D.; Oeser, M. Numerical Investigation of the Temperature Field Effect on the Mechanical Responses of Conventional and Cool Pavements. *Materials* 2022, 15(19), 6813. https://doi.org/10.3390/ma15196813

17. Yu, J.; Zhang, B.; Long, P.; Chen, B.; Guo, F. Optimizing the texturing parameters of concrete pavement by balancing skid-resistance performance and driving stability. *Materials* 2021, 14(20), 6137. https://doi.org/10.3390/ma14206137

18. Guo, F.; Pei, J.; Zhang, J.; Li, R.; Liu, P.; Wang, D. Study on adhesion property and moisture effect between SBS modified asphalt binder and aggregate using molecular dynamics simulation. *Materials* 2022, 15(19), 6912. https://doi.org/10.3390/ma15196912

19. Wei, J.; Shi, S.; Zhou, Y.; Chen, Z.; Yu, F.; Peng, Z.; Duan, X. Research on Performance of SBS-PPA and SBR-PPA Compound Modified Asphalts. *Materials* 2022, 15(6), 2112. https://doi.org/10.3390/ma15062112

20. Fan, Y.; Wu, Y.; Chen, H.; Liu, S.; Huang, W.; Wang, H.; Yang, J. Performance Evaluation and Structure Optimization of Low-Emission Mixed Epoxy Asphalt Pavement. *Materials* 2022, 15(18), 6472. https://doi.org/10.3390/ma15186472

21. Wang, W.; Li, J.; Wang, D.; Liu, P.; Li, X. The Synergistic Effect of Polyphosphates Acid and Different Compounds of Waste Cooking Oil on Conventional and Rheological Properties of Modified Bitumen. *Materials* 2022, 15(23), 8681. https://doi.org/10.3390/ma15238681

Markus Oeser, Michael Wistuba, Pengfei Liu, and Di Wang
Editors

MDPI

Article

Fracture Parameters and Cracking Propagation of Cold Recycled Mixture Considering Material Heterogeneity Based on Extended Finite Element Method

Lei Gao [1], Xingkuan Deng [1,*], Ye Zhang [1], Xue Ji [1] and Qiang Li [2]

[1] Department of Civil and Airport Engineering, Nanjing University of Aeronautics and Astronautics, 29 Jiangjun Road, Nanjing 211106, China; glzjy@nuaa.edu.cn (L.G.); zhangye@nuaa.edu.cn (Y.Z.); jixue001@nuaa.edu.cn (X.J.)
[2] College of Civil Engineering, Nanjing Forestry University, Nanjing 210037, China; liqiang2526@njfu.edu.cn
* Correspondence: dxk777@nuaa.edu.cn

Abstract: Cold recycled mixture (CRM) has been widely used around the world mainly because of its good ability to resist reflection cracking. In this study, mixed-mode cracking tests were carried out by the designed rotary test device to evaluate the cracking resistance of CRM. Through the finite element method, the heterogeneous model of CRM based on its meso-structure was established. The cracking process of CRM was simulated using the extended finite element method, and the influence of different notch lengths on its anti-cracking performance was studied. The results show that the mixed-mode fracture test method can effectively evaluate the cracking resistance of CRM by the proposed fracture parameters. The virtual tests under three of five kinds of mixed-cracking modes have good simulation to capture the cracking behavior of CRM. The effect of notch length on the initial crack angle and the crack propagation process of the CRM is mainly related to the distribution characteristics of its meso-structure. With the increase of the proportion of Mode II cracking, the crack development path gradually deviates, and the failure elements gradually increase. At any mixed-mode level, there is an obvious linear relationship between the peak load, fracture energy, and the notch length.

Keywords: cold recycling mixture; mixed-mode fracture performance; fracture parameters; cracking propagation; Arcan test

Citation: Gao, L.; Deng, X.; Zhang, Y.; Ji, X.; Li, Q. Fracture Parameters and Cracking Propagation of Cold Recycled Mixture Considering Material Heterogeneity Based on Extended Finite Element Method. *Materials* **2021**, *14*, 1993. https://doi.org/10.3390/ma14081993

Academic Editors: Markus Oeser, Michael Wistuba, Pengfei Liu and Di Wang

Received: 12 March 2021
Accepted: 14 April 2021
Published: 16 April 2021

1. Introduction

Cold recycling is a kind of asphalt pavement recycling technology that has the technical characteristics of construction under normal temperature, so it has positive significance for resource conservation and environmental protection. As a result of its low cost and sustainable characteristics, this technology has been widely used around the world. Usually, reclaimed asphalt pavement (RAP), emulsion, water, additional aggregate, and additives are mixed to form rehabilitated pavement without the application of heat [1]. The repaired pavement can well resist the reflection cracking from underlying concrete, which is the key to the success of cold recycled mixture (CRM) [2]. Therefore, it is necessary to understand the cracking behavior and mechanism of CRM.

According to fracture mechanics, the cracking types of asphalt mixture can be divided into Mode I (opening), Mode II (sliding), and their combination. The combination of both is the mixed-mode cracking of asphalt mixture. There have been a lot of research studies on the test methods of asphalt mixture cracking [3–7]. Most of them are focused on Mode I cracking of asphalt mixture, and less attention is paid to Mode II cracking. Gao [8] had designed the Arcan testing device which can simulate five levels of mixed-mode cracking for asphalt concrete. Through a laboratory Arcan test combined with the digital image correlation (DIC) system, the mixed-mode cracking behavior of asphalt mixture can be effectively analyzed. However, due to the time-consuming and special devices needed in the laboratory Arcan test, the numerical simulation method is usually needed to study the mixed-cracking behavior of materials.

A three-dimensional numerical model of asphalt mixture was established by the discrete element method, and the fracture performance of CRM and a hot-mix asphalt mixture was compared [9]. The cohesive zone model (CZM) of cold in-place recycling mixture was established by the finite element method (FEM). The model has the potential capability to obtain the fracture process of Arcan specimen [10]. However, using the CZM model requires defining the cracking propagation path in advance, which doubtlessly has a great influence on the prediction of asphalt mixture crack trajectory. The extended finite element method (XFEM) has an obvious advantage in the study of fracture problems because it does not rely on mesh generation and does not need to define the propagation path of cracks [11]. XFEM has been used to study the crack resistance of a variety of materials [12–14]. In order to study the mechanism of reflective cracks on an asphalt concrete surface, the XFEM model was established, and the effects of temperature and traffic load were considered. The effects of initial cracking lengths and inclined degrees of initial crack on crack initiation and crack propagation were analyzed [15]. The stress distribution, crack initiation temperature, crack opening displacement, and crack propagation path of asphalt pavement with three different overlay thicknesses were studied [16]. Thus, the numerical simulations of mixed-mode cracking for mixtures could be better investigated by XFEM.

In this study, the overall objective is to analyze the mixed-mode cracking parameters and cracking propagation of CRM through the combination of Arcan test and FEM. Furthermore, the specific objectives of this paper are as follows:

(1) Obtain the fracture parameters such as peak load, fracture energy, and crack angle using the Arcan test method to study the cracking resistance of CRM;
(2) Verify the mixed-cracking simulation result of CRM with the fracture parameters from the laboratory Arcan test;
(3) Analyze the cracking process of CRM according to the stress distribution at different times in the virtual test;
(4) Investigate the anti-cracking mechanism of CRM with the effect of different notch length on mixed-cracking resistance of CRM studied by FEM.

2. Materials and Methods

2.1. Materials and Specimen Forming

In this research, CRM were all designed according to the cold in-place recycled specification in Jiangsu Province, China [17]. The mix design results of CRM are shown in Table 1. Portland cement (Wuxi Yangshijin Construction Material Co., Ltd., Wuxi, China) was added at 1.5% by weight of RAP materials. In addition, 3% of mineral filler (Lingshou Runling Mineral Products Trading Co., Ltd., Shijiazhuang, China) was added, and the optimum moisture content was 4.3%. The square specimens of CR-20 mixture were obtained by a rutting test. Then, the high-precision marble cutting machine was used to cut a square surface of the specimens to obtain the Arcan specimen with a smooth surface. The size of the Arcan specimens is 80 mm × 80 mm × 50 mm. Finally, a 40 mm notch was cut in the center of one side of the square specimen to complete the molding of the Arcan specimen. Furthermore, in order to achieve the application of DIC technology to obtain the displacement data that need to be recorded during the test, speckles are formed on the surface of the Arcan specimen by black and white spray paints.

Table 1. Mix design results of the CR-20 mixture.

Mixtures	CR-20	
Optimal Asphalt Content (%)	**3.5(CSS-1)**	
Sieve Size (mm)	**Passing Percent (%)**	**Limits**
26.5	100	100
19	96.7	90–100
16	92.4	-
13.2	84.3	-
9.5	70.1	60–80
4.75	50	35–65
2.36	36	20–50
1.18	22.2	-
0.6	14.5	-
0.3	8.3	3–21
0.15	6.1	-
0.075	3.7	2–8

2.2. Arcan Configuration

The Arcan testing device as shown in Figure 1, by adjusting the angle (90°, 67.5°, 45°, 22.5°, and 0°) between the loading direction and the initial notch, the stress modes of arcan specimens can be divided into Mode A (100% Mode I and 0% Mode II), Mode B (75% Mode I and 25% Mode II), Mode C (50% Mode I and 50% Mode II), Mode D (25% Mode I and 75% Mode II), and Mode E (0% Mode I and 100% Mode II), so as to simulate the stress state of asphalt mixture under different mixed-cracking modes. The Arcan test was carried out by a universal testing machine (Shanghai zhuozhilitian Technology Development Co., Ltd., Shanghai, China). The test temperature was set at −10 °C, and the tensile loading rate was set at 0.5 mm/min. The Arcan test has three key parameters: peak load, crack angle, and fracture energy. The peak load is the maximum load in the process of the Arcan test, which reflects the cracking strength of the asphalt mixture to a certain extent. The crack angle is the acute angle formed by the straight line connecting the crack starting point with the crack end point and the cutting direction after the specimen is damaged. Fracture energy is a comprehensive index to evaluate the anti-cracking performance of CRM. In the Arcan test, the measured displacement data mainly include load line displacement (LLD), crack mouth opening displacement (CMOD), and crack tip opening displacement (CTOD). The fracture energies corresponding to different displacement data are $G_{\text{f-LLD}}$, $G_{\text{f-CMOD}}$, and $G_{\text{f-CTOD}}$. The specific calculation methods of three kinds of displacement fracture energy used in this paper are as follows:

$$G_{f-LLD} = \frac{\int Pdu_1}{A_{\text{lig}}}, \tag{1}$$

$$G_{f-CMOD} = \frac{\int Pdu_2}{A_{\text{lig}}}, \tag{2}$$

$$G_{f-CTOD} = \frac{\int Pdu_3}{A_{\text{lig}}}, \tag{3}$$

where P is the load; A_{lig} is the ligament area; du_1, du_2, and du_3 are changes in *LLD*, *CMOD*, and *CTOD* respectively; $G_{\text{f-LLD}}$ is the fracture energy from the *LLD* measurement; $G_{\text{f-CMOD}}$ is the fracture energy from the *CMOD* measurement, and $G_{\text{f-CTOD}}$ is the fracture energy from the *CTOD* measurement.

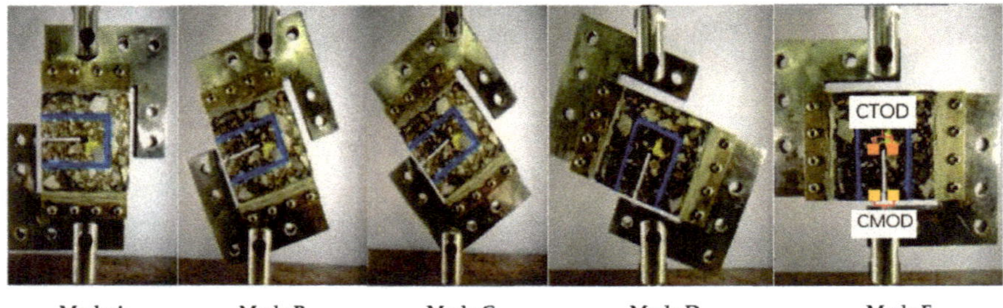

Figure 1. Arcan test configuration with five levels of mixed-mode cracking.

2.3. Numerical Simulation Method

2.3.1. Homogenization Method of Composite Materials

Since the cracking of CRM in low temperature (especially subzero temperature) is mainly brittle, the elastic material characteristics are considered in the finite element simulation. There are two basic methods to homogenize the elastic parameters of composite materials: a series model (Voigt model) [18,19] and parallel model (Reuss model) [20,21]. The parallel model can ensure the deformation compatibility of each component in the material. In addition, the homogenization method of material parameters is needed to determine the element size. The selection of cell size was determined by moving window technology. The images were divided into 100, 256, 400, and 625 windows. The coefficient variation of the percentage of coarse aggregate in the window represents the uniformity of the material within the unit size. As can be seen from Figure 2, the smaller the unit size, the larger the coefficient variation is. Based on the representativeness of material sampling and the accuracy requirements of numerical simulation, this study selected 400 equal parts for element division.

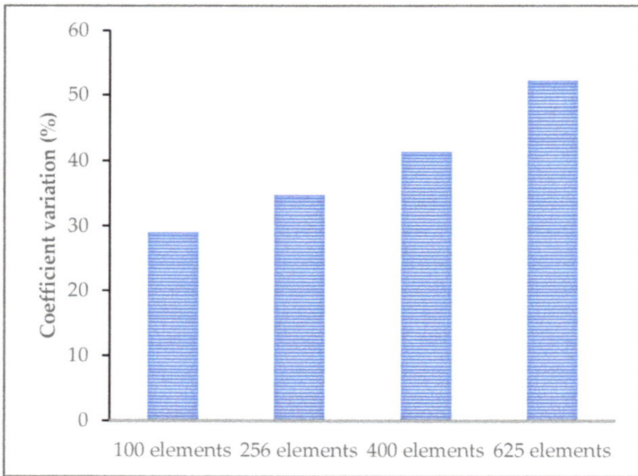

Figure 2. The results of coefficient variation calculated by different element size division.

The equivalent elastic modulus and equivalent shear modulus of each element in the parallel model are defined as follows:

$$\begin{cases} E^* = \sum_{i=1}^{n} C_i E_i \\ G^* = \sum_{i=1}^{n} C_i G_i \end{cases}, \tag{4}$$

where E^* is the equivalent elastic modulus; G^* is the equivalent shear modulus; C_i is the volume fraction of the component i; E_i is the elastic modulus of the component i; and G_i is the shear modulus of the component i.

According to the relationship between shear modulus and Poisson's ratio, the equivalent Poisson's ratio of the element can be deduced as follows:

$$\mu^* = \frac{E^*}{2G^*} - 1 = \frac{\sum_{i=1}^{n} C_i E_i}{\sum_{i=1}^{n} C_i \frac{E_i}{1+\mu_i}} - 1, \tag{5}$$

where μ^* is the equivalent Poisson's ratio, and μ_i is the Poisson's ratio of the component i.

2.3.2. Model Building

According to the stress characteristics and the size of CRM specimen, the finite element model of the testing device was established. The establishment of the digital model of asphalt mixture, and its meso-structure should be considered [22]. Through the homogenization method of composite materials, the heterogeneous model of CRM based on its meso-structure was established. The continuous sectional image of CRM was obtained by CT scanning and digital image processing technology, and the image was divided into 400 elements, as shown in Figure 3a. The volume fraction of each component (emulsified asphalt mortar, aggregate, and voids) along the height direction of the specimen under the area of the element was calculated first, and then, the elastic parameters were determined by the homogenization method of composite materials. In ABAQUS 6.14-2 software, 400 two-dimensional CPE4R elements were established, each element was set as a separate section, and the calculated equivalent elastic parameters were input into the corresponding element. The established heterogeneous model is shown in Figure 3b. By simulating the initial notch with one-dimensional components without material characteristics, the Arcan digital specimens with different notch lengths were established. Through the XFEM method, the cracking process and mixed-mode cracking behavior of CRM under five different types of loads were obtained. According to the experimental results of Paulino, Kim, and Gao [7,23,24], the values of elastic parameters used in this paper are as follows: E = 42.0 GPa, μ = 0.15 for aggregates, and E = 0.185 GPa, μ = 0.25 for emulsified asphalt mortar. The fracture parameters for each element used in the XFEM method were obtained by the Arcan test under Mode A cracking mode, and their values are as follows: the tensile strength of 3.92 MPa and the fracture energy of 275 J/m^2.

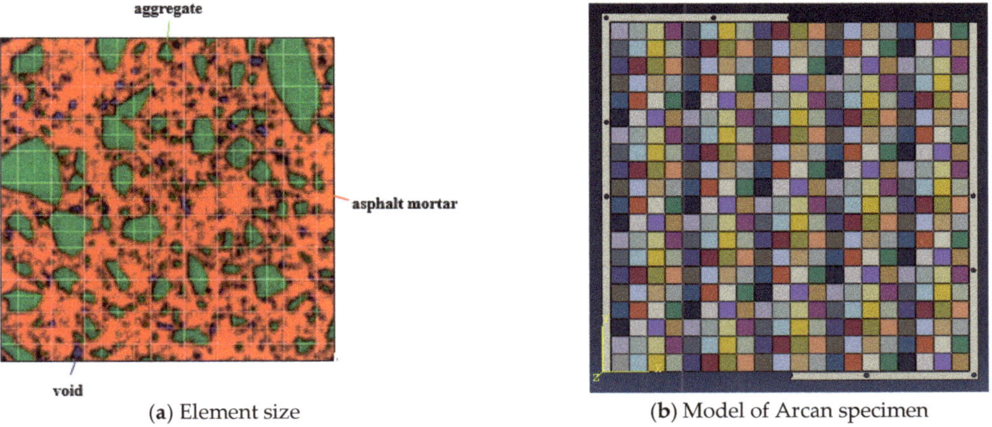

(**a**) Element size (**b**) Model of Arcan specimen

Figure 3. Schematic diagram of FEM model based on homogenization method.

3. Results and Discussion

3.1. Mixed-Mode Cracking Test Results

The load displacement curve of the CR-20 mixture is shown in Figure 4. The ordinate axis corresponds to the tensile load, and the abscissa axis corresponds to the axial load displacement of the Arcan specimen. To some extent, the peak load can reflect the cracking resistance strength of CRM at a certain mixed-mode level. The fracture strength of the CR-20 mixture is affected by the cracking mode. The peak load of the CR-20 mixture reaches the minimum value at the mixed-mode level of Mode C, and it reaches the maximum value at the mixed-mode level of Mode E. Therefore, the crack resistance strength of the CR-20 mixture in the Arcan test is related to the stress mode. Under the mixed-mode level of Mode C, the peak load is the smallest and it cracks the most easily; at the mixed-mode level of Mode E, the peak load is the largest and the most difficult to crack. It can be seen that the fracture process of the CR-20 mixture under the five mixed-cracking modes shows certain softening characteristics. However, the softening characteristics of different loading modes are obviously different. The loading mode of Mode E (pure shear cracking mode) has the maximum peak load. The corresponding fracture curve decreases fastest after the peak load, and the characteristics of brittle fracture are the most obvious. The peak load of the Mode D fracture curve is the second, and the descending speed after the peak load is only next to Mode E. The fracture curve of Mode C has the minimum peak load, the decline rate is the slowest after the peak load, and the softening process is the most obvious in the fracture process. The peak load of the Mode B fracture curve is only greater than that of Mode A, and the decline speed of the fracture curve after the peak load is the second slowest. Therefore, the greater the crack resistance of CRM in a certain mixed-mode level, the greater the possibility of brittle fracture.

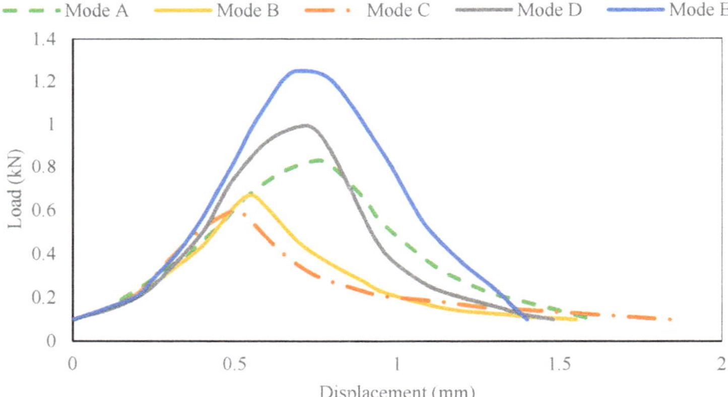

Figure 4. Load–displacement curves of CR-20 mixture in different mixed-mode levels.

Figure 5 summarizes the fracture energy of CR-20 mixture in five different mixed-mode levels. It can be seen that the three kinds of displacement fracture energies of the CR-20 mixture are sensitive to five kinds of mixed-mode cracking levels. In the cracking tests, the three kinds of displacement fracture energies of the CR-20 mixture have the same law: with the proportion of Mode I cracking reduced from 100% to 50%, the fracture energy decreased gradually; but with the proportion of Mode I cracking decreased from 50% to 0%, the fracture energy increased gradually. In addition, three different displacement fracture energies can be used to study the crack resistance of CRM. In any mixed-mode level, the order of displacement fracture energy is as follows: $G_{f\text{-}CMOD} > G_{f\text{-}LLD} > G_{f\text{-}CTOD}$. Therefore, according to the results of $G_{f\text{-}CTOD}$, the CR-20 mixture has the best crack resistance under

Mode A, followed by Mode E, then Mode B and Mode D. Under Mode C, the crack resistance of the CR-20 mixture is the worst.

Fracture energy (J/m²)

	Mode A	Mode B	Mode C	Mode D	Mode E
LLD	301	187	153	170	336
CMOD	343	334	178	195	367
CTOD	227	167	101	110	124

Figure 5. $G_{f\text{-}LLD}$, $G_{f\text{-}CMOD}$, and $G_{f\text{-}CTOD}$ of CR-20 mixture.

The corresponding fracture angles of the CR-20 mixture under different mixed-mode levels (Mode A, Mode B, Mode C, Mode D, and Mode E) are 6°, 15°, 29°, 45°, and 50° respectively. It can be seen that the crack angle of CRM in the Arcan test has a clear relationship with the mixed-cracking level. With the increase of the ratio of Mode II cracking, the crack angle of the CR-20 mixture will increase correspondingly. Therefore, according to the corresponding relationship between the crack angle and the mixed-cracking level in the Arcan test, the types of reflection cracking (tensile, shear, and mixed-mode) of the actual cold recycling pavement can be determined, so as to judge the failure stress mode of the actual pavement.

3.2. *Verification of Numerical Simulation Results*

Figure 6 shows the cracking results of the virtual specimen of the CR-20 mixture with a 40 mm notch under five different mixed-mode levels. The crack angle is the acute angle formed by the straight line connecting the crack starting point with the crack end point and the cutting direction after the specimen is damaged. According to the simulation results, the corresponding crack angles of the CR-20 mixture under five different mixed-mode levels (Mode A, Mode B, Mode C, Mode D, and Mode E) are 3°, 11°, 23°, 38°, and 44° respectively. Compared with the laboratory test results, it can be concluded that the virtual Arcan test has reasonable results.

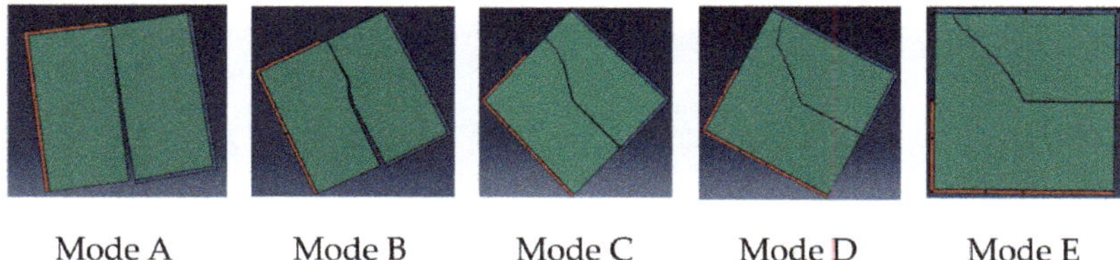

Figure 6. Virtual cracking results of the CR-20 mixture under five different mixed-mode levels.

The LLD of the Arcan test cannot directly reflect the deformation of the specimen, but CMOD and CTOD can directly and accurately determine the deformation of the Arcan specimen and the specific development stage of the Arcan specimen. Therefore, using the P-CMOD curve and P-CTOD curve can better verify the accuracy of simulated results.

Figure 7a–e shows the load–displacement curve comparing the simulated results with the experimental results. Figure 7a corresponds to the Mode A cracking level, in which the simulated results of the P-CTOD curve are closer to the experimental curve, and the peak load of the Arcan test and the development trend of CTOD in the test process are close to the experimental results. The results of the P-CMOD curve for the peak load are consistent, but the calculation results of CMOD in the test process are different from the experimental results. Figure 7b corresponds to the Mode B cracking level; similar to Figure 7a, the simulated result of the P-CTOD curve is better than that of P-CMOD. The main reason for this phenomenon is that CTOD is only affected by the deformation at the notch tip of the Arcan digital specimen, and the factors considered are single. However, CMOD is affected by the deformation of the notch tip and the development of cracks, so the situation of CMOD is more complex, which leads to the deviation of CMOD results of the Arcan virtual test larger than the tip displacement. Figure 7c corresponds to the Mode C cracking level; the peak load result is slightly larger than the laboratory test result, and while the corresponding CMOD and CTOD are close to the experimental result, the overall CMOD and CTOD results are quite different from the experimental curve. Figure 7d corresponds to the Mode D cracking level, which is similar to Figure 7a,b. The similarity of the peak load results is the best, and the similarity between the simulated result of the P-CTOD curve and experimental result is the highest. Figure 7e corresponds to the Mode E cracking level; similar to Figure 7c, its peak load is greater than the experimental results, and the values of CMOD and CTOD are quite different from the laboratory values, so it can not be used as a substitute for a laboratory test.

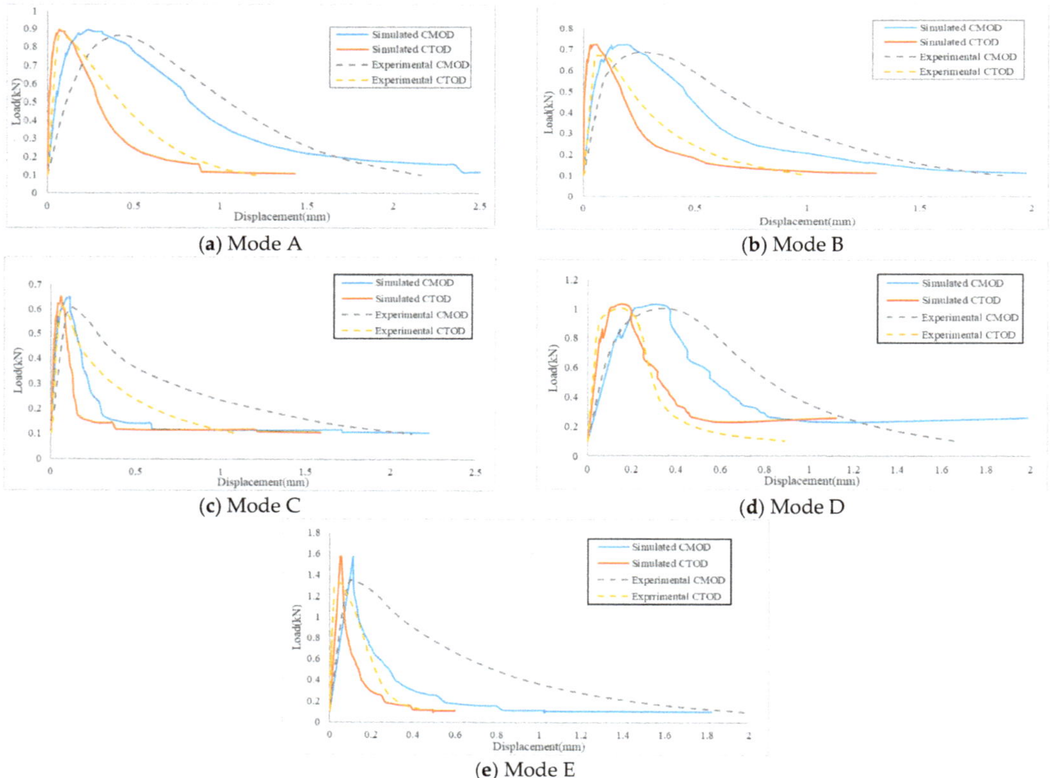

Figure 7. Comparison of simulated results and experimental results of the CR-20 mixture with 40 mm notch under five kinds of mixed-mode levels.

The simulated fracture energy of the Arcan specimen with a 40 mm notch under five different mixed-mode levels was compared with the laboratory results, and the reliability of the simulation results of the virtual test was analyzed. According to the definition and calculation method of the Arcan test fracture energy, the results of $G_{f\text{-}CMOD}$ and $G_{f\text{-}CTOD}$ are shown in Table 2. The results show that the virtual test under Mode A, Mode B, and Mode D has good reliability, and the relative error between simulation results and laboratory results is less than 10%. Among them, the accuracy of the virtual test results is the best under the Mode E cracking level. The simulation results of Mode C and Mode E are quite different from the laboratory results. Therefore, the virtual Arcan test results under the Mode C and Mode E cracking levels cannot be used for analysis.

Table 2. Comparison between experimental fracture energies and the simulated results.

Fracture Energy		Mixed-Mode Level				
		Mode A	Mode B	Mode C	Mode D	Mode E
$G_{f\text{-}CMOD}$ (J/m^2)	Simulated $G_{f\text{-}CMOD}$	365	307	127	207	215
	Experimental $G_{f\text{-}CMOD}$	343	334	178	195	367
	Relative error	6.41%	8.08%	28.65%	6.15%	41.42%
$G_{f\text{-}CTOD}$ (J/m^2)	Simulated $G_{f\text{-}CTOD}$	241	155	73	116	67
	Experimental $G_{f\text{-}CTOD}$	227	167	101	110	124
	Relative error	6.17%	7.18%	27.72%	5.45%	45.97%

3.3. Analysis of Numerical Test Results

In order to analyze the cracking process of the CR-20 mixture, the stress distributions at the cracking initiation, peak load, and failure moment of digital specimen were derived from ABAQUS, as shown in Figure 8. The difference of stress distribution under the three kinds of cracking modes is mainly concentrated on the stress value and crack tip position. In the cracking process of three kinds of cracking modes, stress concentration exists at the crack tip. The partial discontinuity stress gradient was due to the large difference of elastic parameters between some adjacent elements. The peak loads of the three kinds of cracking modes (Mode A, Mode B, and Mode D) were 885, 721, and 1035 N. The load at the initial cracking moment was 754, 586, and 894 N, accounting for 85.2%, 81.3%, and 86.4% of the peak load, respectively. The load at the initial cracking of Mode B accounts for the smallest proportion of the peak load, and the load at the initial cracking of Mode D accounts for the largest proportion of the peak load. Therefore, the Mode B cracking mode is more prone to initial cracking.

In order to further analyze the anti-cracking performance of the CR-20 mixture under three kinds of cracking modes (Mode A, Mode B and Mode D), a numerical Arcan test was carried out on the digital specimens of CRM with different notch lengths (10, 20, 30 and 40 mm). The numerical results of mixed-mode cracking test with different notch lengths are shown in Figure 9. For the Mode A cracking mode, the corresponding crack angles of the virtual test with different notch lengths (40, 30, 20 and 10 mm) are 2.9°, 4.6°, 3.8° and 6.5° respectively. For the Mode B cracking mode, the corresponding crack angles corresponding to different notch lengths are 5.7°, 9.1°, 18.4° and 33.7° respectively. For the Mode D cracking mode, the crack angles corresponding to different notch lengths are 51.3°, 45.0°, 29.2° and 41.6° respectively. For any initial notch length, the order of crack angle is as follows: Mode A < Mode B < Mode D. It is difficult to describe the relationship between the crack angle and the notch length directly through the crack angle calculated from the starting point and the crack end point, which is because the crack propagation path is also related to the distribution characteristics of its meso-structure. In order to further analyze the relationship between the crack angle and notch length, the initial angle of the mixed-mode crack is calculated. Corresponding to the different initial notch lengths (40, 30, 20, 10 mm), the crack initiation angles in Mode A are 2.8°, 8.3°, 10.0° and 10.3°, the crack initiation angles in Mode B are 12.6°, 15.4°, 17.3°, and 18.4°, and the crack initiation angles in Mode D are 40.7°, 41.2°, 45.3° and 59.5°, respectively. It can be found that for the three mixed-cracking modes, the initial angle of mixed-mode crack increases gradually with the decrease of the notch length.

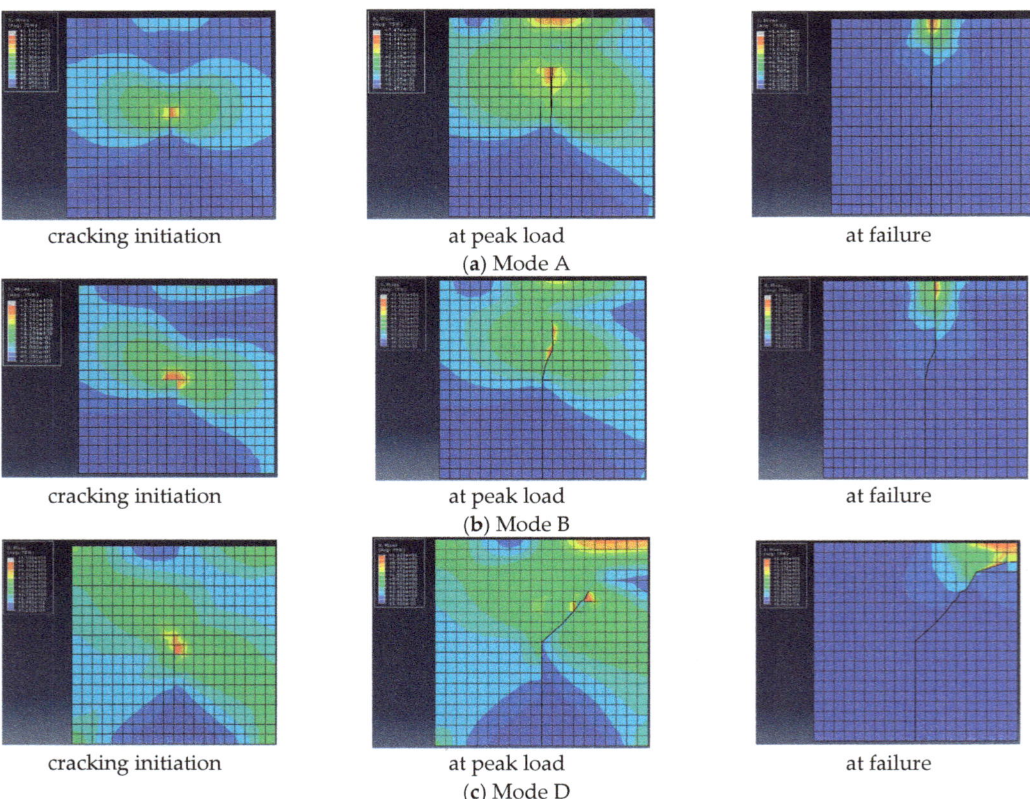

Figure 8. The stress distributions at different stages of loading for three levels of mixed-mode cracking.

Figure 10 shows the peak load results of virtual Arcan tests with different mixed-mode levels and different notch lengths. It can be seen that for the three different mixed-mode levels, the smaller the notch length, the greater the peak load result is. This phenomenon shows that for the three different cracking modes, the larger the notch length, the easier it is for the specimen to crack. For the virtual Arcan test of 20 mm, 30 mm, and 40 mm notch length, the order of peak load of different mixed-mode levels is Mode D > Mode A > Mode B. This is in accordance with the results of the laboratory tests. The order of peak load under the 10 mm notch length is Mode A > Mode B > Mode D. The peak load value of the Mode D cracking mode is the smallest among the three cracking modes under the 10 mm notch length. This is because under the Mode D cracking mode, the crack propagation path of the CRM specimen with the 10 mm notch length is the shortest. The results show that the peak load of the Mode A cracking mode and the 10 mm notch length is the largest, which verifies the previous conjecture. It shows that the peak load results of the test are related to the crack development path of the specimen.

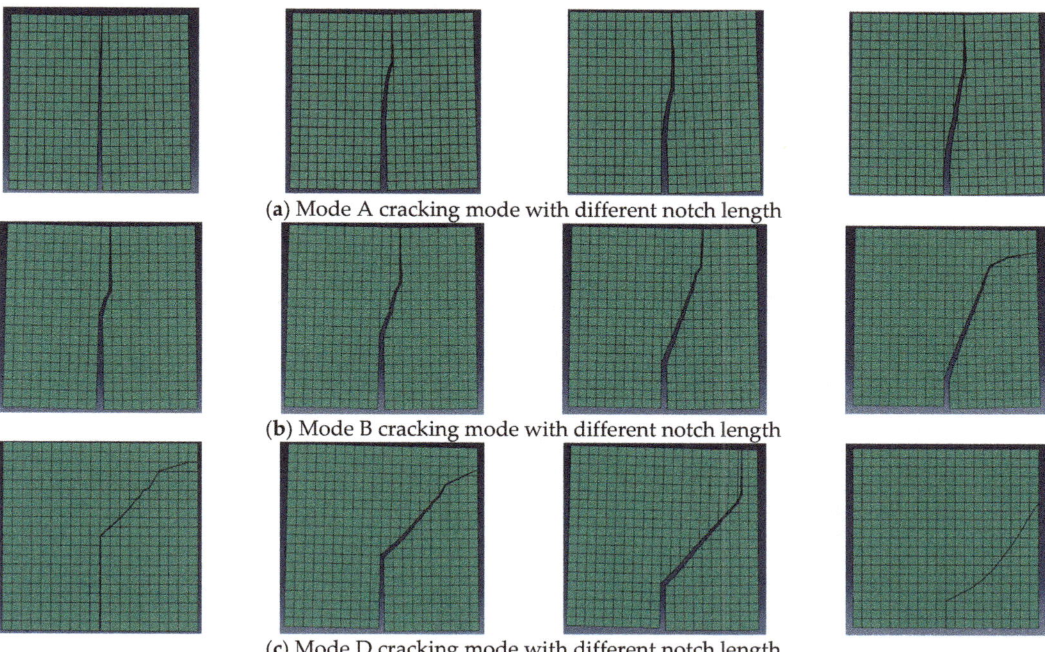

Figure 9. Numerical Arcan test results of mixed-mode cracking with different notch lengths (40, 30, 20 and 10 mm).

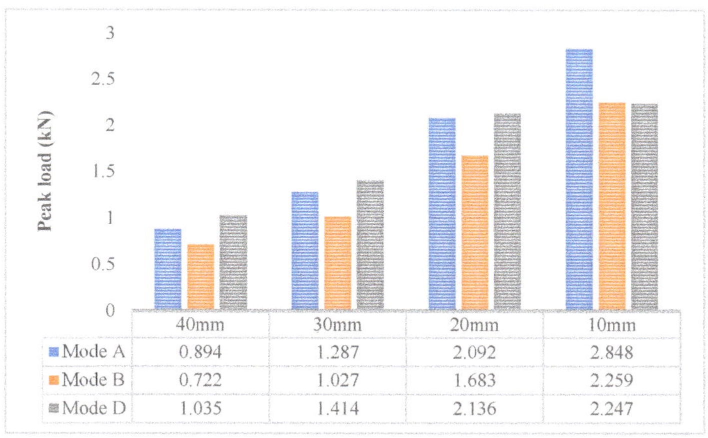

	40mm	30mm	20mm	10mm
Mode A	0.894	1.287	2.092	2.848
Mode B	0.722	1.027	1.683	2.259
Mode D	1.035	1.414	2.136	2.247

Figure 10. The peak load results of the virtual Arcan test with different notch lengths and different mixed-mode levels.

Fracture energy is a reliable and comprehensive parameter to evaluate the crack resistance of materials [22]. $G_{f\text{-}CMOD}$ and $G_{f\text{-}CTOD}$ were used to further evaluate the crack resistance of the CR-20 mixture. The $G_{f\text{-}CMOD}$ and $G_{f\text{-}CTOD}$ results of the virtual Arcan test under different notch lengths and different mixed-mode levels are shown in Figure 11. It can be found that the values of $G_{f\text{-}CMOD}$ and $G_{f\text{-}CTOD}$ increase with the decrease of notch length at any mixed-mode level. The smaller the initial notch length, the greater the fracture energy required for the failure of the CR-20 mixture under the determined mixed-cracking mode. Under the selected initial notch length, the order of $G_{f\text{-}CMOD}$ and $G_{f\text{-}CTOD}$ values is Mode A > Mode B > Mode D, which is consistent with the experimental results. It also reflects that the fracture energy of the Arcan test is a more comprehensive evaluation index, and it is not affected by the length of the notch to a certain extent. Among all the fracture energy results, the fracture energy required for the Arcan specimen with an initial notch length of 10 mm is the largest under the Mode A cracking mode. This is because the Arcan specimen with an initial notch length of 10 mm has the longest crack development path under the Mode A cracking mode, which also confirms the previous peak load results of the virtual Arcan test. The fracture energy of the specimen with an initial notch length of 40 mm is the minimum at the Mode D cracking level. The fracture energy of the specimen with an initial notch length of 40 mm is the minimum at the Mode D cracking level, which is also related to the crack propagation path. It can be concluded that the CR-20 mixture with a 40 mm notch length under the Mode D cracking mode is most likely to crack. Meanwhile, under the Mode A cracking mode, the CR-20 mixture with a 10 mm notch length is the most difficult to crack.

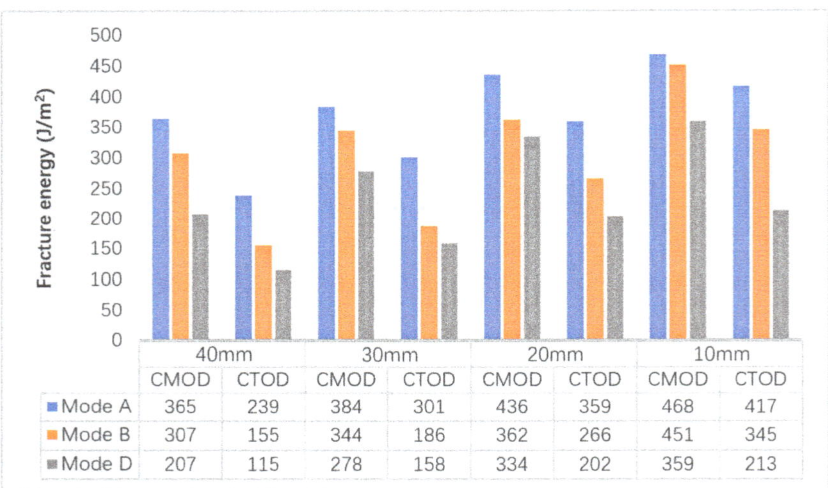

	40mm		30mm		20mm		10mm	
	CMOD	CTOD	CMOD	CTOD	CMOD	CTOD	CMOD	CTOD
Mode A	365	239	384	301	436	359	468	417
Mode B	307	155	344	186	362	266	451	345
Mode D	207	115	278	158	334	202	359	213

Figure 11. The results of $G_{f\text{-}CMOD}$ and $G_{f\text{-}CTOD}$ of a virtual Arcan test with different notch length and different mixed-cracking modes.

4. Conclusions

In this paper, the cracking resistance and cracking behavior of a CR-20 mixture under different mixed-cracking modes were studied by the Arcan test method and FEM. The mixed-mode cracking test of a CR-20 mixture was carried out with an Arcan testing device, and the simulation of the Arcan test of the CR-20 mixture was carried out by FEM. The reliability of the simulation was verified by comparing the simulation results with the experimental results. Based on the results and analysis, the following conclusions can be made:

1. The fracture parameters such as peak load, crack angle, and fracture energy were obtained by Arcan tests. The order of displacement fracture energy is $G_{f\text{-}CMOD}$ > $G_{f\text{-}LLD}$ > $G_{f\text{-}CTOD}$. According to the result of $G_{f\text{-}CTOD}$, the order of crack resistance of the CR-20 mixture under five kinds of mixed-cracking modes is as follows: Mode A > Mode E > Mode B > Mode D > Mode C.
2. The virtual Arcan configuration and two-dimensional digital specimen were set up in the ABAQUS software by using FEM. Considering the material heterogeneity and

combined with the XFEM cracking mechanism, the cracking behavior of the CR-20 mixture under different mixed-mode levels was studied.

3. According to the reliability of the virtual test, Mode A, Mode B, and Mode D cracking has better simulation in five kinds of mixed-mode virtual cracking tests. Furthermore, the relative error of the Mode D cracking mode is the smallest among the three kinds of mixed-cracking modes.
4. With the increase of the proportion of Mode II cracking, the development path of cracks gradually deviates, and the number of failure elements increases. In any mode of cracking process, the stress concentration exists at the crack tip. The cracking propagation of the CR-20 mixture is not only affected by the length of the initial notch but also related to the distribution characteristics of its meso-structure.
5. There is a certain relationship between the initial notch length and the initial crack angle, peak load, and fracture energy. Through two kinds of fracture energy, the anti-cracking performance of the CR-20 mixture with different notch lengths under different mixed-cracking modes was evaluated. The results show that the CR-20 mixture with a 10 mm initial notch length has better cracking resistance under the pure tensile stress mode.

Author Contributions: Conceptualization, L.G. and X.D.; methodology, L.G.; software, X.D.; validation, Y.Z., X.J. and Q.L.; formal analysis, L.G.; investigation, Y.Z.; data curation, X.D.; writing—original draft preparation, Q.L.; writing—review and editing, X.D.; visualization, Y.Z.; supervision, X.J. All authors have read and agreed to the published version of the manuscript.

Funding: This research was funded by National Natural Science Foundation of China (51908286) and Natural Science Foundation of Jiangsu Province (BK20191278).

Institutional Review Board Statement: Not applicable.

Informed Consent Statement: Not applicable.

Data Availability Statement: Data available on request due to restrictions eg privacy or ethical. The data presented in this study are available on request from the corresponding author. The data are not publicly available due to these data are part of ongoing research.

Acknowledgments: The authors would like to thank Zhanqi Wang and Yazhou Zhou of Nanjing University of Aeronautics and Astronautics for their contributions to this work.

Conflicts of Interest: The authors declare no conflict of interest.

References

1. AASHTO-AGC-ARTBA Joint Committee. *Task Force 38-Report on Cold Recycling of Asphalt Pavements*; Ameican Association of State Highway and Transportation Officials: Washington, DC, USA, 1988.
2. Gu, F.; Ma, W.; West, R.C.; Taylor, A.J.; Zhang, Y. Structural performance and sustainability assessment of cold central-plant and in-place recycled asphalt pavements: A case study. *J. Clean. Prod.* **2019**, *208*, 1513–1523. [CrossRef]
3. Ma, T.; Ding, X.; Zhang, D.; Huang, X.; Chen, J. Experimental study of recycled asphalt concrete modified by high-modulus agent. *Constr. Build. Mater.* **2016**, *128*, 128–135. [CrossRef]
4. Jiang, J.; Li, Y.; Zhang, Y.; Bahia, H.U. Distribution of mortar film thickness and its relationship to mixture cracking resistance. *Int. J. Pavement Eng.* **2020**, 1–10. [CrossRef]
5. Sreedhar, S.; Coleri, E.; Haddadi, S.S. Selection of a performance test to assess the cracking resistance of asphalt concrete materials. *Constr. Build. Mater.* **2018**, *179*, 285–293. [CrossRef]
6. Mandal, T.; Ling, C.; Chaturabong, P.; Bahia, H. Evaluation of analysis methods of the semi-circular bend (SCB) test results for measuring cracking resistance of asphalt mixtures. *Int. J. Pavement Res. Technol.* **2019**, *12*, 456–463. [CrossRef]
7. Jiang, J.; Dong, Q.; Ni, F.; Zhao, Y. Effects of loading rate and temperature on cracking resistance characteristics of asphalt mixtures using nonnotched semicircular bending tests. *J. Test. Eval.* **2018**, *47*, 2649–2663. [CrossRef]
8. Gao, L.; Ni, F.; Braham, A.; Luo, H. Mixed-mode cracking behavior of cold recycled mixes with emulsion using Arcan configuration. *Constr. Build. Mater.* **2014**, *55*, 415–422. [CrossRef]
9. Gao, L.; Li, H.; Xie, J.; Yang, X. Mixed-mode fracture modeling of cold recycled mixture using discrete element method. *Constr. Build. Mater.* **2017**, *151*, 625–635. [CrossRef]
10. Zhao, Y.; Ni, F.; Zhou, L.; Gao, L. Three-dimensional fracture simulation of cold in-place recycling mixture using cohesive zone model. *Constr. Build. Mater.* **2016**, *120*, 19–28. [CrossRef]

11. Fries, T.; Belytschko, T. The extended/generalized finite element method: An overview of the method and its applications. *Int. J. Numer. Meth. Eng.* **2010**, *84*, 253–304. [CrossRef]
12. Zhao, Y.; Jiang, J.; Zhou, L.; Ni, F. Improving the calculation accuracy of FEM for asphalt mixtures in simulation of SCB test considering the mesostructure characteristics. *Int. J. Pavement Eng.* **2020**, 1–15. [CrossRef]
13. Pietras, D.; Sadowski, T. A numerical model for description of mechanical behaviour of a Functionally Graded Autoclaved Aerated Concrete created on the basis of experimental results for homogenous Autoclaved Aerated Concretes with different porosities. *Constr. Build. Mater.* **2019**, *204*, 839–848.
14. Chen, K.; Qiu, H.; Sun, M.; Lam, F. Experimental and numerical study of moisture distribution and shrinkage crack propagation in cross section of timber members. *Constr. Build. Mater.* **2019**, *221*, 219–231.
15. Wang, X.; Li, K.; Zhong, Y.; Xu, Q.; Li, C. XFEM simulation of reflective crack in asphalt pavement structure under cyclic temperature. *Constr. Build. Mater.* **2018**, *189*, 1035–1044. [CrossRef]
16. Wang, X.; Zhong, Y. Reflective crack in semi-rigid base asphalt pavement under temperature-traffic coupled dynamics using XFEM. *Constr. Build. Mater.* **2019**, *214*, 280–289.
17. Department of Transportation in Jiangsu Province. *Technical Specification of Cold In-Place Recycling with Emulsions*; Department of Transportation in Jiangsu Province: Jiangsu, China, 2010. (In Chinese)
18. Kima, H.; Hong, S.; Kim, S. On the rule of mixtures for predicting the mechanical properties of composites with homogeneously distributed soft and hard particles. *J. Mater. Process. Technol.* **2001**, *122*, 109–113.
19. Farno, E.; Baudez, J.C.; Eshtiaghi, N. Comparison between classical Kelvin-Voigt and fractional derivative Kelvin-Voigt models in prediction of linear viscoelastic behaviour of waste activated sludge. *Sci. Total Environ.* **2018**, *613–614*, 1031–1036. [CrossRef]
20. Tian, W.; Qi, L.; Su, C.; Liang, J.; Zhou, J. Numerical evaluation on mechanical properties of short-fiber-reinforced metal matrix composites: Two-step mean-field homogenization procedure. *Compos. Struct.* **2016**, *139*, 96–103. [CrossRef]
21. Ye, W.; Li, W.; Shan, Y.; Wu, J.; Ning, H.; Sun, D.; Hu, N.; Fu, S. A mixed-form solution to the macroscopic elastic properties of 2D triaxially braided composites based on a concentric cylinder model and the rule of mixture. *Compos. B Eng.* **2019**, *156*, 355–367. [CrossRef]
22. Ling, C.; Bahia, H. Modeling of aggregates' contact mechanics to study roles of binders and aggregates in asphalt mixtures rutting. *Road Mater. Pavement Des.* **2020**, *21*, 720–736. [CrossRef]
23. Paulino, G.H.; Song, S.H.; Buttlar, W.G. Cohesive zone modeling of fracture in asphalt concrete. In Proceedings of the 5th International RILEM Conference-Cracking in Pavements: Mitigation, Risk Assessment, and Preservation, Limoges, France, 26–30 March 2004; pp. 63–70.
24. Kim, H.; Buttlar, W.G. Discrete fracture modeling of asphalt concrete. *Int. J. Solids Struct.* **2009**, *46*, 2593–2604. [CrossRef]

Article

Cold In-Place Recycling Asphalt Mixtures: Laboratory Performance and Preliminary M-E Design Analysis

Dongzhao Jin [1], Dongdong Ge [1], Siyu Chen [1,2], Tiankai Che [1,3], Hongfu Liu [1], Lance Malburg [4] and Zhanping You [1,*]

[1] Department of Civil and Environmental Engineering, Michigan Technological University, 1400 Townsend Drive, Houghton, MI 49931-1295, USA; dongj@mtu.edu (D.J.); dge1@mtu.edu (D.G.); siychen@mtu.edu (S.C.); tche@mtu.edu (T.C.); hliu16@mtu.edu (H.L.)
[2] School of Transportation, Southeast University, Nanjing 211189, China
[3] School of Transportation and Logistics, Dalian University of Technology, Dalian 116024, China
[4] Dickinson County Road Commission, Iron Mountain, MI 49801, USA; lance@dickinsoncrc.com
* Correspondence: zyou@mtu.edu

Abstract: Cold in-place recycling (CIR) asphalt mixtures are an attractive eco-friendly method for rehabilitating asphalt pavement. However, the on-site CIR asphalt mixture generally has a high air void because of the moisture content during construction, and the moisture susceptibility is vital for estimating the road service life. Therefore, the main purpose of this research is to characterize the effect of moisture on the high-temperature and low-temperature performance of a CIR asphalt mixture to predict CIR pavement distress based on a mechanistic–empirical (M-E) pavement design. Moisture conditioning was simulated by the moisture-induced stress tester (MIST). The moisture susceptibility performance of the CIR asphalt mixture (pre-mist and post-mist) was estimated by a dynamic modulus test and a disk-shaped compact tension (DCT) test. In addition, the standard solvent extraction test was used to obtain the reclaimed asphalt pavement (RAP) and CIR asphalt. Asphalt binder performance, including higher temperature and medium temperature performance, was evaluated by dynamic shear rheometer (DSR) equipment and low-temperature properties were estimated by the asphalt binder cracking device (ABCD). Then the predicted pavement distresses were estimated based on the pavement M-E design method. The experimental results revealed that (1) DCT and dynamic modulus tests are sensitive to moisture conditioning. The dynamic modulus decreased by 13% to 43% at various temperatures and frequencies, and the low-temperature cracking energy decreased by 20%. (2) RAP asphalt incorporated with asphalt emulsion decreased the high-temperature rutting resistance but improved the low-temperature anti-cracking and the fatigue life. The M-E design results showed that the RAP incorporated with asphalt emulsion reduced the international roughness index (IRI) and AC bottom-up fatigue predictions, while increasing the total rutting and AC rutting predictions. The moisture damage in the CIR pavement layer also did not significantly affect the predicted distress with low traffic volume. In summary, the implementation of CIR technology in the project improved low-temperature cracking and fatigue performance in the asphalt pavement. Meanwhile, the moisture damage of the CIR asphalt mixture accelerated high-temperature rutting and low-temperature cracking, but it may be acceptable when used for low-volume roads.

Keywords: cold in-place recycling (CIR); disk-shaped compact tension (DCT); moisture-induced stress tester (MIST); dynamic modulus; dynamic shear rheometer (DSR); asphalt binder cracking device (ABCD)

Citation: Jin, D.; Ge, D.; Chen, S.; Che, T.; Liu, H.; Malburg, L.; You, Z. Cold In-Place Recycling Asphalt Mixtures: Laboratory Performance and Preliminary M-E Design Analysis. *Materials* **2021**, *14*, 2036. https://doi.org/10.3390/ma14082036

Academic Editor: Jeong Gook Jang

Received: 23 March 2021
Accepted: 16 April 2021
Published: 18 April 2021

1. Introduction

Asphalt recycling has increased dramatically in the past several years [1,2]. The rehabilitation of pavement also has lots of techniques, and one eco-friendly methodology is to use a CIR asphalt mixture [3,4]. CIR is a procedure whereby the overlayer of asphalt pavement is milled and mixed with stabilizers like emulsified asphalt [5]. Early applications of CIR focused on low-volume pavements [6]. Cost-effectiveness was the primary factor for using CIR. Bradbury et al. [7] stated it was about one-third cheaper than the original hot asphalt pavement road in Ontario. Scholz et al. [8] paved a 50 mm thick asphalt concrete surface with CIR and saved about 40% of the total cost. Many researchers have focused on the compaction and mixing procedure of CIR mixtures especially relating to optimum moisture content. Anderson et al. [9] discovered that in order to obtain good road performance, especially regarding strength and anti-cracking properties, optimum moisture content was the key factor that affected the on-site compaction degree and even mixing. In general, the Marshall method is always chosen to obtain the optimum moisture content [10], but the relationship between moisture and density could also help obtain it [11]. Curing time was also a vital factor affecting the CIR asphalt mixture, which is typically cured in a 60 °C oven [12]. Woods et al. [13] implied that the criteria content of the moisture is 1.5% based on the moisture sensors at different depths in on-site CIR overlays.

Quantitative researchers have concentrated on the performance characterization test of the CIR asphalt mixture. Wu et al. [14] conducted asphalt concrete APA fatigue tests by using a more traditional theoretical analysis. Marshall stability has traditionally been the main criterion used to select design binder content for bituminous-stabilized CIR mixtures. Yan et al. [15] stated that APA test was conducted by many researchers to estimate anti-rutting performance, and he suggested that Marshall stability should be larger than at 6 kN at 40 °C. Du and Cross [16] used three types of CIR mixture (1.5% asphalt emulsion, 1.5% asphalt emulsion with hydrated lime, 1.5% asphalt emulsion with quick lime) to test rutting performance using APA tests, and the rut depth ranged from 0.37 to 0.67 cm. Indirect Tensile Strength has been used by many people to study CIR asphalt mixture performance. Yan et al. [15] suggested that the peak load at 15 °C should be larger than 0.5 MPa. Kavussi and Modarres [17] conducted an IDT test of CIR mixtures and figured out that the coefficient of variation is no larger than 0.1 in most cases; Thomas et al. [18] tested two types of CIR mixture by an IDT test and by creep compliance and strength to characterize low temperature cracking performance. Based on the above literature review, the performance was one of the vital factors in CIR asphalt mixtures. In this study, all loose materials used were obtained from Dickinson County Road 581, located on the Upper Peninsula of Michigan, where the lowest temperature was below 0 °C for almost half a year, indicating that low-temperature performance is important for road service. Meanwhile, CIR performance at high temperatures is important in summertime [19,20]. A CIR asphalt mixture commonly has higher air voids, which may cause moisture damage. Therefore, it is necessary to investigate the CIR asphalt mixture moisture resistance.

The main purpose of this research is to characterize the effect of moisture on the high-temperature and low-temperature performance of CIR asphalt mixture and to predict pavement distress and deterioration based on a mechanistic–empirical pavement design. The low-temperature cracking performance of the CIR asphalt mixture was estimated by the DCT test. The dynamic modulus was conducted to evaluate the stress and strain response of the CIR asphalt mixture at various temperatures and frequencies. In addition, the Moisture-Induced Stress Tester (MIST) was used to simulate the pore pressure generated in a wet pavement under moving traffic loading. The standard solvent extraction test was used to obtain the reclaimed asphalt pavement (RAP) and CIR asphalt. The high-temperature and fatigue performances were characterized by a dynamic shear rheometer (DSR) equipment, while the low-temperature properties were investigated by the asphalt binder cracking device (ABCD). Then the predicted pavement distresses were estimated based on the M-E inputs.

2. Materials and Methods

2.1. Materials and Mixture Design

All the loose mixtures are obtained from Dickinson County Road 581 in Michigan, USA. The mix design was conducted by the Superpave mix design. The RAP materials obtained from the milled surface asphalt pavement layer were utilized with 2.5% emulsion content (64.8% residue content) and an additional 2% water content based on the mix design results shown in Table 1. The average bulk specific gravity is 2.129. The gradation used in this study is shown in Figure 1. The Superpave gyratory compactor (SGC, manufacturer: PINE test equipment, Grove City, PA, USA) gyration time is 30 for CIR mixture in nearly all US states [21], so the gyration time was 30 and the mass was cured at 60 °C for 48 h until it was constant. The average air voids were 13%. The SGC compaction procedure was followed the AASHTO PP 60-09 specification [22].

Table 1. Mix design for CIR asphalt mixture.

Test Parameters		Test Results		Specification Requirement
Emulsion Content	2.0%	2.5%	3.0%	-
Optimum Water for Mixing	2.0%	2.0%	2.0%	-
Bulk Specific Gravity	2.110	2.125	2.132	-
Maximum Theoretical Specific Gravity	2.443	2.440	2.436	-
Air Voids	13.6%	12.9%	12.5%	-
Marshall Stability (kN)	7.80	7.16	7.28	5.56 kN min
Conditioned Marshal Stability (kN)	5.74	5.47	5.12	-
Retained Stability	74%	76%	70%	70% min
Raveling Test	1.61%	0.30%	0.19%	2% Max

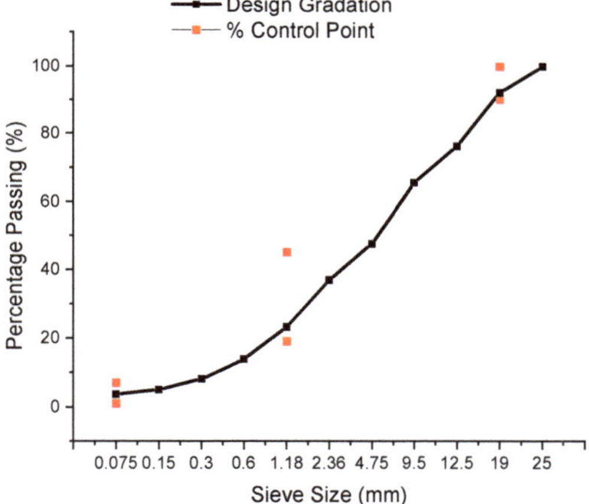

Figure 1. The gradation used in this study.

2.2. Experimental Program

This section concerns the laboratory mixture experiments, which included the moisture conditioning test (MIST), the dynamic modulus test, and the low-temperature DCT test. The laboratory binder experiments, including the standard solvent extraction test, were used to obtain the reclaimed asphalt pavement (RAP) and CIR binder. The dynamic shear rheometer (DSR, manufacturer: Anton Paar, North Ryde, Australia), the asphalt binder cracking device (ABCD) test, and then the prediction pavement distress were estimated by the M-E design. The technical flowchart is shown in Figure 2.

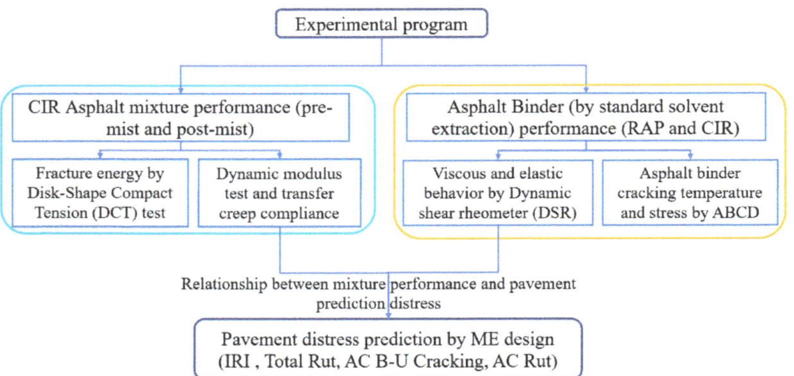

Figure 2. The technical flowchart of this study.

2.3. *Asphalt Mixture Test Program*

2.3.1. Moisture Conditioning via MIST

A CIR asphalt mixture generally shows a high air void, and it causes the asphalt mixture to be more sensitive to moisture damage [23]. Therefore, there is a need to investigate the moisture conditioning performance of the CIR asphalt mixture. The Moisture Induced Stress Tester (MIST, manufacturer: InstroTek, Grand Rapids, MI, USA) is designed to be a quick and logical method for testing the moisture damage susceptibility of asphalt mixture, which is specified in ASTM D7870 [24]. The samples for the dynamic modulus and disk-shaped compact tension tests were conditioned by MIST. It should be noted that the moisture conditioning was set for 500 cycles at 40 °C for the CIR asphalt mixture. Specifically, the CIR asphalt mixture without moisture conditioning was labeled as "pre-mist", while the CIR asphalt mixture under moisture conditioning was labeled as "post-mist".

2.3.2. Low-Temperature Cracking Test

Dickinson County Road 581 is located on the Upper Peninsula of Michigan where the lowest temperature was less than 0 °C for almost half a year, which means that low-temperature performance is vital for road service. The DCT test is used to evaluate the low-temperature cracking resistance of the asphalt mixture and is based on ASTM D7313 [25], "Standard Test Method for Determining Fracture Energy of Asphalt Mixtures Using the Disk-Shaped Compact Tension Geometry". The test is generally used to obtain the fracture energy, peak load, and maximum crack mouth-opening displacement (CMOD). The DCT test temperature for the CIR asphalt mixture was −12 °C based on the PG level. The test was run with a constant CMOD control mode, and the CMOD rate was 1 mm/min. The CMOD after the peak load reflected the propagation of cracking inside the sample during the DCT test. The specimen's geometry for a DCT test is a cylinder with a diameter of 150 mm and height of 45 mm. The details of the DCT test are shown in Figure 3. It should be mentioned that at least three parallel samples were used for each test.

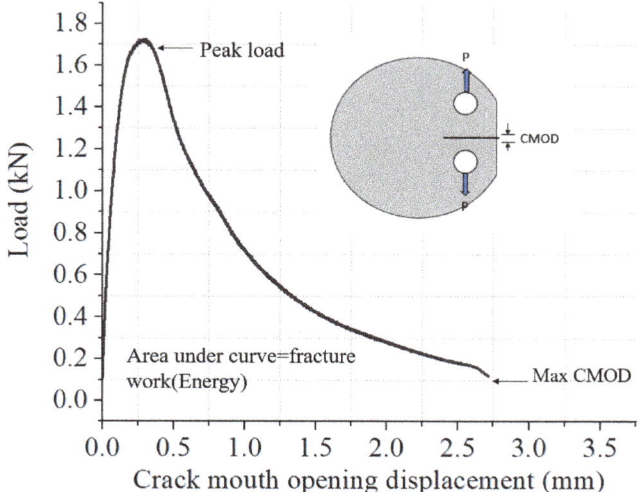

Figure 3. Demonstration of the DCT test procedure.

2.3.3. Dynamic Modulus Test

The dynamic modulus could be used directly to reflect the stress and strain response by specific load [26,27]. The dynamic modulus (E*) test is conducted at temperatures of −10 °C, 10 °C, 21 °C, and 37 °C, and at loading frequencies of 0.1 Hz, 0.5 Hz, 1 Hz, 5 Hz, 10 Hz, and 25 Hz at each temperature. The UTM-100 equipment was used for this test, and it is specified in AASHTO T342. The temperature chamber of the UTM-100 machine can control the temperature from -15 to 60 °C. The diameter and height of the test specimens were 100 mm and 150 mm, respectively. The dynamic modulus (|E*|) reflected the elastic performance of the asphalt mixture, and the phase angle (δ) expressed the gap between the stress and the strain. The rutting parameter (|E*|/sin δ) and fatigue parameter (|E*|·sin δ) could reflect the rutting and fatigue properties of the two types of asphalt mixtures. The master curve of the dynamic modulus was plotted for both the dynamic modulus of asphalt mixture and asphalt binder. It could be used to reveal the high-temperature performance and low-temperature performance. The details of the dynamic modulus master curve are shown in Figure 4. The specimen's geometry for the dynamic modulus test is a cylinder with a diameter of 100 mm and height of 150 mm. It should be mentioned that at least three parallel samples were used for each test.

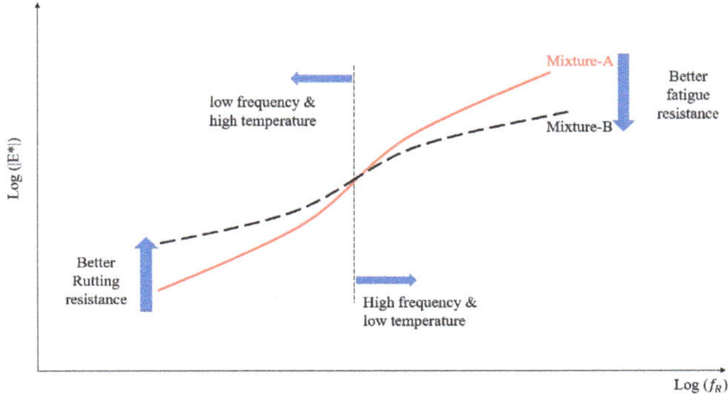

Figure 4. Demonstration of the dynamic modulus master curve.

2.3.4. Creep Compliance

Creep compliance at low temperature could be used to reflect the thermal cracking at the low temperature of the asphalt mixture. It is based on AASHTO T322 (2011) [28] "Determining the Creep Compliance and Strength of Hot Mix Asphalt (HMA) Using the Indirect Tensile Test Device". Creep compliance would be estimated from the |E*| master curve using the procedure developed in 1999 by S.W. Park and R.A. Shapery [29]. The master curve of dynamic modulus |E*| could be used to characterize the linear viscoelastic behavior of asphalt mixtures and conversion between the time and frequency domains.

2.4. *Asphalt Binder Test Program*

2.4.1. Standard Solvent Extraction

The properties of CIR asphalt also show significant variants compared with RAP asphalt from the milled pavement. Therefore, there is a need to characterize the extraction of asphalt binder properties. The automatic asphalt analyzer (manufacturer: CONTROLS Inc., Chicago, IL, USA) was used for the washing of the CIR loose mixture (around 3.5 kg) with trichloroethylene (TCE) solvent, ultrasonic motion, simultaneous heating, and rotation of the drum lined with screening mesh, which is specified in ASTM D8159-19 [30]. The solvent mix incorporated with asphalt binder was moved to the rotary evaporator device to remove the solvents from the asphalt binder by evaporation extracts. Asphalt binder washed out from the mixture with the Automatic Asphalt Analyzer could be separated from the asphalt. In short, the extracted asphalt binder from the RAP loose mixture could be written as RAP, and the extracted asphalt from the CIR loose mixture could be written as CIR.

2.4.2. Dynamic Shear Rheometer (DSR)

The viscous and elastic property of asphalt at medium and high temperatures was characterized by dynamic shear rheometer equipment, which characterized high-temperature and fatigue performance. The high-temperature rheological properties were be conducted at a temperature of 34, 40, 46, 52, 58, 64, 70, 76, and 82 °C at loading frequencies of 0.628, 6.28, 9.99, 18.8, 31.4, and 62.8 rad/s for each temperature. The intermediate temperature was conducted at a temperature of 13, 16, 19, 22, and 25 °C at loading frequencies of 0.628, 6.28, 9.99, 18.8, 31.4, 62.8 rad/s for each temperature. The test procedure was based on the AASHTO 315 specification. At least three parallel samples were used for each test.

2.4.3. Asphalt Binder Cracking Device (ABCD)

The ABCD is a new empirical test for evaluating the low-temperature cracking potential of asphalt binder. The cracking temperature and cracking stress can be obtained from the test, which is specified in AASHTO TP 92(2014) [31]. Strain and temperature are recorded until the cracking starts. The ABCD system consists of an air-cooled environmental chamber that can automatically cool asphalt specimens at a constant rate (20 °C /h) from 25 °C to −60 °C. The ABCD invar rings are equipped with an electrical-strain reading gauge, temperature control sensor, and silicone rubber specimen molds. Typical ABCD setup and test results are shown in Figure 5. At least three parallel samples were used for each test.

(**a**) Asphalt Binder Cracking Device (ABCD) test apparatus

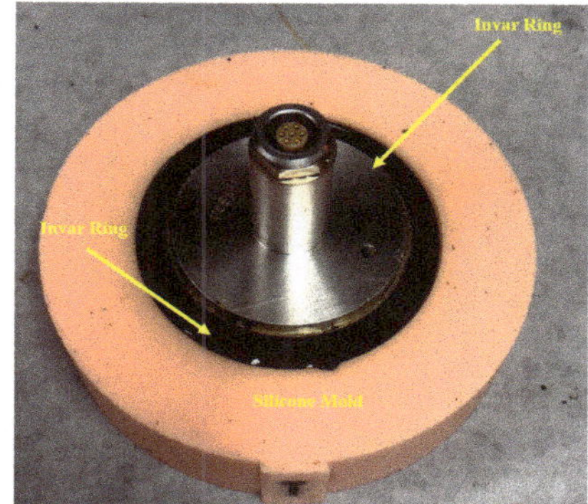

(**b**) Sample in the mold

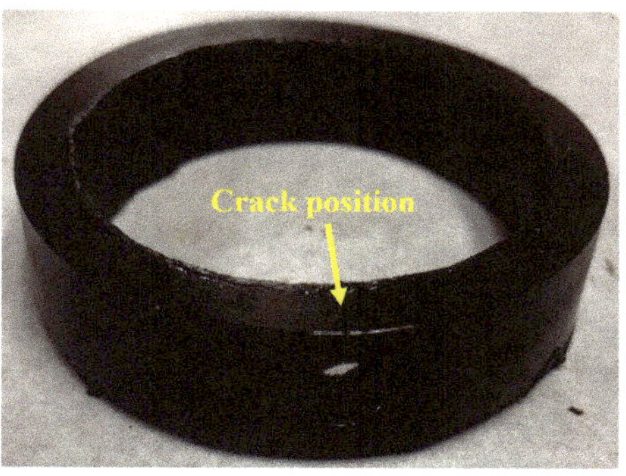

(**c**) Sample with crack after the test

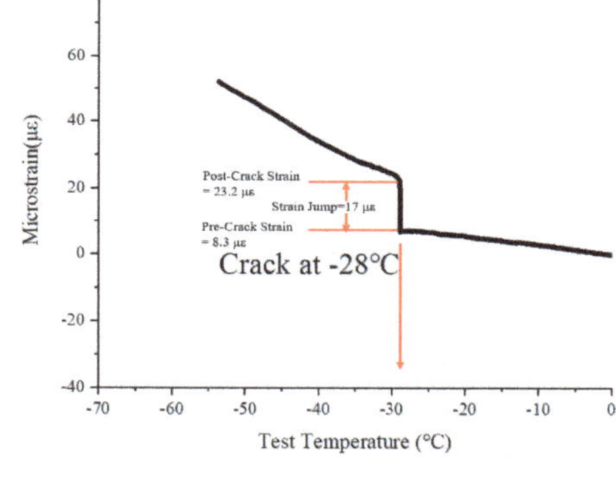

(**d**) Typical ABCD results

Figure 5. The Asphalt Binder Cracking Device (ABCD) test apparatus and procedure: (**a**) Asphalt Binder Cracking Device (ABCD) test apparatus; (**b**) Sample in the mold; (**c**) Sample with crack after the test; (**d**) Typical ABCD results.

2.5. Pavement Distress Prediction by Pavemeent M-E Design Analysis

The M-E Design [29] was used to evaluate the difference in the asphalt binder performance between the RAP and CIR asphalt, especially for the permanent deformation and fatigue cracking. The dynamic modulus pre-mist and post-mist conditions were also used to predict the effect of moisture damage in the CIR asphalt layer on pavement distress. The test parameters were the following: vehicle growth rate was 0.5%, the design life was 20 years, design speed was 90 km/h, the annual average daily traffic (AADT) in 2019 was 778, design function of the traffic volume was compound, and the number of lanes was 2. The traffic value, including single axle, tandem axle, tridem axle, quad axle distribution, was set according to the Buch research report [30]. The climate was the near place according to the M-E used guide recommendation. The specific calibration factor was according to the Michigan DOT User Guide for Mechanistic–empirical Pavement Design 2020, and the pavement structure and thickness used in this road is shown in Figure 6.

Structure	Materials type and thickness
Top course	3.8 cm HMA 5E1, PG 58-34
Levelling course	12.7 cm CIR
Base course	10 cm gravel base
Sand subbase	90 cm gravel subbase
subgrade	Sandy clay subgrade

Figure 6. Demonstration of the structure and materials used in this study.

3. Results and Discussions

3.1. Dynamic Modulus Test Results

Figure 7 shows the dynamic modulus master curve at a reference temperature of 21 °C for the pre-mist and post-mist CIR asphalt mixture. It was determined that the stiffness of all mixtures would decrease after moisture conditioning. As we all know, a lower reduced frequency represents slow traffic speeds or high pavement temperature, and a higher reduced frequency indicates high traffic speeds or low pavement temperature. Both the lower and high reduced frequencies showed that a reduction in stiffness in post-mist CIR asphalt mixture compared with the pre-mist CIR asphalt mixture.

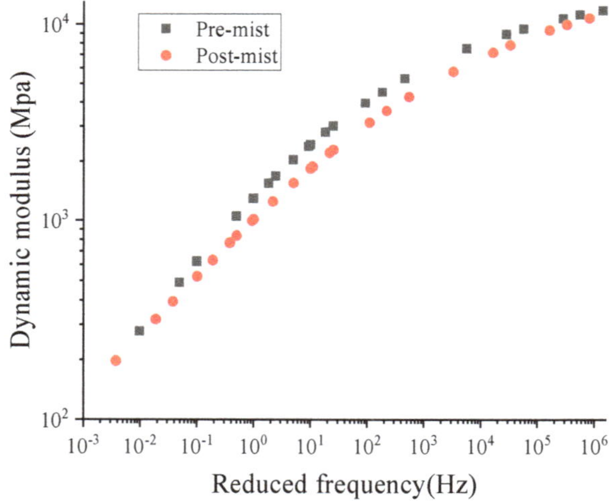

Figure 7. Master curve of dynamic modulus at 21 °C of pre-mist and post-mist CIR asphalt mixture.

The phase-angle master curve at the reference temperature of 21 °C between the pre-mist and post-mist CIR is shown in Figure 8. It was apparent that the phase angle increased after MIST conditioning. At lower frequencies, the phase angle increased as little as 0.2 degrees, while it could be 1 degree at high frequencies. The reason is that after MIST conditioning, the asphalt mixture displayed more viscous action under loading. The results from the dynamic modulus test indicated that after the moisture damage, the asphalt mixture experienced a decrease in modulus and an improvement in phase angle, which meant that the asphalt mixture was softer after the moisture damage.

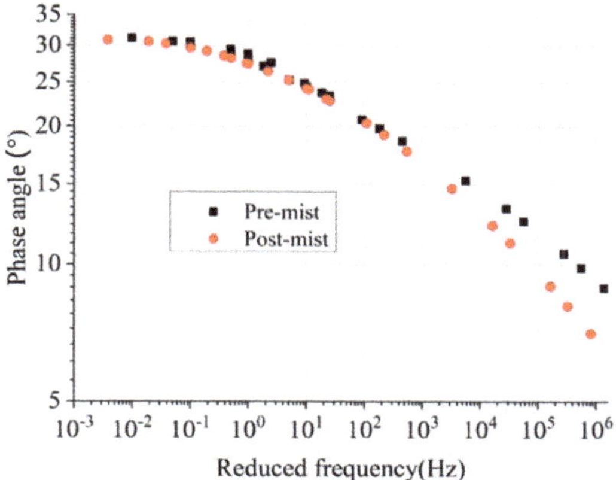

Figure 8. Master curve of phase angle at 21 °C of pre-mist and post-mist CIR asphalt mixture.

The rutting index of the pre-mist and post-mist CIR mixtures at the loading frequency of 10Hz is displayed in Figure 9. The rutting parameter (|E*|/sinδ) quantifies the rutting property of the asphalt mixture. The MIST process reduced the rutting parameter of the mixture. For example, the rutting parameter of the asphalt mixture decreased by 13% at −10 °C and by 41% at 37 °C. This reduction effect was significant at high temperatures, which was very important because at high temperatures rutting was easier to accumulate on the pavement. The stiffness of the asphalt mixture decreased after fiber modification, thus weakening the rutting resistance of the asphalt mixture. It should be mentioned that the high value of error bar at −10 °C may be caused by moisture frozen in the mixture.

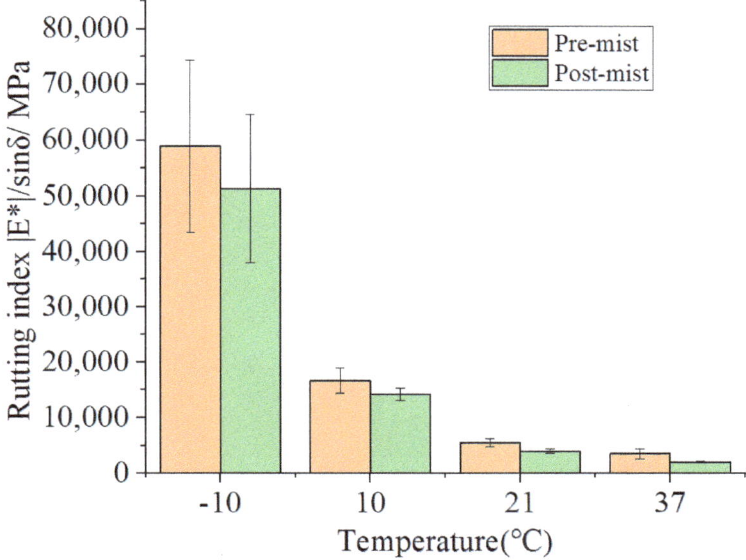

Figure 9. Rutting index at 10 Hz of pre-mist and post-mist CIR asphalt mixture.

The moisture susceptibility of the pre-mist and post-mist CIR mixture was evaluated by the dynamic modulus ratio. For CIR, this was defined by the ratio of the dynamic modulus of pre-mist CIR mixture to that of the post-mist CIR mixture. The results of the various mixtures at different temperatures and frequencies are presented in Figure 10. The lower dynamic modulus ratio values were observed with a decrease in loading frequency and increase in temperature. The reduction in stiffness was more pronounced at lower frequencies, and after conditioning the material experienced a decrease in stiffness at all temperatures. These results indicated that asphalt mixtures were more affected by humidity regulation at lower traffic speeds and higher temperatures.

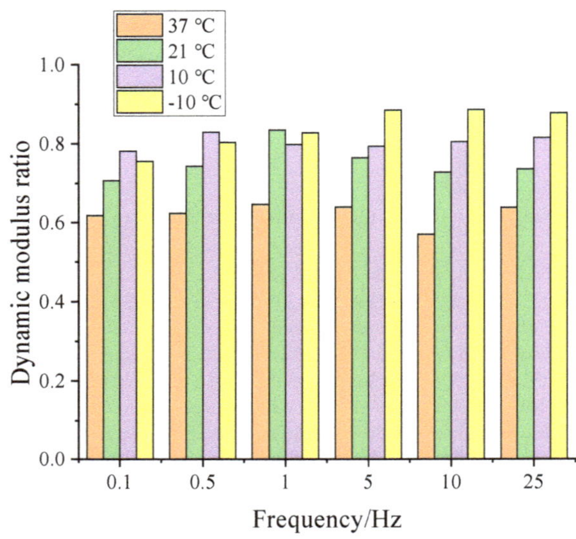

Figure 10. Dynamic modulus ratio of pre-mist and post-mist CIR mixture at 10Hz.

3.2. Creep Compliance

Creep compliance at low temperature was used to determine the cracking resistance for the asphalt mixture [32,33]. Creep compliance revealed low-temperature behavior and predicted the thermally induced cracking in asphalt pavement at low temperature in the U.S. [34]. The results for the pre-mist and post-mist CIR mixtures, are shown in Figure 11. The temperature increases caused the creep compliance to increase as did the duration time increase. It is clear that the post-mist CIR showed higher creep compliance under various temperatures and times. For example, it increased by 55% from 4.396×10^{-6} to 6.85×10^{-6}. This meant that the stiffness of the post-mist CIR mixture decreased and the low temperature cracking resistance decreased compared with the pre-mist CIR mixture.

3.3. DCT Test Results

Low-temperature cracking resistance was evaluated by the DCT test, which investigated the implications of the CIR mixture on low-temperature cracking. The fracture energy directly expressed the cracking performance of the asphalt mixture, and the mixture with the higher fracture energy had a better low-temperature cracking property. The fracture energy of the pre-mist and post-mist CIR mixture is displayed in Figure 12b. Pre-mist CIR mixture showed better low temperature cracking resistance compared with the post-mist CIR mixture. The recommended low-temperature crack resistance threshold determined by MnDOT is 400 J/m^2 [33]. The peak load of the DCT test defines the force needed to initiate the cracking in the asphalt mixture at low temperatures. The peak load represents the cracking propagation possibility of the asphalt mixture; an asphalt mixture with a higher peak load was harder to crack. The peak load of the two asphalt mixtures under different temperatures is shown in Figure 12c. The maximum CMOD represents the deformation that occurs in the asphalt mixture during the DCT test. The asphalt mixture with a higher maximum CMOD had better deformation ability under low temperatures. The maximum CMOD of the pre-mist and post-mist CIR mixture is displayed in Figure 12d. The pre-mist CIR mixture showed better low temperature cracking resistance compared with the post-mist CIR mixture. The results of the DCT test were consistent with the results obtained from creep compliance.

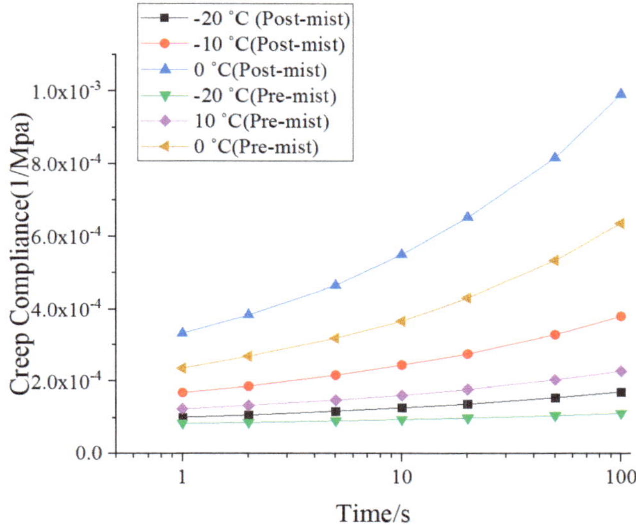

Figure 11. Creep compliance results of pre-mist and post-mist CIR asphalt mixture.

3.4. DSR results

The complex modulus of RAP asphalt and CIR asphalt under six frequencies at 58 °C were displayed in Figure 13a. RAP and CIR asphalt were extracted from the loose materials. Asphalt extracted from the loose materials did not need any aging to simulate the short-term and long-term performance. It was determined that the presence of emulsion asphalt decreased the complex modulus because it made the asphalt softer compared with RAP. Simultaneously, the complex modulus of CIR and RAP asphalt increased with as the frequency increased. For example, the complex modulus of the RAP asphalt binder was 20.95 kPa at 1.00 Hz, and it increased 44.97% and 464.21% to 1.59 Hz and 10.00 Hz, respectively. The reason for this is that asphalt shows higher elastic components and properties with the increased frequency. In contrast, a lower frequency, which represents a longer loading time, generally causes the asphalt to display more viscous properties. This evidence could be used to reveal the reason that rutting distress appears more easily on longitudinal-slope asphalt pavement over long distance. An automobile would generally be driven at low speed over a long time, thus producing low-frequency-related properties.

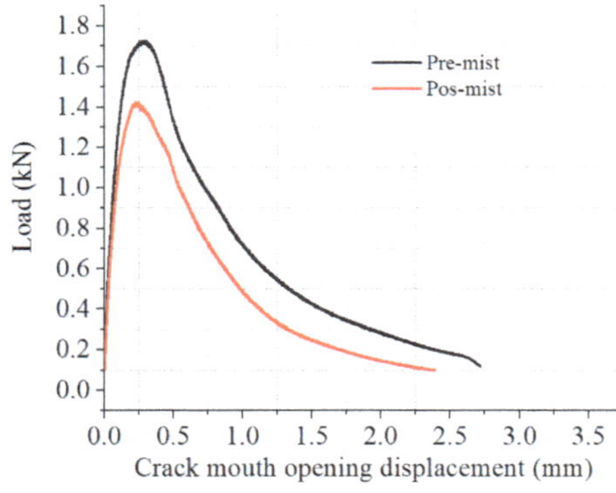

(**a**) CMOD vs. Load between pre-mist and post-mist

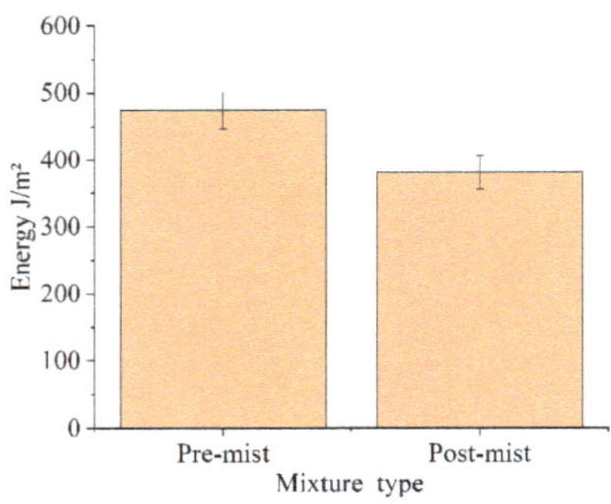

(**b**) Energy difference between pre-mist and post-mist

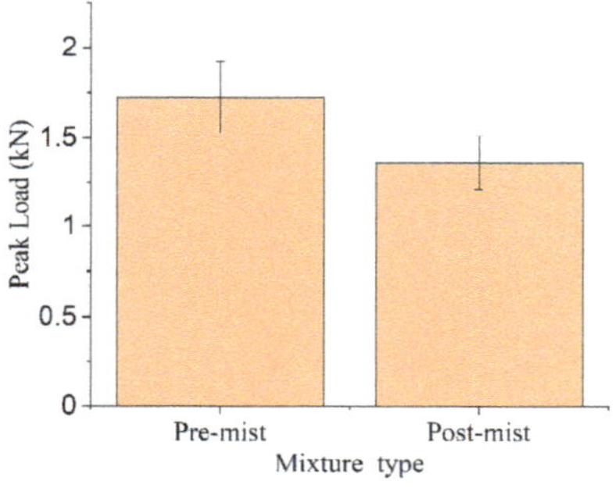

(**c**) peak load difference between pre-mist and post-mist

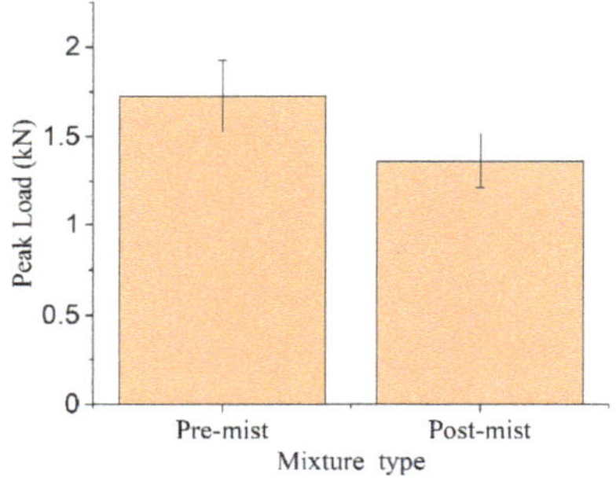

(**d**) Maximum CMOD difference between pre-mist and post-mist

Figure 12. Low-temperature cracking resistance results of pre-mist and post-mist CIR asphalt mixture: (**a**) CMOD vs. Load between pre-mist and post-mist; (**b**) Energy difference between pre-mist and post-mist; (**c**) peak-load difference between pre-mist and post-mist; (**d**) Maximum CMOD difference between pre-mist and post-mist.

Figure 13b illustrates the master curves of phase angles at 58 °C. As the phase angle decreased, the viscous component of asphalt decreased, and the elastic component increased. The addition of emulsified asphalt increased the phase angle of the asphalt. The master curves of the rutting factor for RAP asphalt and CIR asphalt at 58 °C are shown in Figure 13c. It is worth mentioning that low temperature corresponded to the highly reduced frequency. It was revealed that the rutting index increased with increasingly reduced frequency. Meanwhile, the asphalt had more elastic components and a higher resistance to deformation under low-temperature conditions. This was consistent with the fact that a lower temperature indicated a higher rutting coefficient and better resistance to deformation. For example, adding emulsion to the RAP asphalt binder caused the rutting factor value to decrease throughout the reduced frequency, which indicated that the stiffness of the asphalt binder had been reduced compared to that of the RAP asphalt. The master curves of fatigue factor for the two types of asphalt at the reference temperature of 19 °C are presented in Figure 13d. In order to resist fatigue cracking, an asphalt binder should be elastic (able to dissipate energy by rebounding, not cracking) but not too stiff (excessively stiff substances will crack rather than deform and rebound). Therefore, the complex shear modulus viscous portion |G*|×sinδ should be minimal. It is clear that the addition of emulsified asphalt in the RAP asphalt decreased the |G*|×sinδ value of the asphalt, which may improve the resistance of fatigue cracking properties.

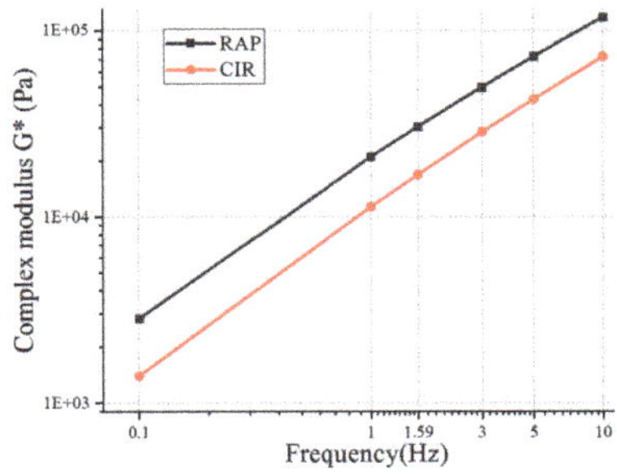

(**a**) The complex modulus of asphalt binder at 58 °C

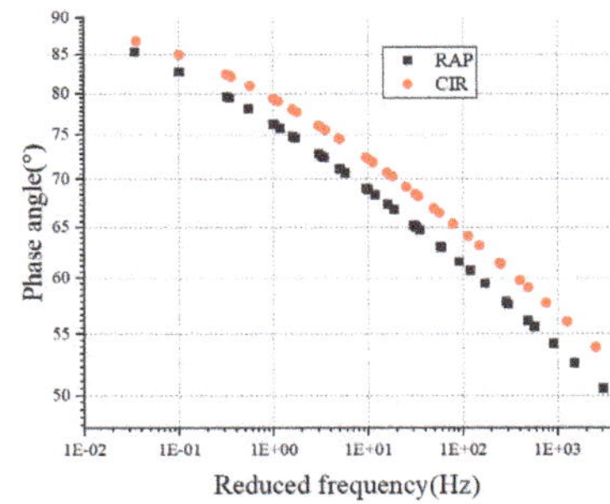

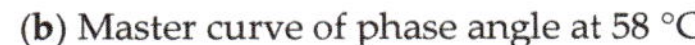

(**b**) Master curve of phase angle at 58 °C

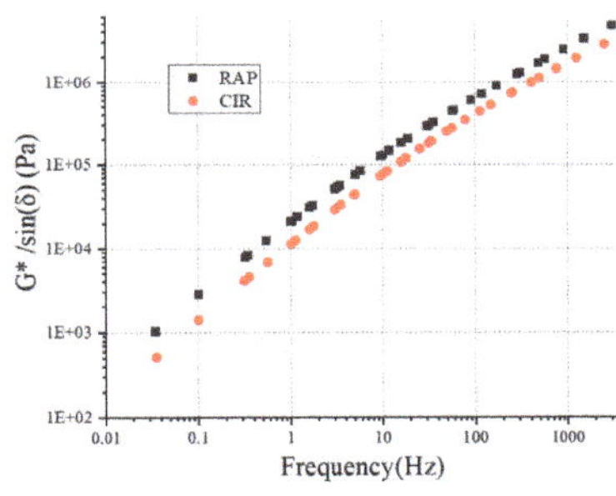

(**c**) Master curve of rutting factor at 58 °C

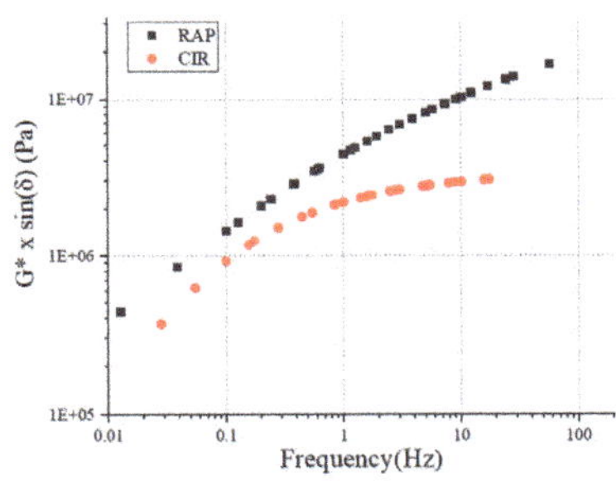

(**d**) Master curve of fatigue index at 19 °C

Figure 13. DSR test results for RAP and CIR asphalt: (**a**) The complex modulus of asphalt binder at 58 °C; (**b**) master curve of phase angle at 58 °C; (**c**) master curve of rutting factor at 58 °C; (**d**) master curve of fatigue index at 19 °C.

3.5. ABCD Results

In this section, the effect of added emulsion on the low-temperature performance of the asphalt binder was evaluated. According to AASHTO TP 92-11, the cracking temperature was determined based on the strain jump when the fracture stress was recorded. The test results are shown in Figure 14. It should be noted that the addition of emulsion improved the low-temperature cracking resistance of asphalt. It not only increased the fracture stress but also decreased the cracking temperature. For example, the cracking temperature of CIR with the added emulsion decreased from −26.45 °C to −29.2 °C. The main reason was that the asphalt emulsion made the CIR asphalt binder softer than the RAP asphalt binder. Since CIR asphalt is softer than RAP asphalt, it showed a lower cracking temperature, while RAP asphalt had higher fracture stress.

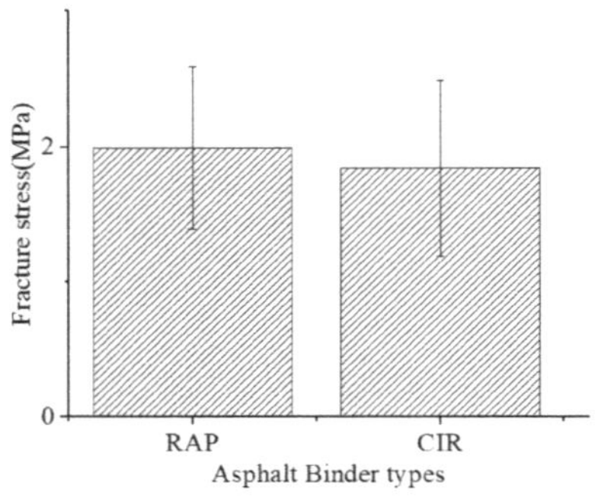

(**a**) Cracking stress between RAP and CIR asphalt

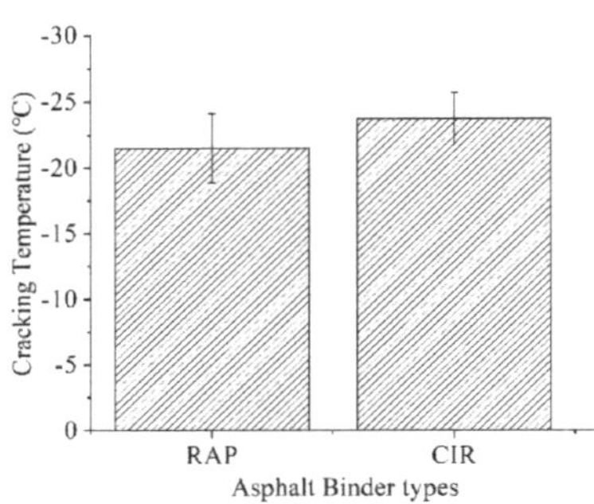

(**b**) Cracking temperature between RAP and CIR asphalt

Figure 14. Asphalt cracking stress and temperature between RAP and CIR: (**a**) Cracking stress between RAP and CIR asphalt; (**b**) Cracking temperature between RAP and CIR asphalt.

3.6. Pavement Distress Prediction by Pavement M-E Design

A pavement M-E design was conducted to predict pavement distress [35,36]. The climate data of the place near Dickinson County was used in this study according to the M-E used-guide recommendation. The dynamic modulus from the asphalt mixture and complex shear modulus from the asphalt binder was the input material property that could be used for distress predictions. As described in the method section, the pavement M-E design was used as the pavement distress predictor in this paper because it uses dynamic modulus values and complex shear modulus values to obtain the master curve for asphalt concrete layers. The dynamic modulus values of pre-mist and post-mist CIR asphalt mixture were measured, and the complex shear modulus (extracted from the RAP and CIR) was measured by DSR equipment used in the M-E pavement analysis. The pavement distress prediction results were analyzed based on the dynamic modulus of CIR (pre-mist and post-mist). The complex shear modulus of CIR asphalt was used to do the M-E analysis.

The pavement distress prediction results of the pre-mist and post-mist CIR mixture dynamic modulus are shown in Figure 14. It should be mentioned that the dynamic modulus of pre-mist and post-mist CIR mixture reflected the effect of moisture damage in the CIR pavement layer on the pavement distress predictions, The M-E input parameters of other pavement layers (top, base, and subbase) were based on the reference value from the report [37]. In reality, the moisture damage not only caused deterioration to the leveling layer, but it also affected the top, base, subbase, and subgrade layers. The distress prediction between pre-mist and post-mist CIR mixture in this study only revealed the maximum moisture damage of the CIR layer. Rut depths of the post-mist CIR pavement layer were 0.34 inches higher compared with the pre-mist mixture, which was caused by a reduction in stiffness. It shows an analogous trend with the rutting-factor results based on asphalt mixture dynamic modulus results. Bottom-up fatigue cracking results of the post-mist CIR pavement layer increased by 6% compared with the pre-mist CIR pavement layer, which implied a similar tendency with the creep compliance and disk-shaped compact tension test results. Furthermore, the international roughness index of the post-mist CIR pavement layer increased by 10% compared with the pre-mist CIR pavement layer. The results of the average dynamic modulus at pre-mist and post-mist conditions are shown in Table 2. Overall, the moisture damage of the CIR asphalt mixture deteriorated high- and low- temperature performance but it was acceptable when used in a low-volume road.

Table 2. Average dynamic modulus value pre-mist and post-mist condition.

	ǀE*ǀ (MPa) Average Value Pre-Mist Condition					
F (Hz)	0.1	0.5	1	5	10	25
T (°C)						
−10	7098.3	8737.7	9418.5	10,529.2	11,181.8	12,379.7
10	2270.4	3125.8	3572.8	4873.0	5404.6	6018.4
21	623.6	956.2	1216.3	2037.7	2505.0	2979.3
37	320.2	475.5	664.4	1047.0	1341.7	1567.5
	ǀE*ǀ (MPa) Average Value Post-Mist Condition					
F (Hz)	0.1	0.5	1	5	10	25
T (°C)						
−10	5363.1	7020.6	7789.8	9318.7	9902.8	10,851.1
10	1773.2	2591.1	2851.7	3864.8	4347.0	4898.0
21	440.6	710.4	1015.3	1556.9	1821.7	2189.7
37	198.0	296.8	429.8	669.4	764.8	999.5

The pavement distress prediction results of RAP and CIR complex shear modulus are shown in Figure 15. It was found that CIR asphalt increased the rut depths of the total pavement and asphalt layer, which was caused by a reduction in stiffness. However, bottom-up fatigue cracking results decreased by 1.34%. This showed a similar trend compared with the rutting and fatigue factors from the DSR results. Moreover, the international roughness index (IRI) of CIR asphalt decreased when compared with the RAP asphalt. It should be mentioned that the results just considered the DSR data (RAP and CIR asphalt) on 12.7 cm CIR layers difference. The input value of other layers was based on the reference value from the report [37,38]. The DSR results of RAP and CIR used in this study are shown in Table 3. The pavement distress prediction results were studied by the complex shear modulus (RAP and CIR). The dynamic modulus of the pre-mist CIR asphalt mixture was used to do the M-E analysis. In fact, the dynamic modulus of the RAP asphalt mixture should be used for the M-E design, so the M-E results in Figure 15 did not completely reflect the predicted distress.

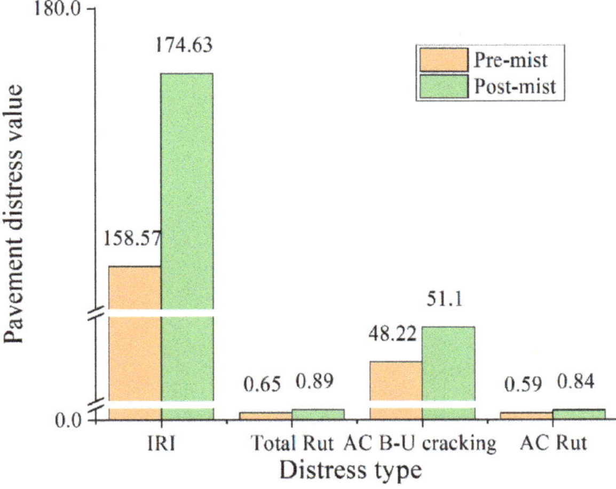

Figure 15. Pavement distress prediction by the dynamic modulus of CIR (pre-mist and post- mist). Note: IRI smoothness (in/mile or 0.016 m/km), Total Rut (in or 0.0254 m), AC B-U(bottom-up) cracking (% lane area), AC Rut (in or 0.0254 m), only consider the difference of 12.7 cm CIR layers.

Table 3. DSR results of RAP and CIR used in this study.

	Frequency @ 10 rad/s for RAP	
T (°C)	∣G*∣ (Pa)	Phase Angle°
13	30,628,000	35
25	70,84,800	46
34	3,888,633	53.6
46	621,153	62.67
58	114,973	70.4
	Frequency @ 10 rad/s for CIR	
T (°C)	∣G*∣ (Pa)	Phase Angle°
13	14,437,000	36
25	56,78,100	45
34	2,281,967	57.3
46	354,476	66.7
58	68,359	73.3

The pavement distress prediction results of the RAP and CIR complex shear modulus are shown in Figure 16. It could be founded that CIR asphalt increased the rut depths of the total pavement and asphalt layer, which was caused by a reduction in stiffness. However, bottom-up fatigue cracking results decreased by 1.34%. It showed a similar trend compared with the rutting and fatigue factors from the DSR results. Moreover, the international roughness index (IRI) of CIR asphalt decreased when compared with the RAP asphalt. It should be mentioned that the results just considered the DSR data (RAP and CIR asphalt) on 12.7 cm CIR layers difference. The input value of other layers was based on the reference value from the report [36]. The pavement distress prediction results were studied by complex shear modulus (RAP and CIR). The dynamic modulus of the pre-mist CIR asphalt mixture was used to do the M-E analysis. In fact, the dynamic modulus of the RAP asphalt mixture should be used for the M-E design, so the M-E results in Figure 16 could not completely reflect the prediction distress.

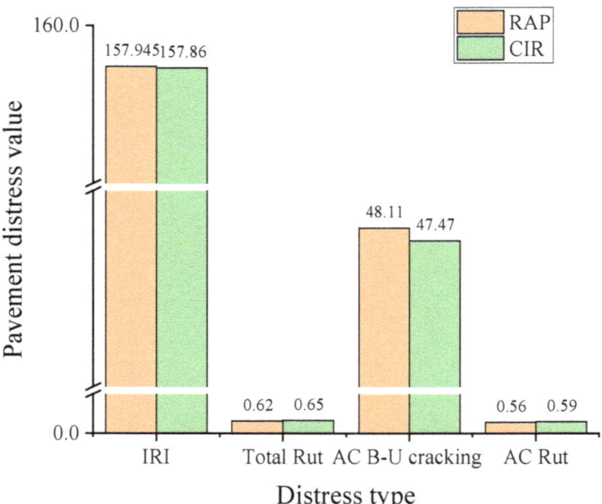

Figure 16. Pavement predicted distresses to complex shear modulus (RAP and CIR). Note: Note: IRI smoothness (in/mile or 0.016 m/km), Total Rut (in or 0.0254 m), AC B-U(bottom-up) cracking (% lane area), AC Rut (in or 0.0254 m), only consider the difference of 12.7 cm CIR layers.

4. Field Construction and Pavement Condition Assessment

A demonstration project was paved by The Dickinson County Road Commission in Dickinson, Michigan, in July 2020. The original roadway condition is shown in Figure 17a,b. It was poor, and the surface layer was milled before the new pavement was placed. The construction procedures in the field are shown in Figure 17c,d. The use of cold in-place recycling restored the old pavement to the desired profile, eliminated existing wheel ruts, restored the crown and cross slope, and eliminate potholes, irregularities, and rough areas. It also eliminated transverse, reflective, and longitudinal cracks. The asphalt emulsion was applied with the milling loose asphalt mixture, and it was placed in the original place as the level layer. After the CIR asphalt pavement layer was compacted, a new top layer was paved over it. The service condition of the road will be estimated in the future.

(**a**) Original road condition at northbound lane station 20+00

(**b**) View of the original roadway condition

(**c**) Construction procedure of CIR layer

(**d**) Compaction of the CIR layer

Figure 17. Original roadway condition and construction procedure: (**a**) Original road condition at northbound lane station 20+00; (**b**) View of the original existing roadway condition; (**c**) Construction procedure of CIR layer; (**d**) Compaction of the CIR layer.

5. Summary and Conclusions

This study concentrated on characterizing the CIR asphalt mixture moisture susceptibility performance and extraction of asphalt binder properties to predict the pavement distress and deterioration based on a mechanistic–empirical pavement design. Asphalt mixture moisture susceptibility performance (pre-mist and post-mist) was estimated by the dynamic modulus test, DCT test, and asphalt-binder performance (RAP and CIR asphalt obtained from the standard solvent extraction process). Higher-temperature and medium-temperature performances were evaluated by DSR equipment and low-temperature properties were estimated by the ABCD. Then the M-E pavement design was conducted to predict the pavement distress. The following conclusions are summarized from this study:

(1) The dynamic modulus test showed that the moisture condition has a significant effect on the CIR asphalt mixture, especially with high temperature and low frequency. The dynamic modulus of the post-mist asphalt mixture decreased 13–43% at various temperatures and frequencies, but pavement M-E results showed that moisture damage

of the CIR layer only increased AC Rut and total Rut by 0.34 inches at low traffic volume. It showed that CIR asphalt pavement may still be used for low-volume roads.

(2) Compliance creep, and disk-shaped compact tension test results showed that MIST conditioning decreased the low temperature cracking resistance of the asphalt mixture, and the low-temperature cracking energy of the CIR asphalt mixture decreased 20% between pre-mist and post-mist condition. The pavement M-E results showed that moisture damage to the CIR layer only increased bottom-up cracking by 2.9%. The lab test results suggest that the performance of CIR materials may have poor performance, but the M-E results showed that it is possible to apply CIR materials on low traffic-volume roads.

(3) The asphalt binder that included emulsion asphalt made the asphalt softer and increased fatigue property and low temperature cracking resistance while decreasing rutting resistance performance.

(4) Pavement M-E analysis based on DSR results of RAP and CIR binder showed that the asphalt binder with emulsion increased the rutting predictions while reducing the IRI, bottom-up fatigue-cracking predictions.

In summary, the implementation of CIR technology in the project improved the low-temperature cracking and fatigue performance in asphalt pavement. Meanwhile, the moisture damage to the CIR asphalt mixture accelerated the high-temperature rutting and low-temperature cracking performance, but it is possibly acceptable when used for low-volume roads.

Author Contributions: Conceptualization, D.J., D.G., Z.Y.; data curation, D.J., T.C, H.L; methodology, D.J., S.C. and H.L., T.C., L.M; resources, L.M.; supervision, Z.Y.; writing—original draft, D.J.; writing—review & editing, D.J., D.G., S.C., T.C., H.L., L.M and Z.Y. All authors have read and agreed to the published version of the manuscript.

Funding: This research received no external funding.

Institutional Review Board Statement: Not applicable.

Informed Consent Statement: Not applicable.

Data Availability Statement: The datasets generated during analyzed during the current study are available from the corresponding author on reasonable request.

Acknowledgments: The work is carried out in cooperation with the Dickinson County Road Commission of Michigan. All the test properties are obtained based on the received field loose mixtures.

Conflicts of Interest: The authors declare no conflict of interest.

References

1. Liu, P.; Xu, H.; Wang, D.; Wang, C.; Schulze, C.; Oeser, M. Comparison of mechanical responses of asphalt mixtures manufactured by different compaction methods. *Constr. Build. Mater.* **2018**, *162*, 765–780. [CrossRef]
2. Chen, S.; Ge, D.; Jin, D.; Zhou, X.; Liu, C.; Lv, S.; You, Z. Investigation of hot mixture asphalt with high ground tire rubber content. *J. Clean. Prod.* **2020**, *277*, 124037. [CrossRef]
3. Li, X.; Lv, X.; Zhou, Y.; You, Z.; Chen, Y.; Cui, Z.; Diab, A. Homogeneity evaluation of hot in-place recycling asphalt mixture using digital image processing technique. *J. Clean. Prod.* **2020**, *258*, 120524. [CrossRef]
4. Wang, C.; Wang, H.; Oeser, M.; Hasan, M.R.M. Investigation on the morphological and mineralogical properties of coarse aggregates under VSI crushing operation. *Int. J. Pavement Eng.* **2020**, 1–14. [CrossRef]
5. Ueckermann, A.; Wang, D.; Oeser, M.; Steinauer, B. Calculation of skid resistance from texture measurements. *J. Traffic Transp. Eng.* **2015**, *2*, 3–16. [CrossRef]
6. Epps, J.A. State-of-the-Art Cold Recycling. *Transp. Res. Rec.* **1980**, *I*, 68–100.
7. BRADBURY, A.; Kazmierowski, T.; Cheng, S.; Raymond, C. Cold in-place recycling in ontario: A case study. In Proceedings of the Thirty-Sixth Annual Conference of Canadian Technical Asphalt Association, Montreal, QC, Canada, April 1991.
8. Scholz, T.V.; Rogge, D.F.; Hicks, R.G.; Allen, D. Evaluation of Mix Properties of Cold In-Place Recycled Mixes. *Transp. Res. Rec.* **1991**, *1317*, 77–89.

9. Anderson, D.A.; Luhr, D.R.; Lahr, M. *Cold In-Place Recycling of Low-Volume Roads in Susquehanna County Volume I, Technical Report*; Vol. 1, no. FHWA-PA-84-020 Final Rpt.; FHWA: Washington, DC, USA, 1985.
10. Carter, A.; Feisthauer, B.; Lacroix, D.; Perraton, D. *Comparison of cold In-Place Recycling and Full-Depth Reclamation Materials*; No. 10-1325; Transportation Research Board: Washington, DC, USA, 2010.
11. Kim, Y.; Lee, H. "David" Development of Mix Design Procedure for Cold In-Place Recycling with Foamed Asphalt. *J. Mater. Civ. Eng.* **2006**, *18*, 116–124. [CrossRef]
12. Cross, S.A. *Determination of Ndesign for CIR Mixtures Using the Superpave Gyratory Compactor*; University of Kansas Center for Research: Kansas City, KS, USA, 2002.
13. Woods, A.; Kim, Y.; Lee, H. Determining Timing of Overlay on Cold In-Place Recycling Layer: Development of New Tool Based on Moisture Loss Index and in Situ Stiffness. *Transp. Res. Rec. J. Transp. Res. Board* **2012**, *2306*, 52–61. [CrossRef]
14. Wu, H.; Huang, B.; Shu, X. Characterizing Fatigue Behavior of Asphalt Mixtures Utilizing Loaded Wheel Tester. *J. Mater. Civ. Eng.* **2014**, *26*, 152–159. [CrossRef]
15. Yan, J.; Ni, F.; Tao, Z.; Jia, J. Development of asphalt emulsion cold in-place recycling specifications. In *Asphalt Material Characterization, Accelerated Testing, and Highway Management: Selected Papers from the 2009 GeoHunan International Conference*; ASCE: Reston, VA, USA, 2009; pp. 49–55.
16. Du, J.-C.; Cross, S.A. Cold in-place recycling pavement rutting prediction model using grey modeling method. *Constr. Build. Mater.* **2007**, *21*, 921–927. [CrossRef]
17. Kavussi, A.; Modarres, A. A model for resilient modulus determination of recycled mixes with bitumen emulsion and cement from ITS testing results. *Constr. Build. Mater.* **2010**, *24*, 2252–2259. [CrossRef]
18. Thomas, T.; Kadrmas, A.; Huffman, J. Cold In-Place Recycling on US-283 in Kansas. *Transp. Res. Rec. J. Transp. Res. Board* **2000**, *1723*, 53–56. [CrossRef]
19. Kim, Y.; Lee, H.D.; Heitzman, M. Dynamic Modulus and Repeated Load Tests of Cold In-Place Recycling Mixtures Using Foamed Asphalt. *J. Mater. Civ. Eng.* **2009**, *21*, 279–285. [CrossRef]
20. Graziani, A.; Mignini, C.; Bocci, E.; Bocci, M. Complex Modulus Testing and Rheological Modeling of Cold-Recycled Mixtures. *J. Test. Eval.* **2020**, *48*, 120–133. [CrossRef]
21. Cox, B.C.; Howard, I.L. *Cold In-Place Recycling Characterization Framework and Design Guidance for Single or Multiple Component binder Systems*; A Dissertation in Mississippi State University: Starkville, MS, USA, 2015.
22. AASHTO. *Preparation of Cylindrical Performance Test Specimens Using the Superpave Gyratory Compactor (SGC)*; AASHTO PP 60-09; AASHTO: Washington, DC, USA, 2009.
23. Che, T.; Pan, B.; Sha, D.; Zhang, Y.; You, Z. Relationship between Air Voids and Permeability: Effect on Water Scouring Resistance in HMA. *J. Mater. Civ. Eng.* **2021**, *33*, 04021022. [CrossRef]
24. ASTM Standard D7870/7870M. *Standard Practice for Moisture Conditioning Compacted Asphalt Mixture Specimens by Using Hydrostatic Pore Pressure*; ASTM: West Conshohocken, PA, USA, 2013.
25. ASTM D-7313. *Standard Test Method for Determining Fracture Energy of Asphalt-Aggregate Mixtures Using the Disk-Shaped Compact Tension Geometry*; American Society for Testing and Materials: West Conshohocken, PA, USA, 2007.
26. Gharaibeh, H.M. Managing the Cost of Power Transmission Projects: Lessons Learned. *J. Constr. Eng. Manag.* **2013**, *139*, 1063–1067. [CrossRef]
27. Si, C.; Zhou, X.; You, Z.; He, Y.; Chen, E.; Zhang, R. Micro-mechanical analysis of high modulus asphalt concrete pavement. *Constr. Build. Mater.* **2019**, *220*, 128–141. [CrossRef]
28. American Association of State Highway and Transportation Officials (AASHTO) Standard T 322. Determining the creep compliance and strength of hot-mix asphalt (HMA) using the indirect tensile test device. In *Standard Specifications for Transportation Materials and Methods of Sampling and Testing*, 25th ed.; AASHTO: Washington, DC, USA, 2011.
29. Park, S.; Schapery, R. Methods of interconversion between linear viscoelastic material functions. Part I—A numerical method based on Prony series. *Int. J. Solids Struct.* **1999**, *36*, 1653–1675. [CrossRef]
30. A.S.T.M. Standard. *ASTM D8159: Standard Test Method for Automated Extraction of Asphalt Binder from Asphalt Mixtures*; ASTM 2019 International: West Conshohocken, PA, USA, 2019.
31. American Association of State and Highway Transportation Officials. *AASHTO TP 92-11 Determining the Cracking Temperature of Asphalt Binder using the Asphalt Binder Cracking Device*; AASHTO: Washington, DC, USA, 2011.
32. Zofka, A.; Marasteanu, M.; Turos, M. Investigation of Asphalt Mixture Creep Compliance at Low Temperatures. *Road Mater. Pavement Des.* **2008**, *9*, 269–285. [CrossRef]
33. Zofka, A.; Marasteanu, M.O.; Turos, M. Determination of Asphalt Mixture Creep Compliance at Low Temperatures by Using Thin Beam Specimens. *Transp. Res. Rec. J. Transp. Res. Board* **2008**, *2057*, 134–139. [CrossRef]
34. Ge, D.; You, Z.; Chen, S.; Liu, C.; Gao, J.; Lv, S. The performance of asphalt binder with trichloroethylene: Improving the efficiency of using reclaimed asphalt pavement. *J. Clean. Prod.* **2019**, *232*, 205–212. [CrossRef]
35. Dave, E.V. *Moisture Susceptibility Testing for Hot Mix Asphalt Pavements in New England*; Final Report for New England Transportation Consortium, Project (2018) 15-3; NETC: Durham, NH, USA, 2018.
36. Olidid, C.; Hein, D. Guide for the mechanistic-empirical design of new and rehabilitated pavement structures. In Proceedings of the 2004 Annual Conference and Exhibition of the Transportation Association of Canada-Transportation Innovation-Accelerating the Pace, Quebec City, QC, Canada, September 2004.

37. You, Z.; Yang, X.; Hiller, J.; Watkins, D.; Dong, J. *Improvement of Michigan Climatic Files in Pavement ME Design*; Michigan Technological University: Houghton, MI, USA, 2015.
38. Buch, N.; Haider, S.W.; Brown, J.; Chatti, K. *Characterization of Truck Traffic in Michigan for the New Mechanistic Empirical Pavement Design Guide*; Michigan. Dept. of Transportation, Construction and Technology Division: Lansing, MI, USA, 2009.

 materials

Article

Reducing Compaction Temperature of Asphalt Mixtures by GNP Modification and Aggregate Packing Optimization

Tianhao Yan *, Mugurel Turos, Jia-Liang Le and Mihai Marasteanu *

Department of Civil, Environmental and Geo- Engineering, University of Minnesota, Twin Cities, Minneapolis, MN 55455, USA
* Correspondence: yan00004@umn.edu (T.Y.); maras002@umn.edu (M.M.)

Abstract: Compaction of hot mix asphalt (HMA) requires high temperatures in the range of 125 to 145 °C to ensure the fluidity of asphalt binder and, therefore, the workability of asphalt mixtures. The high temperatures are associated with high energy consumption, and higher NOx emissions, and can also accelerate the aging of asphalt binders. In previous research, the authors have developed two approaches for improving the compactability of asphalt mixtures: (1) addition of Graphite Nanoplatelets (GNPs), and (2) optimizing aggregate packing. This research explores the effects of these two approaches, and the combination of them, on reducing compaction temperatures while the production temperature is kept at the traditional levels. A reduction in compaction temperatures is desired for prolonging the paving window, extending the hauling distance, reducing the energy consumption for reheating, and for reducing the number of repairs and their negative environmental and safety effects, by improving the durability of the mixtures. A Superpave asphalt mixture was chosen as the control mixture. Three modified mixtures were designed, respectively, by (1) adding 6% GNP by the weight of binder, (2) optimizing aggregate packing, and (3) combining the two previous approaches. Gyratory compaction tests were performed on the four mixtures at two compaction temperatures: 135 °C (the compaction temperature of the control mixture) and 95 °C. A method was proposed based on the gyratory compaction to estimate the compaction temperature of the mixtures. The results show that all the three methods increase the compactability of mixtures and thus significantly reduce the compaction temperatures. Method 3 (combining GNP modification and aggregate packing optimization) has the most significant effect, followed by method 1 (GNP modification), and method 2 (aggregate packing optimization).

Keywords: temperature reduction; compaction; asphalt mixture; Graphite Nanoplatelets (GNP); aggregate packing

Citation: Yan, T.; Turos, M.; Le, J.-L.; Marasteanu, M. Reducing Compaction Temperature of Asphalt Mixtures by GNP Modification and Aggregate Packing Optimization. *Materials* **2022**, *15*, 6060. https://doi.org/10.3390/ma15176060

Academic Editors: Markus Oeser, Michael Wistuba, Pengfei Liu, Di Wang and Simon Hesp

Received: 7 July 2022
Accepted: 26 August 2022
Published: 1 September 2022

1. Introduction

High temperature is required for the construction of hot mix asphalt (HMA), which is typically 145 to 165 °C for producing (mixing) asphalt mixtures, and 125 to 145 °C for paving and compacting. The high temperatures lower the viscosity of asphalt binders and thus ensure the workability of asphalt mixtures [1,2]. However, high production temperature increases the fuel consumption for heating, the emission of greenhouse gases and pollution gases, and the aging of asphalt binders. High compaction temperature shortens the hauling distance, the time window for compaction, and the paving season every year [3].

The most widely used temperature reduction technology is the Warm Mix Asphalt (WMA), which reduces the production and compaction temperatures of HMA by 20–40 °C [3,4]. The temperature reduction of WMA is typically achieved by one of the following approaches: (1) the addition of organic additives, (2) the addition of chemical additives, or (3) the foaming processes [5,6]. Organic additives are mainly waxes. They melt below the melting point of the binder, thus reducing the viscosity of binder during mixing and compaction which improves the coating and workability of the mix [4]. Chemical additives are in the form of emulsions and surfactants, which work at the microscopic interface of the binder and aggregates to reduce the frictional forces and facilitate lubrication between the binder and aggregate during mixing and compaction [7,8]. Foaming technologies involve the introduction of small amounts of water into the asphalt binder. The water evaporates during the mixing process with the hot binder and, therefore, the steam remains entrapped, generating a large volume of foam. The foam causes a temporary viscosity reduction in the mix which results in improved aggregate coating and easier compaction of the asphalt mix at lower temperatures [5,9]. Numerous studies have shown the benefits of WMA on mitigating the issues caused by the high mixing and compaction temperatures [10–16].

In previous studies, the authors have developed two different approaches for improving the compactability of asphalt mixtures, which are (1) the addition of Graphite Nanoplatelets (GNPs) [17–20] and (2) the optimization of aggregate packing [21]. In these studies, the authors demonstrated that both methods increase the compactability of asphalt mixtures, while retaining or improving relevant mechanical properties of asphalt mixtures, such as resistance to rutting, cracking, and fatigue [18,21]. Given the improvement in the compactability of asphalt mixtures, it is expected that these two methods can reduce the compaction temperature of asphalt mixtures.

The objective of this study is to explore the effects of the two methods (adding GNP and optimizing aggregate packing) and the combination of them on reducing compaction temperatures. It is important to note that, although the GNP modification could also lead to a reduction of the mixing temperature, the focus of this study is only on the reduction of compaction temperature. Therefore, the mixing temperature is maintained at the traditional level in this study. The produced mixtures are therefore in between HMA and WMA, which can be classified as the hot-mixed warm-compacted (HMWC) mixtures, meaning that the mixtures are mixed at the same temperature as HMA but compacted at a lower temperature than HMA [22]. Similar to WMA, the HMWC mixtures have shown many advantages over HMA due to the reduction of compaction temperature, for example, (1) extending the hauling distance (beneficial for paving at remote locations and emergency paving after natural disasters) [22,23], (2) facilitating cold-weather paving [24], (3) extending the paving window [25], and (4) reducing the energy consumption for reheating. Moreover, in some cases, the mixing temperatures of mixtures are predominantly governed by the compaction temperature and the required hauling distance and paving window [26]. Under such circumstances, the reduction of compaction temperature helps reducing the mixing temperature as well.

In addition, a new method is proposed to determine the compaction temperature. Most of previous methods assume that compaction is governed by the viscosity of asphalt binder and therefore determine the compaction temperature by the viscosity–temperature relationship of asphalt binder [27,28]. For polymer-modified binders (typically non-Newtonian), the viscosity is shear rate-dependent, so it has been argued that viscosity should be measured at the shear rate representative for compaction of asphalt mixtures. However, the representative shear rate is still under debate [1,2,27–30]. These methods use the binder viscosity measured by a Rotational Viscometer (RV). The Dynamic Shear Rheometer (DSR) has also been used for measuring the binder viscosity and determining the compaction temperature [31]. In addition to the viscosity-based methods, phase angle of asphalt binder (measured by DSR) was also used as an index for determining compaction temperature [32]; it is assumed that the change in phase angle characterizes the solid to liquid transition of asphalt binder. It is important to note that all these methods are based on rheological properties (viscosity or phase angle) of asphalt binder. However, many studies have questioned this approach, since the compactability of asphalt mixtures is also dependent on many other factors, such as binder content, lubrication of binder, aggregate gradation, and angularity [33–35].

In this study, a new method for determining the compaction temperature is proposed based on characterizing the compactability of mixtures by the gyratory compaction. The density at 30 gyrations (ϕ_{30}) is selected as compactability index, since previous studies have shown that the ϕ_{30} approximately represents the field density of the mixture [36,37]. Given the desired field density, the compaction temperature is interpolated by gyratory compaction results at different temperatures.

The paper is structured as follows. First, the two methods for improving compactability are introduced in Section 2. An experimental study is then performed to investigate the effects of the two methods and their combination on reducing the compaction temperature. The material information and experimental plan of gyratory compaction are detailed in Section 3, and the results of gyratory compaction and the method for estimating compaction temperatures are presented and discussed in Section 4.

2. Two Approaches for Improving Compactability

2.1. GNP Modification

The GNPs are nano-discs with a sub-micrometer diameter and a thickness on the order of nanometer. The GNPs could be produced from either graphene or natural graphite. If the GNPs are prepared directly from graphene, each platelet typically consists of several layers of graphene sheets, each of which is a single layer of carbon atoms. Depending on its type and carbon purity, the cost of GNPs can be as low as $3 per pound, which is comparable to some existing asphalt modifiers such as the SBS.

Recent studies have shown that addition of small amount (3% to 6% by weight of the binder) of GNPs to asphalt binder can result in 20% to 40% reduction in the compaction effort required for achieving the target air voids, 130% increase in the flexural strength of asphalt binders, and almost 100% increase in the fracture energy of asphalt mixtures. Meanwhile, other mechanical properties of binders and mixtures, e.g., rutting, stiffness, fatigue, and rheology, are not adversely affected by the GNP modification [17,18].

The physical mechanisms of GNPs on improving the compactability of asphalt mixtures have been investigated in the past years [19,20]. It was found that the GNP addition increases the viscosity of the binder, which implies that the improved compactability cannot be attributed to the viscosity of asphalt binder. A tribological test was developed measuring the lubricating effect of binder between rough surfaces [19]. The test fixture is shown in Figure 1. As shown in Figure 1a, the tribological test fixture is developed based on the AR 2000 TA Dynamic Shear Rheometer (DSR) equipment (TA Instruments, New Castle DE, USA). As shown in Figure 1b, the fixture contains five different components: a lower cup, three steel plates, a steel ball, a shaft, and a ring to attach the ball to the shaft. In the lower cup, there are three plates with an angle of 45° with respect to the horizontal plane and the asphalt sample. The steel ball is attached to the shaft, which then gets attached to the DSR head. To better simulate the surface of the aggregates, the surfaces of the ball and the plates were roughened by immersing the ball and the plates in hydrochloric acid (HCl) for a period of time. Hydrochloric acid corroded the surfaces of the parts and made them rough and looking like an orange skin. The roughness increases with the corroding period. It was found that a corroding period of three days best simulates the roughness of aggregate surfaces. Therefore, the corroding period was strictly controlled at three days, to achieve consistent roughness for all the testing balls and plates. Figure 1c–f compares the original smooth ball and plate and the ball and plate after they were roughened by the acid. During the tribological test, the axial force was kept constant and equal to 10 N, whereas the rotational speed was increased in logarithmic steps from 0.1 to 1433 rpm. The friction coefficient was measured as the function of the rotational speed. The results showed that the addition of GNPs lowers the friction coefficient between rough surfaces and therefore explained the effect of GNPs on improving the compactability of mixtures.

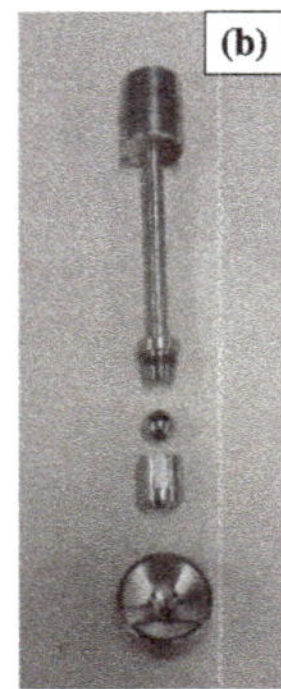

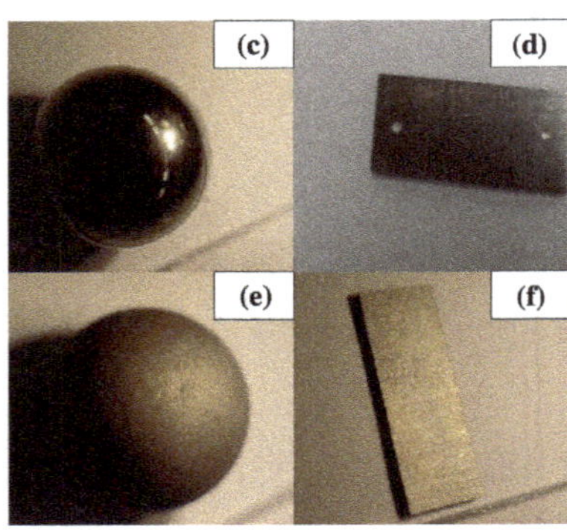

Figure 1. Tribological test fixture: (**a**) general view, (**b**) components of the fixture, (**c**) smooth ball, (**d**) smooth plate (used), (**e**) rough ball, (**f**) rough plate.

Microscopically, it is believed that GNPs occupy the space between the asperities of the aggregates as shown in Figure 2, which improve the lubricating effect of asphalt binders and, thus, the compactability of asphalt mixtures.

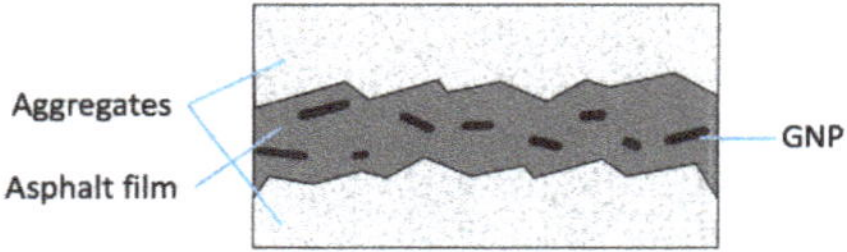

Figure 2. Scheme of the GNP-modified asphalt film between aggregate surfaces.

2.2. *Aggregate Packing Optimization*

In a recent study on developing the Superpave 5 asphalt mix designs in Minnesota [21], a method was developed to improve the compactability of asphalt mixtures by optimizing the aggregate packing. In this method, aggregate sources are divided into the coarse and fine aggregates. Then, the mass ratio between the coarse and fine aggregates (C/F ratio) is adjusted to maximize the packing density of the mixture. This method is based on the binary aggregate packing theory which shows that when mixing two aggregate sources of different sizes, there exists an optimum C/F ratio which raises the maximum packing density for the mixture [38–44].

One mixture from a previous study [21] is used to demonstrate this mix design method. The mixture was originally designed as a traditional Superpave mixture (4% air voids at N_{design} = 60) and was modified to a Superpave 5 mixture (5% air voids at N_{design} = 30) by this method. The mixture contains five aggregate sources and two reclaimed asphalt pavements (RAP). The aggregate gradations of the aggregate sources are plotted in Figure 3. As shown, aggregate sources are separated into the coarse aggregates (C-1 and C-2) and fine aggregates (F-1, F-2, and F-3) and RAP's (R-1 and R-2).

Trial blends were designed by adjusting the C/F ratio of the aggregates while keeping the binder content (5.6%) and the RAP contents (20% for R-1 and 10% for R-2) unchanged. The aggregate gradations of the Trial blends are shown in Figure 4a. It is seen that as the C/F ratio increases, the aggregate gradation changes from fine-graded to coarse-graded. Densities of mixtures (ϕ) at 30 gyrations, ϕ_{30}, are plotted in Figure 4b. It is seen that the Trial 2 (C/F = 1.33) has the densest aggregate packing and thus achieves the maximum ϕ_{30}, 95.25 %G_{mm}, which represents a 1.08% increase compared to the original mix design. Therefore, the aggregate gradation of the Trial 2 (C/F = 1.33) was selected for the modified mix design.

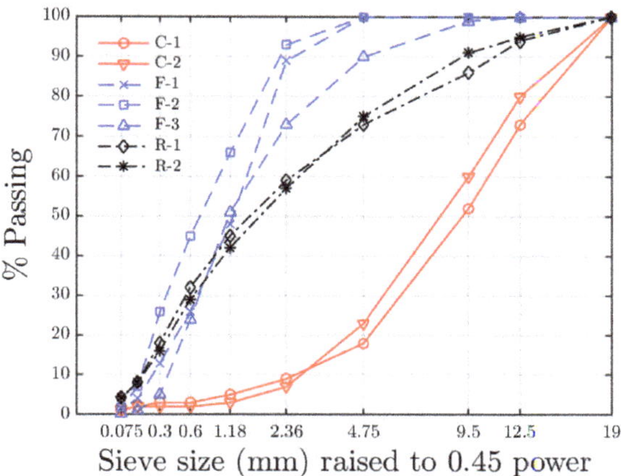

Figure 3. Gradation of aggregate sources. The line colors and styles indicate the type of aggregate.

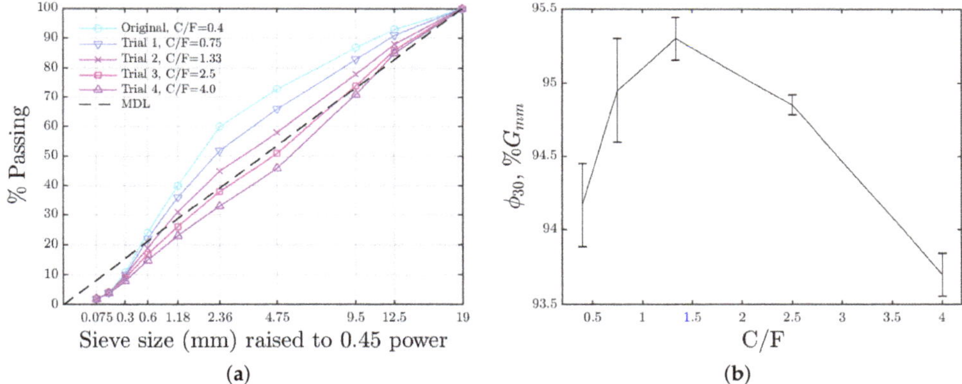

Figure 4. (**a**) Aggregate gradations of the trial blends. (**b**) Densities of the trial blends at 30 gyrations in gyratory compaction. The error bar represents the difference between the two replicates.

Laboratory performance tests were conducted to compare the mechanical properties of the modified and the original mixtures. They are the Semi-Circular Bending (SCB) [45], Flow Number [46], and Diametral dynamic modulus [47]. Test specimens are prepared by gyratory compaction to 30 gyrations. The average densities of the modified mixture and the original mixture are 94.17 $\%G_{mm}$ and 95.25 $\%G_{mm}$, respectively. The performance test results are shown in Figure 5. It is seen that the modified mixture (Trial 2) had higher fracture energy (at both −12 and −20 °C), flow number, and dynamic modulus than the original mixture, indicating that the modified mixture would have better performance in the field than the original mixture.

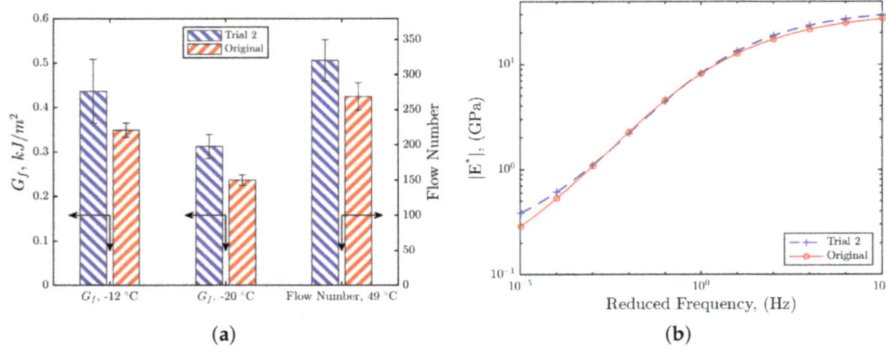

Figure 5. Performance tests results of the original and modified mixtures. (**a**) SCB fracture energy and flow number results. The error bar indicates the standard error of three replicates. The arrows indicate the corresponding axes for the data. (**b**) Dynamic modulus master curves where the reference temperature is 12 °C.

3. Experimental Study

3.1. Materials Information

The traditional Superpave asphalt mixture mentioned in Section 2.2 was used as the control mixture (Mix-C) in this study. Mix-C was used for the wearing course of a traffic level 3 (20-year design Equivalent Single Axle Loads (ESAL's) are between 1 to 3 million) project in Minnesota in 2020. It was designed at 4% air voids with the N_{design} = 60. The aggregate gradation of the Mix-C is shown in Figure 4a as the original gradation. More information of the Mix-C is listed in Table 1.

Table 1. Material information of the Mix-C.

Mix ID	NMAS (mm)	Binder PG	% AC	% Virgin AC	% RAP	FAA (%)	CAA1 (%)	CAA2 (%)
Mix-C	12.5	58S-28	5.6	4.2	30	42	99	93

Note: NMAS = nominal maximum aggregate sizes, PG = performance grade, AC = asphalt content, RAP = reclaimed asphalt pavement, FAA = fine aggregate angularity, CAA1 = coarse aggregate angularity of one fractured surface, CAA2 = coarse aggregate angularity of two fractured surfaces.

Based on the raw materials of the Mix-C, three modified mixtures were designed by (1) GNP modification, (2) optimizing aggregate packing, and (3) a combination of GNP modification and optimizing aggregate packing. The first modified mixture is denoted as Mix-GNP. It modified Mix-C by adding 6% GNPs by the weight of binder to the mixture. The GNPs used in this study are made of a synthetic graphite material with 99.66% carbon and 0.34% ash, characterized by an enhanced surface area equal to 250 m^2/g. The same GNPs have been used in previous studies [17–20]. In this study, the GNPs were added into the mixture during the mixing process after aggregates and binder were thoroughly mixed for 5 minutes by a mechanical mixer at 145 °C.

The second modified mixture was designed by optimizing aggregate packing. Details of this approach is described in Section 2.2. This modified mixture is denoted as Mix-Agg. The aggregate gradation of Mix-Agg is shown in Figure 3a as that of the Trial 2. Other properties of Mix-Agg are the same as Mix-C as listed in Table 1.

The third modified mixture was designed by combining the two former approaches. It contains 6% GNPs while it has the optimized aggregate gradation of the Mix-Agg. This modified mixture is denoted as Mix-GNP&Agg.

3.2. Experimental Plan

Laboratory gyratory compaction tests (AASHTO T312) [48] were performed for the four mixtures to 100 gyrations at two compaction temperatures: 135 °C and 95 °C. The 135 °C was chosen since it is the recommended compaction temperature for Mix-C. Since the focus of this study is on the reduction of compaction temperature, the mixing temperature was kept constant at 145 °C for all mixtures. For each mixture, two samples were compacted. Based on the gyratory compaction results at the two reference temperatures, the compaction temperatures of the modified mixtures are estimated by the direct characterization of the compactability of mixtures by gyratory compaction.

4. Results and Discussion

4.1. Gyratory Compaction Results

The gyratory compaction curves of the four mixtures at the two compaction temperatures (135 °C and 95 °C) are shown in Figure 6. The compaction curves show the average of two replicates, while the error bars show the differences between the two replicates. It is seen that the three modified mixtures have higher compactability than the control mixture (Mix-C) at both compaction temperatures. Among the modified mixtures, Mix-GNP&Agg is the most compactable, followed by Mix-GNP and Mix-Agg. The compaction curves of Mix-GNP and Mix-Agg are very close to each other at both temperatures, but it is seen that Mix-GNP is slightly more compactable than Mix-Agg at the starting phase of gyratory compaction while Mix-GNP is surpassed by Mix-Agg at the later phase of gyratory compaction. It is also noticed that the variability of compaction curves is higher at 95 °C than at 135 °C, which indicates that lower compaction temperature may increase the variability of field density.

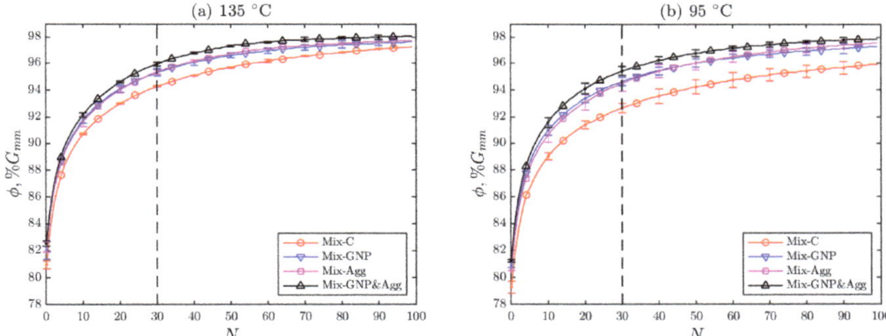

Figure 6. Gyratory compaction results at (**a**) 135 °C and (**b**) 95 °C.

Previous studies have shown that the field compaction effort of current asphalt pavement projects is approximately equivalent to 30 gyrations in the gyratory compaction [36,37]. Therefore, we use the density of mixtures at 30 gyrations, ϕ_{30}, to characterize the compactability of the mixture. The higher the ϕ_{30}, the more compactable the mixture is. The results of ϕ_{30} for the four mixtures at the two temperatures are shown in Figure 7, where the heights of the bars show the average ϕ_{30} of the two replicates, and the error bars show the difference of ϕ_{30} between the two replicates.

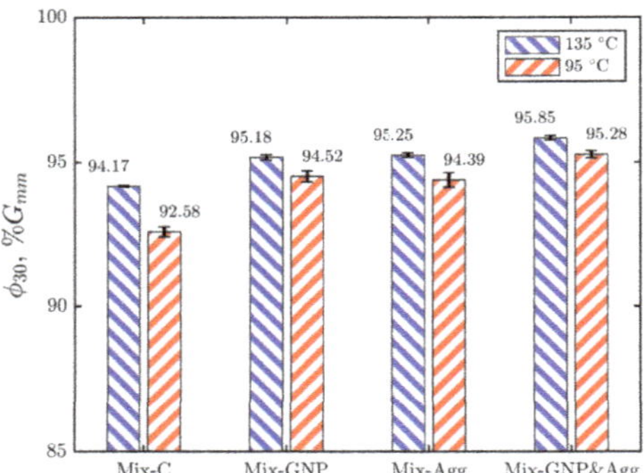

Figure 7. Bar chart of ϕ_{30} at the two compaction temperatures.

As shown in Figure 7, as expected, it is seen that mixtures are more compactable at 135 °C than at 95 °C, which can be attributed to the lower viscosity of binder at higher temperatures. The comparison between ϕ_{30} at 135 °C and 95 °C shows that the effect of temperature on compactability is more significant for the control mixture (Mix-C) than the modified mixtures. Moreover, it is seen again that the variability of compactability is higher at the lower compaction temperature.

Another way to evaluate the compactability of mixtures is by the corresponding number of gyrations needed for achieving a certain density level. In this study, two density levels are of particular interest: (1) the average field density of the control mixture (Mix-C) and (2) the required field density level of the Superpave 5 mix design. The average field density of the Mix-C was 93.9 $\%G_{mm}$, which was obtained from the quality control and quality assurance (QC/QA) data of the project. The required field density level of Superpave 5 mixtures is 95 $\%G_{mm}$, which is about 1–2% increase compared with the traditional Superpave mixtures [21,49]. Based on the gyratory compaction results, the numbers of gyrations needed for achieving the two density levels were calculated. An example of such procedure is shown in Figure 8 for Mix-C. The results of all mixtures are shown in Figure 9, where the heights of the bars show the average equivalent number of gyrations of the two replicates, and the error bars show the difference of equivalent number of gyrations between the two replicates.

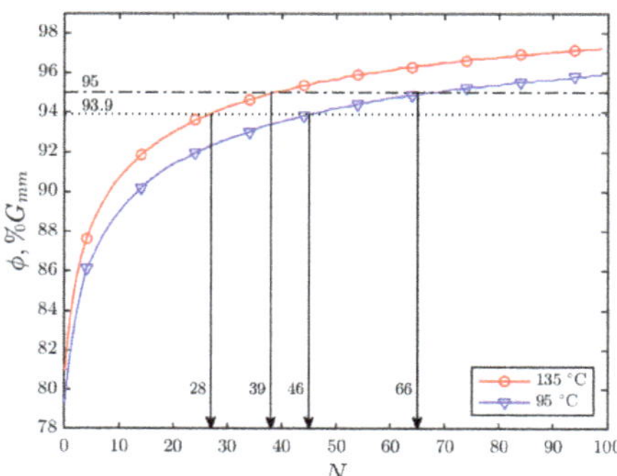

Figure 8. Calculation of the number of gyrations needed for achieving the two density levels for Mix-C.

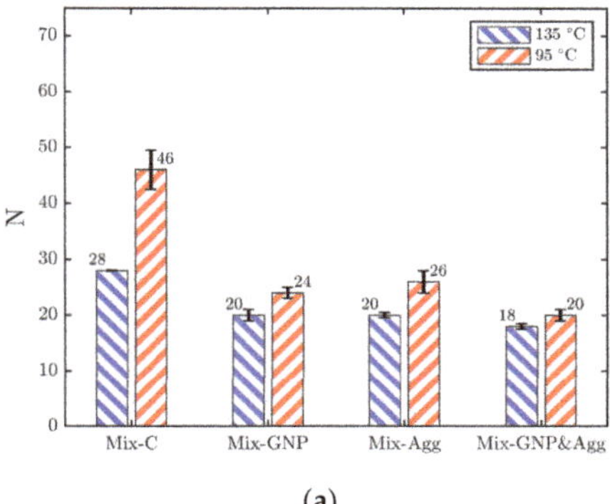

(a)

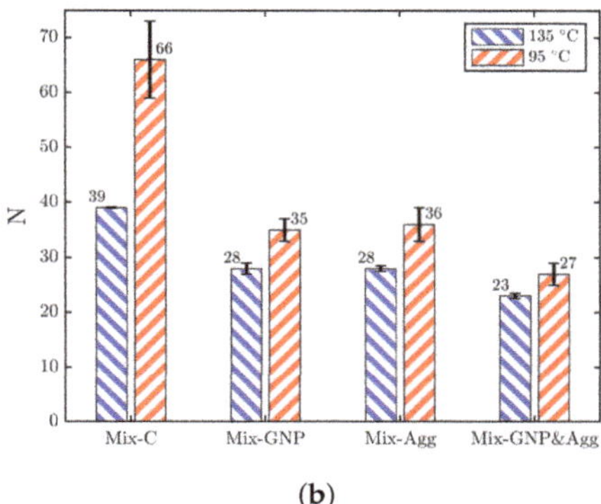

(b)

Figure 9. The numbers of gyrations needed for achieving the two density levels: (**a**) 93.9 $\%G_{mm}$ and (**b**) 95 $\%G_{mm}$.

As shown in Figure 9a , for achieving the average field density (93.9 $\%G_{mm}$), Mix-C needs 28 gyrations at the recommended compaction temperature (135 °C), which is consistent with previous studies [36,37], showing that the field compaction effort is approximately equivalent to 30 gyrations in gyratory compaction. Because of the improved compactability, the modified mixtures achieved 93.9 $\%G_{mm}$ at much lower compaction effort at 135 °C, which are 20, 20, and 18 gyrations for Mix-GNP, Mix-Agg, and Mix-GNP&Agg, respectively. At the lower compaction temperature (95 °C), due to the increase in binder viscosity, more compaction effort is needed than at 135 °C. The increase in number of gyrations is the most significant for Mix-C, which increased by 64% (from 28 to 46), while for the modified mixtures, such increase becomes less significant. It is important to note that, even at the lower compaction temperature 95 °C, all the modified mixtures need less compaction effort (25, 26, and 20 gyrations for Mix-GNP, Mix-Agg, and Mix-GNP&Agg, respectively) than the original compaction effort of Mix-C at 135 °C (28 gyrations), which implies that the improved compactability of the modified mixtures would enable them to be compacted at lower temperatures than 95 °C for achieving the density level of 93.9 $\%G_{mm}$. More details about the estimation of compaction temperatures will be discussed in Section 4.2.

Figure 9b shows the numbers of gyrations needed for achieving 95 $\%G_{mm}$, the density required by the Superpave 5 mix design. Similarly, it is seen that more compaction effort is needed at 95 °C than at 135 °C. The comparison between Figure 9a,b shows that, as expected, more compaction effort is needed for achieving 95 $\%G_{mm}$ than 93.9 $\%G_{mm}$. The increase is quite consistent, which is about 30% to 40% for all mixtures at both compaction temperatures. It is important to note that, at 135 °C, all the modified mixtures require less or equal compaction effort (28, 28, and 23 gyrations for Mix-GNP, Mix-Agg, and Mix-GNP&Agg, respectively) to achieving 95 $\%G_{mm}$ than the original compaction effort of Mix-C (28 gyrations). This result shows that, at 135 °C, all the modified mixtures would be able to achieve higher field densities than 95 $\%G_{mm}$ (requirement of the Superpave 5 mix design) with the regular field compaction effort. It is seen from Figure 9a,b again that the variability of compactability is higher at the lower compaction temperature.

4.2. Compaction Temperature Estimation

A method is proposed for estimating the compaction temperature, which is based on the direct characterization of the compactability of asphalt mixtures by the gyratory compaction. A schematic diagram of this method is depicted in Figure 10. In this method, results of gyratory compaction curves at multiple ($\geq$2) temperatures are required. Previous studies have shown that the field density is typically achieved at about 30 gyrations in the gyratory compaction [36,37]. Therefore, we use the density of mixtures at 30 gyrations, ϕ_{30}, to estimate the field density level of the mixtures. For each mixture, the values of ϕ_{30}

are obtained at the multiple temperatures, as shown in Figure 10a. Then, based on the ϕ_{30} results at the these temperatures, the compaction temperature T^* is interpolated given the desired field density level ϕ^*_{30}, as shown in Figure 10b.

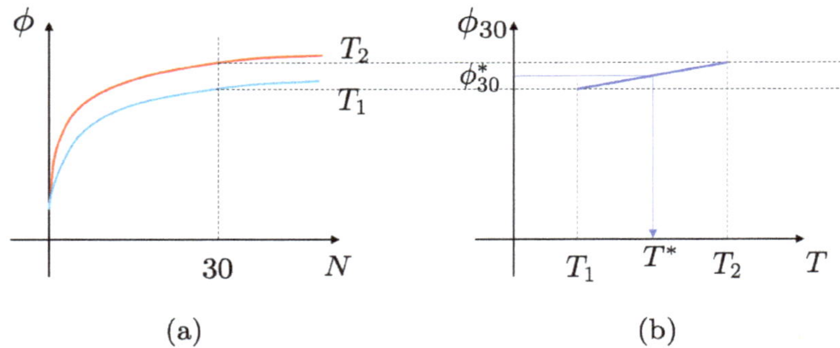

Figure 10. A schematic diagram for compaction temperature estimation: (**a**) Gyratory compaction curves at multiple temperatures, (**b**) compaction temperature interpolation.

In this study, gyratory compaction tests were performed at two temperatures: 95 °C and 135 °C. The ϕ_{30} results at the two temperatures are shown in Figure 7. As seen in Figure 10b, a model for the relationship between T and ϕ_{30} is needed for the interpolation of compaction temperature. Previous studies showed that the effect of temperature on compactability is nonlinear [50–54]. The change of compactability with temperature is more significantly at low temperatures than at high temperatures [52]. At extremely high temperatures (over 150 °C), temperature can even have a negative effect on compactability [50,51,54]. However, to the best knowledge of the authors, there is no generally accepted model for the relationship between compaction temperature and the compactability of asphalt mixtures. For simplicity and due to the limited number of data points, we employ a linear function as the first-order approximation to the nonlinear relationship between T and ϕ_{30}.

$$\phi_{30} = \alpha T + \beta, \tag{1}$$

where α and β are model coefficients. It is important to note that this linear approximation model is acceptable only when compaction temperature is close to the experimental compaction temperatures, while it is expected that the linear approximation would become increasingly inaccurate as compaction temperature gets away from the experimental compaction temperatures. The linear model is applied to the four mixtures. The model coefficients are obtained and listed in Table 2.

Table 2. Coefficients of the linear model between T and ϕ_{30}.

Coefficients	Mix-C	Mix-GNP	Mix-Agg	Mix-GNP&Agg
α, %/°C	10.42	4.32	5.63	3.74
β, %	71.97	85.97	83.24	87.89

The linear approximations of the relationships between T and ϕ_{30} are shown in Figure 11 for the four mixtures. Again, it is seen that the compactability of Mix-C is the most temperature-sensitive, followed by Mix-Agg, Mix-GNP, and Mix-GNP&Agg.

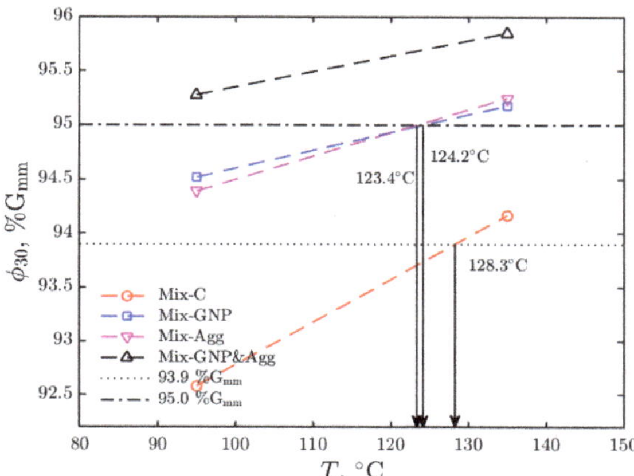

Figure 11. Estimation of the compaction temperatures by the linear model.

Based on the obtained linear models between T and ϕ_{30}, we can then estimate the compaction temperatures of the mixtures for achieving the desired field density level ϕ^*_{30}. In this study, we consider two different ϕ^*_{30} levels: 93.9 %G_{mm} and 95 %G_{mm}, which is the average field density of the control mixture (Mix-C) and the required field density level of the Superpave 5 mix design, respectively. As shown in Figure 11, for ϕ^*_{30} = 93.9 %G_{mm}, the compaction temperature for Mix-C is interpolated as 128.3 °C. For the other mixtures, it is clear that the compaction temperatures are lower than 95 °C. However, their compaction temperatures cannot be accurately estimated due to the limitation of the linear model. For ϕ^*_{30} = 95 %G_{mm}, the compaction temperatures for Mix-GNP and Mix-Agg are interpolated as 124.2 °C and 123.4 °C, respectively. The compaction temperature for Mix-C is clearly higher than 135 °C and the compaction temperature for Mix-GNP&Agg is clearly lower than 95 °C. However, only the estimation of the bounds can be made instead of the exact values, because of the limitation of the linear model. The estimated compaction temperatures for the four mixtures are listed in Table 3.

Table 3. Estimation of compaction temperatures.

ϕ^*_{30}, %G_{mm}	Mix-C	Mix-GNP	Mix-Agg	Mix-GNP&Agg
93.9	128.3 °C	<95 °C †	<95 °C †	<95 °C †
95.0	>135 °C †	124.2 °C	123.4 °C	<95 °C †

Note: † a bound is given for the compaction temperature, because the exact compaction temperature cannot be estimated due to the limitation of the linear approximation.

As shown in Figure 11, the modified mixtures can significantly reduce the compaction temperature of the control mixture. For both the two field density levels, Mix-GNP&Agg would need the lowest compaction temperature followed by Mix-GNP, Mix-Agg, and Mix-C. As shown in Table 3, for achieving the field density level of 93.9 %G_{mm}, the compaction temperatures are lower than 95 °C for the modified mixtures, which means the compaction temperature can be reduced by more than 30% compared to the original compaction temperature (135 °C). For achieving the field density level of 95 %G_{mm}, Mix-C need a compaction temperature higher than 135 °C, while the compaction temperatures are reduced to lower than 135 °C for the modified mixtures. In particular, the compaction temperature for the most compactable mixture, Mix-GNP&Agg, is reduced to even lower than 95 °C. These results show that the three methods can effectively reduce the compaction temperature of mixtures, with their effects comparable to the temperature reduction effect of WMA.

It is worthwhile to discuss the potential ways for improving the accuracy of compaction temperature estimation. One possible way is by increasing the number of experimental compaction temperatures, which would enable the characterization of the nonlinear relationship between T and ϕ_{30}. Another way is by developing more realistic theoretical models for the T-ϕ_{30} relationship.

5. Conclusions and Recommendations

In this paper, two approaches for improving the compactability of asphalt mixtures were introduced. One is the addition of Graphite Nanoplatelets (GNPs), and the other is the optimization of aggregate packing. An experimental study was performed to estimate the effects of the two approaches and their combination on reducing compaction temperatures. A method for determining compaction temperature was proposed based on the gyratory compaction. The main conclusions of this research are summarized as follows.

1. GNP modification, aggregate packing optimizing, and their combination can significantly improve the compactability of mixtures and thus reduce the compaction temperatures.
2. The combination of GNP modification and aggregate packing optimization has the most significant temperature reduction effect, followed by GNP modification, and aggregate packing optimization.
3. The three methods can reduce the compaction temperature of the original mixture by more than 30%, which is comparable to the temperature reduction effect of WMA.
4. The proposed method for compaction temperature estimation is based on the direct characterization of the compactability of asphalt mixtures. It can take into account many affecting factors which were not possible to consider with the traditional viscosity-based method, for example, the properties of aggregates and the lubricating effect of binder. Thus, the method can potentially be used in practice for estimating the compaction temperatures for asphalt mixtures.

While the results presented in this proof-of-concept study are based on limited experimental data, it is reasonable to conclude that the proposed two methods (GNP modification and Aggregate packing optimization), and their combination, have good potential as practical methods for reducing the compaction temperatures of asphalt mixtures. It was also observed that the variability of the compactability of mixtures increased with the decrease in the compaction temperature, which implies that compaction temperature would influence the reliability of field compaction. Such an effect was not considered in the compaction temperature estimation in this study. It is recommended that future studies can consider the effect of compaction temperature on the reliability of field compaction. Moreover, it is seen from this study that one of the main difficulties for the compaction temperature estimation is the lack of knowledge on the relationship between the compaction temperature and the compactability of mixtures. To improve the accuracy of compaction temperature estimation, future research is suggested to explore the relationship between the compaction temperature and the compactability of mixtures, using a combination of experimental investigation and theoretical modeling.

Author Contributions: Conceptualization, M.M., J.-L.L., M.T. and T.Y.; experiments, M.T. and T.Y.; formal analysis, T.Y. and M.T.; writing—original draft preparation, T.Y., M.M. and J.-L.L.; writing—review and editing, M.M., T.Y. and J.-L.L. All authors have read and agreed to the published version of the manuscript.

Funding: The authors gratefully acknowledge the financial support provided by the Minnesota Department of Transportation (grant: 1003325 WO#106) and the University of Minnesota, Center for Transportation Studies (grant: CTS #2018076).

Data Availability Statement: Data are available from the corresponding authors by request.

Conflicts of Interest: The authors declare no conflict of interest.

References

1. Bahia, H.U.; Fahim, A.; Nam, K. Prediction of compaction temperatures using binder rheology. In *Transportation Research Circular, E-C105*; Transportation Research Board: Washington, DC, USA, 2006; pp. 3–17.
2. Yildirim, Y.; Ideker, J.; Hazlett, D. Evaluation of Viscosity Values for Mixing and Compaction Temperatures. *J. Mater. Civ. Eng.* **2006**, *18*, 545–553. [CrossRef]
3. D'Angelo, J.; Harm, E.; Bartoszek, J.; Baumgardner, G.; Corrigan, M.; Cowsert, J.; Harman, T.; Jamshidi, M.; Jones, W.; Newcomb, D.; et al. *Warm-Mix Asphalt: European Practice*; Technical Report, United States; Federal Highway Administration: Washington, DC, USA, 2008.
4. Rubio, M.C.; Martínez, G.; Baena, L.; Moreno, F. Warm Mix Asphalt: An overview. *J. Clean. Prod.* **2012**, *24*, 76–84. [CrossRef]
5. Cheraghian, G.; Falchetto, A.C.; You, Z.; Chen, S.; Kim, Y.S.; Westerhoff, J.; Moon, K.H.; Wistuba, M.P. Warm mix asphalt technology: An up to date review. *J. Clean. Prod.* **2020**, *268*, 122128. [CrossRef]
6. Caputo, P.; Abe, A.A.; Loise, V.; Porto, M.; Calandra, P.; Angelico, R.; Rossi, C.O. The Role of Additives in Warm Mix Asphalt Technology: An Insight into Their Mechanisms of Improving an Emerging Technology. *Nanomaterials* **2020**, *10*, 1202. [CrossRef]
7. Li, X.; Wang, H.; Zhang, C.; Diab, A.; You, Z. Characteristics of a surfactant produced warm mix asphalt binder and workability of the mixture. *J. Test. Eval.* **2015**, *44*, 2219–2230. [CrossRef]
8. Kakar, M.; Hamzah, M.; Akhtar, M.; Woodward, D. Surface free energy and moisture susceptibility evaluation of asphalt binders modified with surfactant-based chemical additive. *J. Clean. Prod.* **2016**, *112*, 2342–2353. [CrossRef]
9. Hasan, M.R.M.; You, Z.; Yin, H.; You, L.; Zhang, R. Characterizations of foamed asphalt binders prepared using combinations of physical and chemical foaming agents. *Constr. Build. Mater.* **2019**, *204*, 94–104. [CrossRef]
10. Rubio, M.C.; Moreno, F.; Martínez-Echevarría, M.J.; Martínez, G.; Vázquez, J.M. Comparative analysis of emissions from the manufacture and use of hot and half-warm mix asphalt. *J. Clean. Prod.* **2013**, *41*, 1–6. [CrossRef]
11. Thives, L.P.; Ghisi, E. Asphalt mixtures emission and energy consumption: A review. *Renew. Sustain. Energy Rev.* **2017**, *72*, 473–484. [CrossRef]
12. Sharma, A.; Lee, B.K. Energy savings and reduction of CO_2 emission using $Ca(OH)_2$ incorporated zeolite as an additive for warm and hot mix asphalt production. *Energy* **2017**, *136*, 142–150. [CrossRef]
13. Wang, F.; Li, N.; Hoff, I.; Wu, S.; Li, J.; Maria Barbieri, D.; Zhang, L. Characteristics of VOCs generated during production and construction of an asphalt pavement. *Transp. Res. Part D Transp. Environ.* **2020**, *87*, 473–484. [CrossRef]
14. Gandhi, T.; Akisetty, C.; Amirkhanian, S. Laboratory evaluation of warm asphalt binder aging characteristics. *Int. J. Pavement Eng.* **2009**, *10*, 353–359. [CrossRef]
15. Hofko, B.; Cannone Falchetto, A.; Grenfell, J.; Huber, L.; Lu, X.; Porot, L.; Poulikakos, L.D.; You, Z. Workability of hot-mix asphalt. *Road Mater. Pavement Des.* **2017**, *18*, 108–117. [CrossRef]
16. Poulikakos, L.D.; Falchetto, C.A.; Wang, D.; Porot, L.; Hofko, B. Impact of asphalt aging temperature on chemo-mechanics. *RSC Adv.* **2019**, *9*, 11602–11613.
17. Le, J.-L.; Marasteanu, M.; Turos, M. *Graphene Nanoplatelet (GNP) Reinforced Asphalt Mixtures: A Novel Multifunctional Pavement Material (NCHRP IDEAProject 173)*; Technical Report; Transportation Research Board: Washington, DC, USA, 2016.
18. Le, J.-L.; Marasteanu, M.; Turos, M. Mechanical and compaction properties of graphite nanoplatelet-modified asphalt binders and mixtures. *Road Mater. Pavement Des.* **2020**, *21*, 1799–1814. [CrossRef]
19. Yan, T.; Ingrassia, L.P.; Kumar, R.; Turos, M.; Canestrari, F.; Lu, X.; Marasteanu, M. Evaluation of graphite nanoplatelets influence on the lubrication properties of asphalt binders. *Materials* **2020**, *13*, 772. [CrossRef]
20. Ingrassia, L.P.; Lu, X.; Marasteanu, M.; Canestrari, F. Tribological Characterization of Graphene Nano-Platelet (GNP) Bituminous Binders. In *Airfield and Highway Pavements 2019: Innovation and Sustainability in Highway and Airfield Pavement Technology*; American Society of Civil Engineers: Reston, VA, USA, 2019; pp. 96–105.
21. Yan, T.; Marasteanu, M.; Le, J.-L.; Turos, M.; Cash, K. *Development of Superpave 5 Asphalt Mix Designs for Minnesota Pavements*; Technical Report; Minnesota Department of Transportation: St. Paul, MN, USA, 2022.
22. Howard, I.L.; Baumgardner, G.L.; Jordan, W.S.; Menapace, A.M.; Mogawer, W.S.; Hemsley, J.M. Haul time effects on unmodified, foamed, and additive-modified binders used in hot-mix asphalt. *Transp. Res. Rec.* **2013**, *2347*, 88–95. [CrossRef]
23. Howard, I.L.; Doyle, J.D.; Hemsley, J.M.; Baumgardner, G.L.; Cooley, L.A. Emergency paving using hot-mixed asphalt incorporating warm mix technology. *Int. J. Pavement Eng.* **2014**, *15*, 202–214. [CrossRef]
24. Goh, S.W.; Z., Y. Warm Mix Asphalt Using Sasobit in Cold Region. In *Cold Regions Engineering 2009*; Amercian Society of Civil Engineers: Reston, VA, USA, 2009; pp. 288–298.
25. Kheradmand, B.; Muniandy, R.; Hua, L.T.; Yunus, R.B.; Solouki, A. An overview of the emerging warm mix asphalt technology. *Int. J. Pavement Eng.* **2014**, *15*, 79–94. [CrossRef]
26. Abed, A.; Thom, N.; Grenfell, J.R.A. Evaluation of mixing temperature impact on warm mix asphalt performance. In *Bearing Capacity of Roads, Railways and Airfields*; CRC Press: Boca Raton, FL, USA, 2017; pp. 251–258.
27. Yildirim, Y.; Solaimanian, M.; Kennedy, T.W. *Mixing and Compaction Temperatures for Hot Mix Asphalt Concrete*; Technical Report; Texas Department of Transportation: Austin, TX, USA, 2000.
28. Bahia, H.U.; Hanson, D.I.; Zeng, M.; Zhai, H.; Khatri, M.A.; Anderson, R.M. *Characterization of Modified Asphalt Binders in Superpave Mix Design (NCHRP Report 459)*; Technical Report; Transportation Research Board: Washington, DC, USA, 2001.

29. Khatri, A.; Bahia, H.U.; Hanson, D. Mixing and compaction temperature for modified binders using the Superpave gyratory compactor. *J. Assoc. Asph. Paving Technol.* **2001**, *70*, 368–402.
30. Shenoy, A. Determination of the temperature for mixing aggregates with polymer-modified asphalts. *Energy* **2001**, *2*, 33–47. [CrossRef]
31. Reinke, G. Determination of Mixing and Compaction Temperature of PG Binders Using a Steady Shear Flow Test. 2003. Available online: https://engineering.purdue.edu/~spave/old/Technical%20Info/Meetings/Binder%20ETG%20Sept%2003%20Las%20Vegas,%20NV/Reinke_MIX%20AND%20COMPACTINO%20INFO%20FOR%20ETG%209-15-03.pdf (accessed on 1 July 2022).
32. West, R.C.; Watson, D.E.; Turner, P.A.; Casola, J.R. *Mixing and Compaction Temperatures of Asphalt Binders in Hot-Mix Asphalt. (NCHRP Report 648)*; Technical Report; Transportation Research Board: Washington, DC, USA, 2010.
33. Gudimettla, J.M.; Cooley, L.A.; Brown, E.R. Workability of hot-mix asphalt. *Transp. Res. Rec.* **2004**, *1891*, 229–237. [CrossRef]
34. Delgadillo, R.; Bahia, H.U. Parameters to define the laboratory compaction temperature range of hot-mix asphalt. *J. Assoc. Asph. Paving Technol.* **1998**, *67*, 125–152.
35. Canestrari, F.; Ingrassia, L.; Ferrotti, G.; Lu, X. State of the Art of Tribological Tests for Bituminous Binders. *Constr. Build. Mater.* **2017**, *157*, 718–728. [CrossRef]
36. Yan, T.; Turos, M.; Bennett, C.; Garrity, J.; Marasteanu, M. Relating Ndesign to Field Compaction: A Case Study in Minnesota. *Transp. Res. Rec.* **2022**, *2676*, 192–201. [CrossRef]
37. Yan, T.; Le, J.-L.; Marasteanu, M.; Turos, M. A probabilistic model for field density distribution of asphalt pavements. *Int. J. Pavement Eng.* 2022, *under revision*.
38. Aïm, R.B.; Goff, P.L. Effet de paroi dans les empilements désordonnés de sphères et application à la porosité de mélanges binaires. *Powder Technol.* **1968**, *1*, 281–290. [CrossRef]
39. Toufar, W.; Born, M.; Klose, E. Beitrag zur Optimierung der Packungsdichte Polydisperser körniger Systeme. In *Freiberger Forschungsheft A 558*; VEB Deutscher Verlag für Grundstoffindustrie: Leipzig, Germany, 1976; Volume 72, pp. 29–44.
40. Goltermann, P.; Johansen, V.; Palbøl, L. Packing of Aggregates: An Alternative Tool to Determine the Optimal Aggregate Mix. *ACI Mater. J.* **1997**, *94*, 435?–443.
41. Johansen, V.; Andersen, P. Particle Packing and Concrete Properties. In *Materials Science of Concrete II*; American Ceramic Society: Columbus, OH, USA, 1991; pp. 111–149.
42. De Larrard, F. *Concrete Mixture Proportioning, A Scientific Approach*; CRC Press: London, UK, 1999.
43. Olard, F.; Perraton, D. On the optimization of the aggregate packing characteristics for the design of high-performance asphalt concretes. *Road Mater. Pavement Des.* **2010**, *11*, 145–169. [CrossRef]
44. Pouranian, M.R.; Haddock, J.E. A new framework for understanding aggregate structure in asphalt mixtures. *Int. J. Pavement Eng.* **2019**, *22*, 11602–11613. [CrossRef]
45. *AASHTO TP105*; Standard Method of Test for Determining the Fracture Energy of Asphalt Mixtures Using the Semicircular Bend Geometry (SCB). American Association of State Highway and Transportation Officials: Washington, DC, USA, 2020.
46. *AASHTO T378*; Standard Method of Test for Determining the Dynamic Modulus and Flow Number for Asphalt Mixtures Using the Asphalt Mixture Performance Tester (AMPT). American Association of State Highway and Transportation Officials: Washington, DC, USA, 2017.
47. Kim, Y.; Seo, Y.; King, M.; Momen, M. Dynamic modulus testing of asphalt concrete in indirect tension mode. *Transp. Res. Rec.* **2004**, *1891*, 163–173. [CrossRef]
48. *AASHTO T312*; Standard Method of Test for Preparing and Determining the Density of Asphalt Mixture Specimens by Means of the Superpave Gyratory Compactor. American Association of State Highway and Transportation Officials: Washington, DC, USA, 2019.
49. Hekmatfar, A.; McDaniel, R.S.; Shah, A.; Haddock, J.E. *Optimizing Laboratory Mixture Design as It Relates to Field Compaction to Improve Asphalt Mixture Durability*; Technical Report; Purdue University: West Lafayette, IN, USA, 2001.
50. Pérez-Jiménez, F.; Martínez, A.H.; Miró, R.; Hernández-Barrera, D.; Araya-Zamorano, L. Effect of compaction temperature and procedure on the design of asphalt mixtures using Marshall and gyratory compactors. *Constr. Build. Mater.* **2014**, *65*, 264–269. [CrossRef]
51. Rahmat, N.A.; Hassan, N.A.; Jaya, R.P.; Satar, M.K.I.M.; Azahar, N.M.; Ismail, S.; Hainin, M.R. Effect of compaction temperature on the performance of dense-graded asphalt mixture. *IOP Conf. Ser. Earth Environ. Sci.* **2019**, *244*, 012012. [CrossRef]
52. Delgadillo, R.; Bahia, H.U. Effects of Temperature and Pressure on Hot Mixed Asphalt Compaction: Field and Laboratory Study. *J. Mater. Civ. Eng.* **2008**, *20*, 440–448. [CrossRef]
53. Azari, H.; McCuen.; H., R.; Stuart, K.D. Optimum Compaction Temperature for Modified Binders. *J. Transp. Eng.* **2003**, *129*, 531–537. [CrossRef]
54. Bennert, T.; Reinke, G.; Mogawer, W.; Mooney, K. Assessment of workability and compactability of warm-mix asphalt. *Transp. Res. Rec.* **2010**, *2180*, 36–47. [CrossRef]

MDPI

Article

A Study on Heat Storage and Dissipation Efficiency at Permeable Road Pavements

Ching-Che Yang [1], Jun-Han Siao [1], Wen-Cheng Yeh [1] and Yu-Min Wang [2,*]

[1] Department of Civil Engineering, National Pingtung University of Science and Technology, Pingtung 91201, Taiwan; a000550@oa.pthg.gov.tw (C.-C.Y.); p10833002@g4e.npust.edu.tw (J.-H.S.); weyeh@mail.npust.edu.tw (W.-C.Y.)

[2] General Research Service Center, National Pingtung University of Science and Technology, Pingtung 91201, Taiwan

* Correspondence: wangym@mail.npust.edu.tw; Tel.: +886-8-7703202 (ext. 6394)

Citation: Yang, C.-C.; Siao, J.-H.; Yeh, W.-C.; Wang, Y.-M. A Study on Heat Storage and Dissipation Efficiency at Permeable Road Pavements. *Materials* **2021**, *14*, 3431. https://doi.org/10.3390/ma14123431

Academic Editors: Markus Oeser, Michael Wistuba, Pengfei Liu and Di Wang

Received: 19 May 2021
Accepted: 16 June 2021
Published: 21 June 2021

Abstract: The main contributing factor of the urban heat island (UHI) effect is caused by daytime heating. Traditional pavements in cities aggravate the UHI effect due to their heat storage and volumetric heat capacity. In order to alleviate UHI, this study aims to understand the heating and dissipating process of different types of permeable road pavements. The Ke Da Road in Pingtung County of Taiwan has a permeable pavement materials experiment zone with two different section configurations which were named as section I and section II for semi-permeable pavement and fully permeable pavement, respectively. The temperature sensors were installed during construction at the depths of the surface course (0 cm and 5 cm), base course (30 cm and 55 cm) and subgrade (70 cm) to monitor the temperature variations in the permeable road pavements. Hourly temperature and weather station data in January and June 2017 were collected for analysis. Based on these collected data, heat storage and dissipation efficiencies with respect to depth have been modelled by using multi regression for the two studied pavement types. It is found that the fully permeable pavement has higher heat storage and heat dissipation efficiencies than semi-permeable pavement in winter and summer monitoring period. By observing the regressed model, it is found that the slope of the model lines are almost flat after the depth of 30 cm. Thus, from the view point of UHI, one can conclude that the reasonable design depth of permeable road pavement could be 30 cm.

Keywords: porous asphalt concrete; permeable road pavement; temperature distributions; urban heat island effect

1. Introduction

With the development of cities, the population requires more facilities, which leads to more roads and building construction [1]. Urban heat island (UHI) is a commonly observed phenomenon worldwide [2]. The formation of UHI can be mainly ascribed to the increased absorption and trapping of solar radiation in built-up urban fabrics, and to other factors, including population density of built-up areas and vegetation fractions [3]. It is an urban area where temperatures are significantly higher than those in the surrounding areas [2]. Therefore, urban environmental problems will inevitably become more serious and important [4]. Previous studies about the UHI effect showed that daytime heating is the main contributing factor [5,6]. Some studies have also found that the high temperature in the summer caused by UHI [7], that will cause public health problems such as the increase of heat strokes during the day [8], and an increase in the number of tropical nights leads to an increase in the morbidity and mortality of the elderly or people with chronic diseases [9,10].

Pavement as a key contributor to higher surface and air temperatures have been pointed by researchers [6,11–14]. Many references mentioned that traditional pavement had high thermal conductivity properties tend to absorb large amounts of solar radiation during daytime and to release it to the cooler surrounding atmosphere subsequently at night, further the formation of the UHI phenomenon [10,14–19]. Thus, the Environmental Protection Agency (EPA) and many researches have recommend permeable pavements to mitigate UHI [3,19–23], and, all around the world, many research results have been presented and compared temperatures of traditional pavements and various permeable pavements whether their tests in fields or in laboratories as shown next the paragraph.

In Taiwan, the in-situ study by Cheng et al. [18] and Paramitha et al. [24] utilized T-type thermocouples to obtain temperature fluctuations in the diurnal cycle of dense-graded asphalt concrete, permeable asphalt, permeable concrete, and permeable interlocking concrete block pavements in summer. From temperature fluctuations in diurnal cycle results by Cheng et al. [18] pointed out that, due to solar radiation can penetrate deeply into the structural porosity of porous asphalt concrete pavement, lead to the peak temperatures at near-surface of it was higher than that of dense-graded asphalt concrete pavement. In addition, the results by Paramitha et al. [24] found that temperature between pavements different, especially at noon. Permeable pavements had lower temperature than impermeable pavement at noon. In Japan, Tokyo, from Asaeda and Ca's field experiment results [25] found that the porous concrete pavement had higher maximum temperatures, but lower minimum temperatures relative to the traditional concrete pavement in summer. In a parking lot of Iowa, Kevern et al. [26] also utilized temperature sensors to obtain temperature behaviors inside of traditional portland cement concrete and portland cement pervious concrete. The research results conducted that portland cement pervious concrete was hotter than the traditional portland cement concrete during daytime. However, the portland cement pervious concrete was cooler than traditional concrete pavement during nighttime. In Phoenix, Arizona, Stempihar et al. [23] also conducted temperatures comparison of dense graded asphalt, porous asphalt, and porland cement concrete pavements. It was observed that porous asphalt had higher daytime surface temperatures and lowest nighttime temperatures than other materials. In addition to the field tests, the researchers also obtained similar results in the laboratory.

In the laboratory study by Hassne et al. [19], temperature of asphalt slabs with different air voids contents had been measured by J-type thermocouples. The outcome was found that the asphalt mixture slab with high air content is easier to reach a stable temperature than that the low air content. It also be found that asphalt mixture with high air voids content heating and dissipation rates higher than that with the low air voids content. Similarly, the laboratory-simulated study by Wu et al. [22] also showed that the porous surface of the open-graded specimen had lower temperatures than air temperatures during nighttime. Open-graded mixture also had better dissipation effect and declined faster than the dense-graded specimen, especially under wind convection conditions. According to the reference, the wind convection and evaporation in the environment can help to temperatures of permeable pavements decline rapidly and increase dropping rate [22]. Haselbach et al. [27] showed that after a rainfall, porous concrete cooled more rapidly than conventional concrete, thus indicating that evaporative dissipation was occurring. As anticipated, ability of porous asphalt to exhibit cool temperatures at nighttime by quickly dissipating high daytime temperatures makes it an excellent tool to mitigate the UHI effect.

In addition, a one-dimensional pavement model by Stempihar et al. [23] to measure surface temperatures of porous asphalt, dense-graded asphalt, and portland cement concrete. Results showed that porous asphalt had higher surface temperatures than others in daytime. However, porous asphalt had a lower surface temperature than others at nighttime. Even though temperatures of the porous asphalt pavement were rapidly increasing as air temperatures increase, but the temperature of the porous asphalt pavement was rapidly decreasing as air temperatures decrease. Consistent with results of Cheng et al. [28], porous pavements had lower temperature than those for traditional pavements on very hot days. Thus, the large-scale applications of porous pavements could help mitigate urban heat island impacts.

As mentioned in front, permeable pavements could effective to aid UHI effect mitigation. According to the Stempihar et al. [23] study, it becomes important to evaluate the entire pavement structure and material properties when selecting pavement materials to mitigate urban heat island. In the present study constructed two kinds of porous asphalt road pavements, called that fully permeable and semi-permeable. These pavements differ between their materials under the subgrade. Moreover, the temperature sensors were embedded in two permeable road pavements for monitoring the temperature behaviors. In the past, these the storage and dissipation efficiencies of permeable road pavements have not been given much attention during hotter and cool period and are not easily available in the literature. Thus, with the objective of understanding the heating and dissipating process of permeable road pavements. The heat storage and dissipation efficiencies of pavements were calculated by formulas and using regression models to find out the relationships of pavements depths verse those. To further recommend the best permeable road pavement for mitigating UHI effect.

2. Materials and Methods

2.1. Study Site

The study site, Ke Da Road, is approximately 10 km in south of Pingtung City, Taiwan. The Ke Da Road is about 5.4 km in length, starting from the Formosa Highway and towards the National Pingtung University of Science and Technology (NPUST) Campus. The experiment zone in the present study is 0.2 km long as shown in Figure 1. Two different kinds of permeable road pavements were constructed by measuring 100 m in length * 8.5 m in width with a surface slope of 2%. The pavements were diverse such that section I is semi-permeable while section II is fully permeable. As Figure 2 presented, two type permeable road pavements had the same material configuration at surface course, base course, and subgrade. The material is Porous Asphalt Concrete (PAC) at the surface course, Permeable Concrete (PC) at the base course, and filter layer at the subgrade, respectively. The material physical property of semi and fully permeable pavements is listed in Table 1. In addition, under subgrade configuration where there is an impermeable cloth in section I while a geotextile in section II. In general, high density polyethylene geomembrane (HDPE) impermeable cloth is used in the slopeland engineering more frequently than used in the road engineering. The HDPE impermeable cloth is used to block the infiltration of water. However, the geotextile in Section II, which was woven with polyolefin fibers, and has good water conduction and filtration. The main purpose of using the HDPE impermeable cloth and the geotextile in permeable road pavements in this study is to compare the difference for heat storage and dissipation capabilities under permeable and non-permeable conditions. These geotechnical material specifications are listed in Tables 2 and 3.

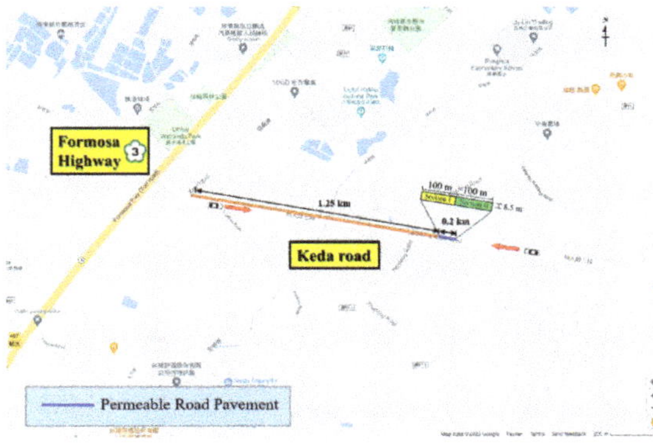

Figure 1. Location of the study area.

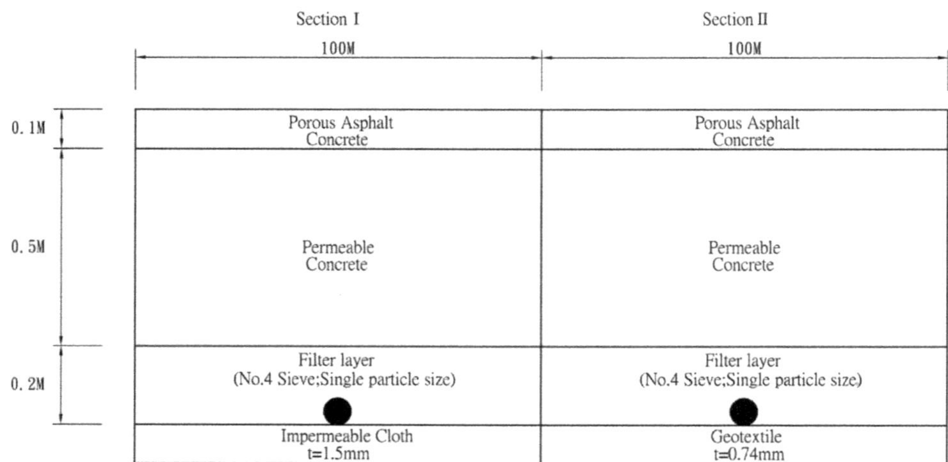

Figure 2. The pavement designs of this study.

Table 1. Material Physical property of semi-permeable and fully permeable pavements.

Course	Physical Property of Materials	Section I	Section II
Surface Course		Porous Asphalt Concrete (PAC)	
	Asphalt Contents (%)	4.8	
	Void ratio (%)	19.5	
	Thickness (cm)	10	
Base Course		Permeable Concrete (PC)	
	Water–Cement Ratio (%)	0.54	
	Amount of Cement(kg)	286	
	Aggregate Volume (m^3)	0.712	
	Amount of Aggregate (kg)	1920	
	Material Quality (kg)	2335	
	Porosity (%)	24.1	
	Thickness (cm)	50	
Subgrade Course		Filter Layer No. 4 Sieve; Single particle size	
	Thickness (cm)	20	
Type of Geosynthetic	Material	Impermeable Cloth	Geotextile
	Thickness (mm)	1.5	0.74

Table 2. Material specification of H.D.P.E impermeable cloth in Section I.

Physical Property	Test Method	Specification
Thickness (mm)	ASTM D5199	2.0 ± 10%
Density (g/cm^3)	ASTM D1505/792	≥0.90

Table 3. Material specification of geotextile in Section II.

Physical Property	Test Method	Specification
Positive permeable rate (1/S)	ASTM-D4491 or CNS 13298	≥0.6
Apparent opening size (mm)	ASTM-D4751 or CNS 14262	≤0.4

2.2. Temperature Data Collection

In order to observe the temperature behaviors of permeable road pavements, five type K thermocouple sensors (from Campbell Scientific, Inc.®, Logan, UT, USA and with measurement temperature range from −55 °C to 125 °C) were embedded into the pavements. As shown in Figure 3, the sensors were embedded within pavement at depths of 0, 5, 30, 55 and 70 cm. The first temperature sensor was installed on the surface for monitoring the surface temperature. The second and fifth temperature sensors were embedded in the middle of the materials for monitoring the temperatures of the porous asphalt concrete and the filter layer, respectively. In addition, the third and fourth temperature sensors were arranged at equal distances depths, 25 cm and 50 cm, are apart from the second sensor to investigate the heat conduction effect of various materials going forward. On the other hand, a long term investigation was carried out from January to December 2017. Temperature data of the different permeable road pavements were automatically recorded and collected by the CR1000 data logger (from Campbell Scientific, Inc.®) at hourly intervals. The air temperatures of the environment were logged in NPUST weather station. The analysis covers days with the negligible precipitation and relatively higher air temperatures in 2017, which are extreme weather conditions for UHI impact. Based on many studies, the UHI effect has been investigated during the summer [14,18,21,23–27]. However, researchers [21,23] suggested that cooler climates should be considered. Thus, temperature distributions of permeable road pavements in January and in June are taken into account in this study. Using data with no rainfall in January and June to represent the variation of temperatures along different pavement depths in winter and summer.

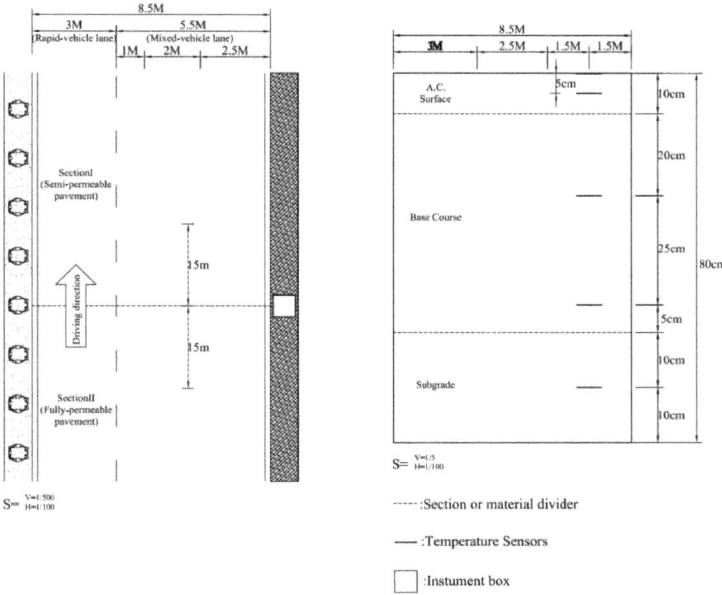

Figure 3. Sketch for the sensor locations in the study.

2.3. Calculation of Heat Storage and Dissipation Efficiency

In the present study, Equation (1) is adopted to evaluate the heat storage efficiency and heat dissipation efficiency. The heat storage efficiency and heat dissipation efficiency were defined as the average gradient of temperature with respect to corresponding time intervals. The rising and dropping limbs of the temperature profile in Figures 4–7 are used in the Equation (1) to estimate the heat storage efficiency and heat dissipation efficiency, respectively. Equation (1) is used to calculate heat storage efficiency and heat dissipation efficiency for the pavements in this study. Heat storage efficiency is that the highest temperature of the day subtracts the lowest temperature, then the result is divided by time intervals of the rising limb. On the other hand, heat dissipation efficiency is the lowest temperature minus the highest temperature of the day, then the result is divided by time intervals of dropping limbs.

$$\text{Heat storage efficiency and Heat dissipation efficiency} = \frac{T_{max} - T_{min}}{\Delta t}, \quad (1)$$

where T_{max} is the maximum temperature, °C; $Tmin$ is the minimum temperature, °C; Δt is the corresponding time intervals, h.

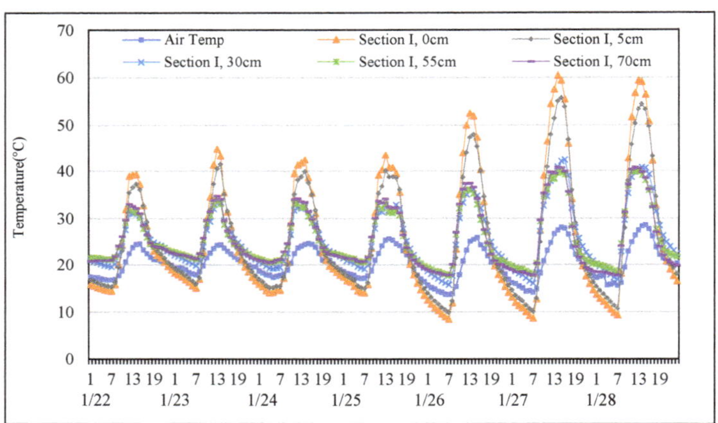

Figure 4. Temperature distribution at each depth of Section I from the 22nd to the 28th of January.

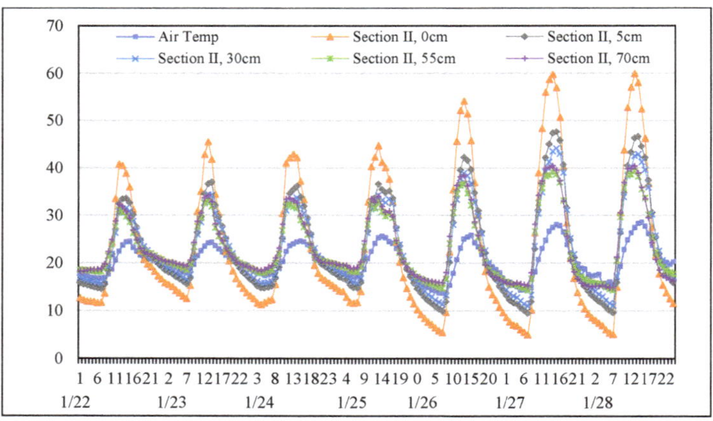

Figure 5. Temperature distribution at each depth of Section II from the 22nd to the 28th of January.

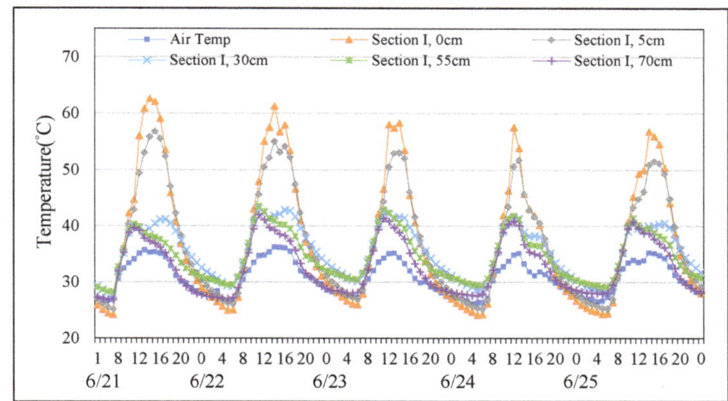

Figure 6. Temperature distribution at each depth of Section I from the 21st to the 25th of June.

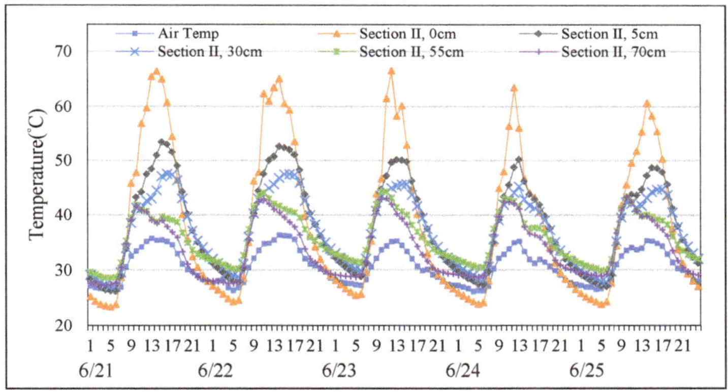

Figure 7. Temperature distribution at each depth of Section II from the 21st to the 25th of June.

2.4. Regression Analysis of Permeable Road Pavement Temperatures with Respect to Depths

In order to understand the heating and dissipating capacities of permeable road pavements, heat storage and dissipation efficiencies obtained from Equation (1) were employed as a basis for modelling, the method is established by regression analysis using the heat storage and dissipation efficiencies for various depths of 22nd to 28th in January and 21st to 25th in June of 2017 at the experiment zone as dependent variables.

3. Results and Discussion

3.1. Temperature Distribution of Permeable Road Pavements in January

This study agrees with previous studies [17,23,26,28,29] which found that the temperature at each depth changed basically synchronously with the air temperature are shown in Figures 4 and 5. Amplitudes of pavements temperature variations decreased while the depth increased. Results showed that temperatures of sections I and II always warmer than air temperature during the hottest period of the day. Due to heat absorption capacities of permeable road pavements leads to their surface temperatures are higher than air temperature. Both sections subsequently cooled than air temperatures in the early morning. In addition, the results also showed that the mean surface temperature of section II approximately 1 °C was warmer than that section I. Although the mean surface temperature of section II was higher than that of section I during the day, the mean surface temperature of section II approximately 3 °C cooled than that of section I early in the morning. It is found that the dissipation capacity of fully permeable road pavement is better than that of semi-permeable road pavement in winter.

3.2. Heat Storage and Dissipation Efficiency in January

As listed, Table 4 is a summary of heat storage and dissipation efficiencies at each depth for permeable road pavements, from 22nd to 28th in January. The results showed that heat storage efficiencies of section I were 5.3, 4.6, 2.4, 2.6, and 2.8 °C/h at depths of 0, 5, 30, 55, and 70 cm, respectively. It should be noted that the heat storage efficiency of the surface was higher than that of other depths in section I. However, the heat storage efficiencies gradually decreased as the depth increases until the depth of 30 cm. Under the continuous pavement heat conduction, the heat storage efficiencies at depths of 55 cm to 70 cm in section I slightly increased. The results also showed that heat dissipation efficiencies at the surface of section I had higher about 2 °C/h than other depths. Below the surface course, heat dissipation efficiencies almost the same in the base course and the subgrade with heat dissipation efficiencies about 1 °C/h.

Table 4. Heat storage and dissipation efficiencies in 22nd–28th January.

	Course		**Surface Course**		**Base Course**		**Subgrade**
Section	**Item**	**Air Temp.**	**0 cm**	**5 cm**	**30 cm**	**55 cm**	**70 cm**
I	Time interval (h)	–	7	7	7	6	6
	Max. Temp. at rising limb (°C)	26.0	48.9	45.3	35.7	35.5	36.5
	Min. Temp. at rising limb (°C)	16.1	12.0	13.2	18.1	19.7	19.5
	Heat storage efficiency (°C/h)	–	5.3	4.6	2.4	2.6	2.8
	Time interval (h)	–	17	17	17	18	18
	Max. Temp. at dropping limb (°C)	26.0	47.2	43.8	35.0	34.7	35.8
	Min. Temp. at dropping limb (°C)	16.1	11.6	12.8	18.0	19.4	19.3
	Heat dissipation efficiency (°C/h)	–	2.1	1.8	1.0	0.9	0.9
II	Time interval (h)	–	7	7	7	6	6
	Max. Temp. at rising limb (°C)	26.0	49.7	40.0	37.2	34.9	36.2
	Min. Temp. at rising limb (°C)	16.1	9.0	12.7	14.0	16.6	17.4
	Heat storage efficiency (°C/h)	–	5.8	3.9	3.3	3.1	3.1
	Time interval (h)	–	17	17	17	18	18
	Max. Temp. at dropping limb (°C)	26.0	48.0	38.9	36.2	34.2	35.5
	Min. Temp. at dropping limb (°C)	16.1	8.5	12.4	13.7	16.4	17.2
	Heat dissipation efficiency (°C/h)	–	2.3	1.4	1.3	1.0	1.0

On the other hand, it was observed that section II had high heat storage efficiency about 5.8 °C/h at surface. However, the heat storage efficiency at the depth of 5 cm was lower about 2 °C/h than the heat storage efficiency at the surface. Although the surface heat storage efficiency of section II was very high, the heat storage efficiency at the depth of 5 cm in section II decreased a lot. Subsequently, the heat storage efficiency gradually becomes the same as the depth increases. In terms of the heat dissipation efficiency of section II, the heat dissipation efficiency was 2.3 °C/h at the surface, which was higher about 1 °C/h than other depths. However, the heat dissipation efficiencies at depths of 5 cm to 70 cm in section II almost no difference.

In comparison with two sections, the heat storage and dissipation efficiencies at depths of 0 cm, 30 cm, 55 cm, and 70 cm of section II were higher than that of section I. Hence, it is found that heat storage and dissipation capacities of fully permeable road pavement performs is better than that of semi-permeable road pavement in winter.

3.3. Temperature Distribution of Permeable Road Pavements in June

Using consecutive five days data with no rainfall in June to represent temperature variation along pavement depths in summer. As shown in Figures 6 and 7, the temperature at each depth changed basically synchronously with the air temperature from June the 21st to the 25th. The results showed that temperatures of sections I and II always warmer than air temperature during the hottest period of the day. However, temperatures of two permeable road pavements were similar with air temperatures in the morning. In addition, the mean surface temperature of section II was warmer approximately 5 °C than of section I. Similar findings have also documented by previous research [30]. Meanwhile, the mean surface temperature of section II cooled approximately 1 °C than that of section I in the morning. In summer, fully permeable road pavement had better heat dissipation comparing with semi-permeable road pavement.

3.4. Heat Storage and Dissipation Efficiency in June

The summary of heat storage and dissipation efficiencies of two sections from 21st to 25th in June are listed in Table 5. The results showed that the heat storage efficiency of section I was 4.3 °C/h originally decreased to 2.1 °C/h as the depth increased. In addition, the results also showed that the heat dissipation efficiencies at the depth of 0 cm of section I was similar with the heat dissipation efficiencies at the depth of 5 cm in section I. However, below the surface course, the same heat dissipation efficiencies of 0.7 °C/h in section I.

Table 5. Heat storage and dissipation efficiencies in 21th–25th June.

	Course		**Surface Course**		**Base Course**		**Subgrade**
Section	**Item**	**Air Temp.**	**0 cm**	**5 cm**	**30 cm**	**55 cm**	**70 cm**
I	Time interval (h)	–	8	8	7	6	6
	Max. Temp. at rising limb (°C)	35.5	59.3	53.6	41.7	42.1	40.8
	Min. Temp. at rising limb (°C)	26.5	24.7	25.8	28.8	29.4	27.5
	Heat storage efficiency (°C/h)	–	4.3	3.5	2.3	2.1	2.1
	Time interval (h)	–	16	16	17	18	18
	Max. Temp. at dropping limb (°C)	35.5	60.0	54.2	41.7	42.1	41.0
	Min. Temp. at dropping limb (°C)	26.5	24.9	25.9	29.1	29.4	27.7
	Heat dissipation efficiency (°C/h)	–	2.2	1.8	0.7	0.7	0.7
II	Time interval (h)	–	8	8	7	6	6
	Max. Temp. at rising limb (°C)	35.5	64.5	51.0	44.6	43.4	42.7
	Min. Temp. at rising limb (°C)	26.5	24.2	27.5	28.6	30.2	28.3
	Heat storage efficiency (°C/h)	–	5.0	2.9	2.3	2.4	2.4
	Time interval (h)	–	16	16	17	18	18
	Max. Temp. at dropping limb (°C)	35.5	65.4	51.6	46.4	43.4	42.5
	Min. Temp. at dropping limb (°C)	26.5	24.4	27.8	28.8	30.6	28.6
	Heat dissipation efficiency (°C/h)	–	2.6	1.5	1.1	0.7	0.8

On the other hand, it was observed that section II had high heat storage efficiency of about 5.0 °C/h on surface. Although the surface heat storage efficiency of section II was very high, the heat storage efficiency at the depth of 5 cm in section II decreased a lot by comparing the surface. Then, the heat storage efficiency gradually became the same as the depth increased. It was also observed that the heat dissipation efficiency of section II gradually decreased as the depth increased.

In comparison with two sections, the heat storage efficiency at the depth of 0 cm in section II was higher than section I. However, the heat storage efficiencies at the depth of 5 cm in section II was lower than that in section I. As the depths increased, the heat storage efficiencies of section II almost the same with section I. Thus, more heating store on the surface of fully permeable road pavement than semi-permeable road pavement in summer. However, section II has higher heat dissipation efficiencies comparing with section I. The dissipation capacity of fully permeable road pavement is better than in semi-permeable road pavement in summer.

3.5. Modelling Permeable Road Pavement Temperature with Respect to Depth

The relationship of heat storage efficiencies and depths in semi and fully permeable road pavements are illustrated in Figure 8a,b. From the trend, it shows that the heat storage efficiency decreases while the depth increases. It can be clearly seen from the figures that the heat storage efficiencies of the upper course in semi and fully permeable pavements are higher than that in the lower courses for winter and summer. It is also found that the slope of the regressed model line is almost flat after depth equal to 30 cm. In addition, the regressed model line for winter is higher than that for summer. Also, the permeable road pavements storage less heat in summer comparing with in winter. Meanwhile, multi regression models were developed for estimating heat storage efficiencies at different depths of semi and fully permeable road pavements as presented by Equations (2) and (3), respectively. It can be seen that these two models will have good performance because their coefficients of determination (R^2) are 0.94 and 0.97.

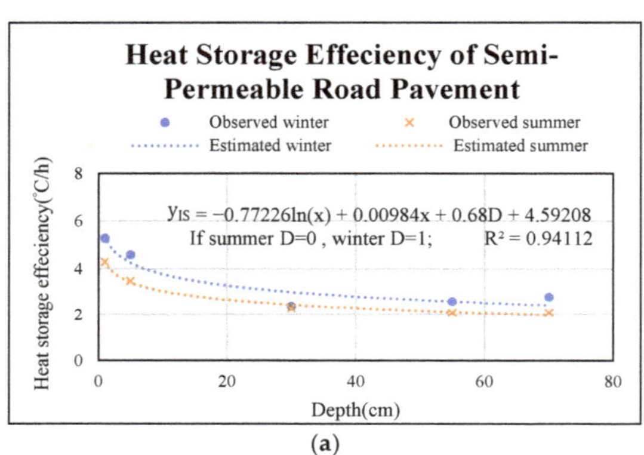

(a)

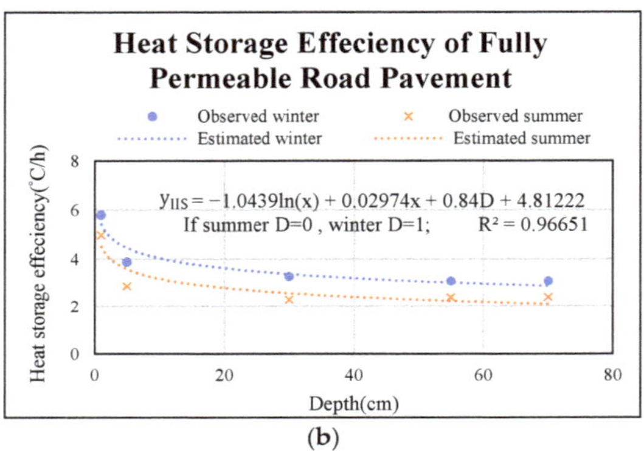

(b)

Figure 8. Relationships between heat storage efficiency verse permeable road pavements depth (**a**) Model for semi-permeable road pavement. (**b**) Model for fully permeable road pavement.

The model predicting heat storage efficiency at specified depth can be presented by the following equation:

$$y_{IS} = -0.77226\ln(x) + 0.00984x + 0.68D + 4.59208, \quad (2)$$

$$y_{IIS} = -1.0439\ln(x) + 0.02974x + 0.84D + 4.81222, \quad (3)$$

where y_{IS} is the heat storage efficiency of section I for winter and summer, °C/h; y_{IIS} is the heat storage efficiency of section II for winter and summer, °C/h; x is the specified pavement depth, cm; D is dummy variable that is 1 for winter and 0 for summer.

On the other hand, the relationship between heat dissipation efficiency and depth of semi and fully permeable road pavements are shown in Figure 9a,b. It is observed that heat dissipation efficiencies of permeable road pavements decrease while the depth increases. Obviously, the heat dissipation efficiencies in the surface course of semi and fully permeable pavements are larger than in the lower courses of semi and fully permeable pavements in winter and summer. It is also found that the slope of the regressed model line is almost flat after depth equal to 30 cm. One can see in these Figures, the regressed model line for winter are slightly higher than that for summer in each permeable road pavement. It is clear that no matter which season, the heat dissipation efficiencies of permeable road pavements are similar. In addition, the heat dissipation efficiency verse depth of fully permeable road pavement performs better than that of semi-permeable road pavement. For example, when the dissipation efficiency is 1 °C/h for semi-permeable road pavement at 25 cm depth, however the same dissipation efficiency rate is found at 30 cm depth in fully permeable road pavement, as shown in Figure 9a,b. By introducing dummy variable for winter and summer seasons, multi-regression models have been developed to form heat dissipation efficiencies at different depths of semi-permeable and fully permeable pavements as shown by Equations (4) and (5), respectively. It is observed that R^2 are 0.95 and 0.91, indicating that two models will have good performance.

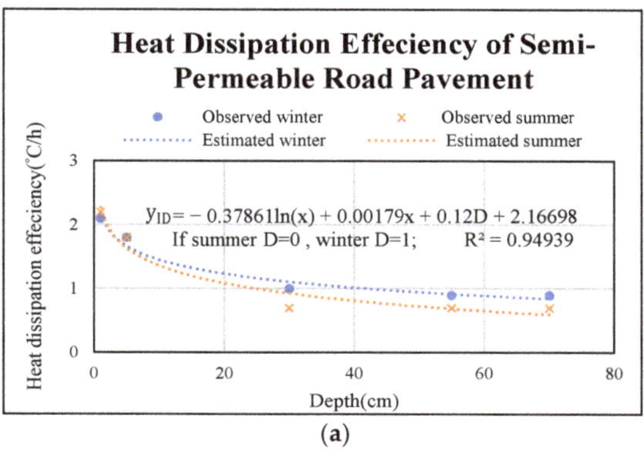

(a)

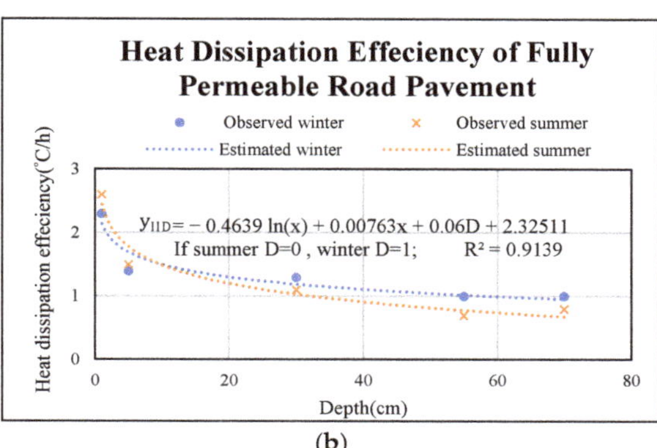

(b)

Figure 9. Relationships between heat dissipation efficiency verse permeable road pavements depth (**a**) Model for semi-permeable road pavement. (**b**) Model for fully permeable road pavement.

The model predicting heat dissipation efficiency at specified depth can be presented by the following equation:

$$y_{ID} = -0.37861\ln(x) + 0.00179x + 0.12D + 2.16698, \quad (4)$$

$$y_{IID} = -0.4639\ \ln(x) + 0.00763x + 0.06D + 2.32511, \quad (5)$$

where y_{ID} is the heat dissipation efficiency of section I for winter and summer, °C/h; y_{IID} is the heat dissipation efficiency of section II for winter and summer, °C/h; x is the specified pavement depth, cm; D is dummy variable that is 1 for winter and 0 for summer.

4. Conclusions

In this paper, two permeable road pavements, including semi and fully permeable road pavements, were monitored for temperatures at the depths of 0, 5, 30, 55, and 70 cm in January and June. Based on the collected temperature data, heat storage and dissipation efficiencies with respect to depth have been studied and modelled by using multi regression for the studied pavements. Through the collected and modelled results, the following conclusions are obtained:

- From temperature results in the two seasons, the maximum surface temperatures of permeable road pavements are higher than those of air temperatures in daytime. Meanwhile, the minimum surface temperatures of permeable road pavements are lower than that of the air temperature at early morning. It is also found that temperature variations of permeable road pavements have the same trend as air temperatures.
- According to the calculated heat storage and dissipation efficiencies, it is revealed that the heat storage and dissipation efficiencies of fully permeable road pavement are higher than those of semi-permeable road pavement in both January and June. No matter which season, it is found that heat storage and dissipation capacities with respect to depth in fully permeable road pavement performs better than that in semi-permeable road pavement for UHI.
- Through the results of modelling, urban heat island effect is easily to occur in summer. Reduction rates in depth of heat storage and the dissipation efficiencies in fully

permeable road pavement has better capacities in comparison with those of semi-permeable road pavement in summer. Thus, this further proves that using fully permeable road pavement is better than semi-permeable road pavement in hotter environments to aid urban heat island effect.

- Multi regression models were developed to relate the heat storage and dissipation efficiencies and depth for semi and fully permeable road pavements in this study. It is found that the slope of the regressed model lines is almost flat after the depth of 30 cm. Thus, from the view point of UHI, one can conclude that a reasonable design depth of permeable road pavement could be 30 cm.

Author Contributions: Conceptualization, Y.-M.W.; formal Analysis, C.-C.Y. and J.-H.S.; methodology, C.-C.Y. and J.-H.S.; resources, C.-C.Y.; Investigation, J.-H.S.; writing-original draft, J.-H.S.; writing-review & editing, W.-C.Y. and Y.-M.W.; supervision, Y.-M.W. All authors have read and agreed to the published version of the manuscript.

Funding: This study was received partly funding from Pingtung county government, Taiwan, R.O.C.

Conflicts of Interest: The authors declare no conflict of interest.

References

1. Rodriguez-Hernandez, J.; Fernández-Barrera, A.H.; Andrés-Valeri, V.C.; Vega-Zamanillo, A.; Castro-Fresno, D. Relationship between urban runoff pollutant and catchment characteristics. *J. Irrig. Drain. Eng.* **2013**, *139*, 833–840. [CrossRef]
2. Lee, K.; Kim, Y.; Sung, H.C.; Ryu, J.; Jeon, S.W. Trend Analysis of Urban Heat Island Intensity According to Urban Area Change in Asian Mega Cities. *Sustainability* **2020**, *12*, 112. [CrossRef]
3. Grimmond, S. Urbanization and global environmental change: Local effects of urban warming. *Geogr. J.* **2007**, *173*, 83–88. [CrossRef]
4. Huang, J.-M.; Chen, L.-C. A Numerical Study on Mitigation Strategies of Urban Heat Islands in a Tropical Megacity: A Case Study in Kaohsiung City, Taiwan. *Sustainability* **2020**, *12*, 3952. [CrossRef]
5. Haselbach, L.M.; Gaither, A. Preliminary field testing: Urban heat island impacts and pervious concrete. In Proceedings of the NRMCA 2008 Concrete Technology Forum: Focus on Sustainable Development, Denver, CO, USA, 20–22 May 2008.
6. Gui, J.; Phelan, P.E.; Kaloush, K.E.; Golden, J.S. Impact of pavement thermophysical properties on surface temperatures. *J. Mater. Civ. Eng.* **2007**, *19*, 683–690. [CrossRef]
7. Benrazavi, R.S.; Dola, K.B.; Ujang, N.; Benrazavi, N.S. Effect of pavement materials on surface temperatures in tropical environment. *Sustain. Cities Soc.* **2016**, *22*, 94–103. [CrossRef]
8. Higashiyama, H.; Sano, M.; Nakanishi, F.; Takahashi, O.; Tsukuma, S. Field measurements of road surface temperature of several asphalt pavements with temperature rise reducing function. *Case Stud. Constr. Mater.* **2016**, *4*, 73–80. [CrossRef]
9. Royé, D. The effects of hot nights on mortality in Barcelona, Spain. *Int. J. Biometeorol.* **2017**, *61*, 2127–2140. [CrossRef] [PubMed]
10. Scott, A.A.; Waugh, D.W.; Zaitchik, B.F. Reduced Urban Heat Island intensity under warmer conditions. *Environ. Res. Lett.* **2018**, *13*, 064003. [CrossRef]
11. Doulos, L.; Santamouris, M.; Livada, I. Passive cooling of outdoor urban spaces. The role of materials. *Sol. Energy* **2004**, *77*, 231–249. [CrossRef]
12. Golden, J.S.; Kaloush, K.E. Mesoscale and microscale evaluation of surface pavement impacts on the urban heat island effects. *Int. J. Pavement Eng.* **2006**, *7*, 37–52. [CrossRef]
13. Oke, T.; Johnson, G.; Steyn, D.; Watson, I. Simulation of surface urban heat islands under 'ideal'conditions at night part 2: Diagnosis of causation. *Bound. Layer Meteorol.* **1991**, *56*, 339–358. [CrossRef]
14. Mohajerani, A.; Bakaric, J.; Jeffrey-Bailey, T. The urban heat island effect, its causes, and mitigation, with reference to the thermal properties of asphalt concrete. *J. Environ. Manage.* **2017**, *197*, 522–538. [CrossRef] [PubMed]
15. Musco, F. *Counteracting Urban Heat Island Effects in a Global Climate Change Scenario*; Springer International Publishing: Cham, Switzerland, 2016; ISBN 978-3-319-10424-9.
16. Deilami, K.; Kamruzzaman, M.; Liu, Y. Urban heat island effect: A systematic review of spatio-temporal factors, data, methods, and mitigation measures. *Int. J. Appl. Earth Obs. Geoinf.* **2018**, *67*, 30–42. [CrossRef]
17. Yang, H.; Yang, K.; Miao, Y.; Wang, L.; Ye, C. Comparison of potential contribution of typical pavement materials to heat island effect. *Sustainability* **2020**, *12*, 4752. [CrossRef]
18. Hsu, C.-Y.; Chen, S.-H.; Lin, J.-D. The In-Situ Temperature Evaluations of Permeable Pavements in Summer. *J. Mar. Sci. Technol.* **2015**, *23*, 288–292.
19. Hassn, A.; Aboufoul, M.; Wu, Y.; Dawson, A.; Garcia, A. Effect of air voids content on thermal properties of asphalt mixtures. *Constr. Build Mater.* **2016**, *115*, 327–335. [CrossRef]
20. United States Environmental Protection Agency (USEPA). *Reducing Urban Heat Islands: Compendium of Strategies-Cool Pavements*; 2012; Draft.

21. Santamouris, M. Using cool pavements as a mitigation strategy to fight urban heat island—A review of the actual developments. *Renew. Sust. Energ. Rev.* **2013**, *26*, 224–240. [CrossRef]
22. Wu, H.; Sun, B.; Li, Z.; Yu, J. Characterizing thermal behaviors of various pavement materials and their thermal impacts on ambient environment. *J. Clean. Prod.* **2018**, *172*, 1358–1367. [CrossRef]
23. Stempihar, J.J.; Pourshams-Manzouri, T.; Kaloush, K.E.; Rodezno, M.C. Porous Asphalt Pavement Temperature Effects for Urban Heat Island Analysis. *Transp. Res. Rec.* **2012**, *2293*, 123–130. [CrossRef]
24. Paramitha, P.A.; Djakfar, L.; Lin, J.-D.; Hsu, C.-Y. The Effect of Permeable Pavement Use on Air Temperature of Surrounding Building. *Rekayasa Sipil* **2014**, *8*, 1–6.
25. Asaeda, T.; Ca, V.T. Environment, Characteristics of permeable pavement during hot summer weather and impact on the thermal environment. *Build Environ.* **2000**, *4*, 363–375. [CrossRef]
26. Kevern, J.T.; Schaefer, V.R.; Wang, K. Temperature behavior of pervious concrete systems. *J. Transp. Res. Rec.* **2009**, *2098*, 94–101. [CrossRef]
27. Haselbach, L.; Boyer, M.; Kevern, J.T.; Schaefer, V.R. Cyclic heat island impacts on traditional versus pervious concrete pavement systems. *J. Transp. Res. Rec.* **2011**, *2240*, 107–115. [CrossRef]
28. Cheng, Y.-Y.; Lo, S.-L.; Ho, C.-C.; Lin, J.-Y.; Yu, S.L. Field testing of porous pavement performance on runoff and temperature control in Taipei City. *Water* **2019**, *11*, 2635. [CrossRef]
29. Kevern, J.T.; Haselbach, L.; Schaefer, V.R. Hot weather comparative heat balances in pervious concrete and impervious concrete pavement systems. *Int. J. Heat Technol.* **2012**, *7*, 2012.
30. Li, H.; Harvey, J.T.; Holland, T.J.; Kayhanian, M. The use of reflective and permeable pavements as a potential practice for heat island mitigation and stormwater management. *Environ. Res. Lett.* **2013**, *8*, 015023. [CrossRef]

Article

Noise Reduction Characteristics of Macroporous Asphalt Pavement Based on A Weighted Sound Pressure Level Sensor

Feng Lai [1], Zhiyong Huang [2,*] and Feng Guo [3]

1 School of Highway, Chang'an University, Middle Section of South Second Ring Road, Xi'an 710064, China; 2017021068@chd.edu.cn
2 School of Civil Engineering and Transportation, South China University of Technology, Wushan Road, Tianhe District, Guangzhou 510641, China
3 Department of Civil and Environmental Engineering, University of South Carolina, Columbia, SC 29201, USA; fengg@email.sc.edu
* Correspondence: 201910101471@mail.scut.edu.cn; Tel.: +86-138-0887-5012

Abstract: Based on the manual of macroporous noise-reducing asphalt pavement design, the indoor main drive pavement function accelerated loading test system was applied to investigate the impact of speed, loading conditions (dry and wet) and structural depth on the noise reduction of macroporous Open Graded Friction Course (OGFC) pavement, as well as its long-term noise reduction. Combined with the noise spectrum of the weighted sound pressure level, the main components and sensitive frequency bands of pavement noise under different factors were analyzed and compared. According to experimental results, the noise reduction effect of different asphalt pavements from strong to weak is as follows: OGFC-13 > SMA-13 > AC-13 > MS-III. The noise reduction effect of OGFC concentrates on the frequency of 1–4 kHz when high porosity effectively reduces the air pump effect. As the effect of wheels increases and the depth of the road structure decreases, the noise reduction effect of OGFC decreases. It indicates the noise reduction performance attenuates at a later stage, similar to the noise level of densely graded roads.

Keywords: asphalt pavement; accelerated loading test; weighted sound pressure level; pavement noise; noise spectrum

Citation: Lai, F.; Huang, Z.; Guo, F. Noise Reduction Characteristics of Macroporous Asphalt Pavement Based on A Weighted Sound Pressure Level Sensor. *Materials* **2021**, *14*, 4356. https://doi.org/10.3390/ma14164356

Academic Editors: Markus Oeser, Michael Wistuba, Pengfei Liu and Di Wang

Received: 1 July 2021
Accepted: 26 July 2021
Published: 4 August 2021

1. Introduction

With the economic development, the number of vehicles has increased rapidly. Concomitant with this, the traffic noise pollution caused by vehicles has become more and more serious, not only bringing negative impacts on the surrounding ecological environment but also endangering human health [1–4]. Traffic noise mainly includes vehicle noise and pavement noise, and the latter is caused by the friction between tires and the road during acceleration, deceleration and braking of the vehicle, which is greatly affected by traffic conditions, road type, road gradient, driving speed, etc. [5,6]. The shaping factors of pavement noise are air pumping effect, tire vibration and aerodynamic noise [7]. According to previous studies, when the speed of a car is over 30 km/h and the speed of a truck or bus is over 50 km/h, pavement noise becomes the main component of traffic noise [8].

The influence factors of tire or pavement noise can be divided into tire factors and road factors. The former factor mainly includes tire structure and tread pattern design, and the latter factor includes the surface structure characteristics of the road surface and its own acoustic impedance (or sound absorption) characteristics, etc. [9–11]. Asphalt pavement is designed to provide a safe [12], high performance [13–16], long service life [17–20] and comfortable driving with low road noise [4,7,8]. The road industry usually designs low-noise pavement to reduce tire pavement noise, and the representative pavement is porous asphalt pavement (OGFC asphalt pavement) [21]. According to the previous study [22,23], the noise generated by OGFC with a thickness of 40–50 mm is 3–6 dB lower than that of ordinary asphalt concrete pavement, which is equivalent to half of the traffic volume. The research and application of macroporous asphalt pavement started early in China. However, it is limited to a large-scale application due to its short service life. The connecting pores of macroporous asphalt pavement are easily blocked, leading to excessively rapid degradation of drainage and noise reduction performance. As a result, how to balance the durability of the macroporous asphalt mixture with the anti-skid and noise reduction function to optimize the design of the mixture ratio has attracted great attention.

Currently, the method of outdoor pavement noise testing is relatively mature, which mainly uses special vehicles or trailers to collect noise on the test road at different driving speeds, expressed by sound pressure levels [24,25]. To follow the design manual, it is necessary to carry out a large number of designs of noise-reducing asphalt mixtures and to perform amount-of-noise tests [26,27]. Due to limited indoor fields and the size of the test piece, it is difficult to use the vehicle driving mode to conduct tests. Chang'an University has developed an indoor test method, which simulates the noise generated by the interaction between tires and the road surface when the vehicle is running, making the test tire roll freely from the track with a certain slope and dive onto the road slab test piece. Additionally, it can analyze the noise characteristics of the moment when the tire is in contact with the road slab test piece [28]. This method greatly reduces the cost of paving the test road, but it has high requirements for background noise control, and it is difficult to control the tire rolling speed during the test [29,30].

In order to accurately and robustly test the driving noise between the tire and different pavement, this study adopts the main driving road function accelerated loading test system, which is an all-weather road function accelerated loading simulation test system and can realize the real simulation of the interaction between the tire and the road. This system can be used to control the variable parameters, such as test temperature, humidity, ultraviolet aging, rainfall, tire axle load, the number of actions, etc. The noise law and noise reduction mechanism of different asphalt pavements are comprehensively analyzed to identify the main influence factors of pavement noise. Besides, the study of the attenuation law of pavement noise reduction performance under long-term action was conducted, aiming to provide a reference for the optimization design of durable noise reduction pavement.

1.1. Tire Noise Generation Mechanism

Tire or pavement noise can be divided into tire vibration noise and aerodynamic noise, and the former one is caused by the impact of tread patterns on the road surface, vibration due to an uneven road surface, the friction between tires and road surface, etc. [5]. As shown in Figure 1, during the process from tire tread touching the road to leaving, the tread pattern impacts on the road surface and the deformation of tread cause the vibration along the radial direction of tread, and this induces the whole tread and sidewall vibrations. The noise pressure mainly depends on the tire rubber material, tread pattern design and road smoothness characteristics. Vibration noise is one of the sources of high-frequency part of tire or pavement noise (noise frequency is greater than 2 kHz) [7].

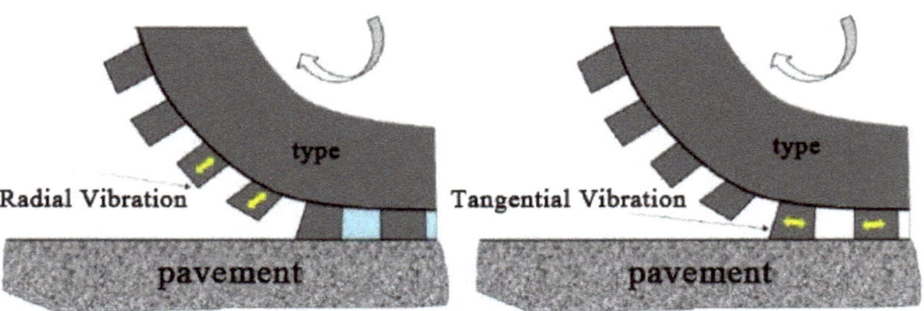

Figure 1. Tire vibration and noise.

Aerodynamic noise includes air pumping effect, air turbulence effect, acoustic tube resonance effect and Helmholtz resonance effect, among which air pumping noise contributes the most to aerodynamic noise [31]. When the tire is rolling on the pavement, once the tread of the contact part is deformed, there will be many small cavities between the tread patterns and the pavement. The tire is in close contact with the pavement during its advancement, the air in the small cavity is squeezed, part of the air is discharged to form a partial unstable air flow, and the remaining part forms a larger pressure air mass in the cavity. When the tire leaves the pavement, the volume of the compressed small cavity suddenly increases, forming a partial vacuum, and the air is sucked in abruptly. This phenomenon is called the "air pumping effect", as shown in Figure 2. Air pumping noise frequency is around 1 kHz in the middle frequency range, which is the main source of high-speed driving noise [32].

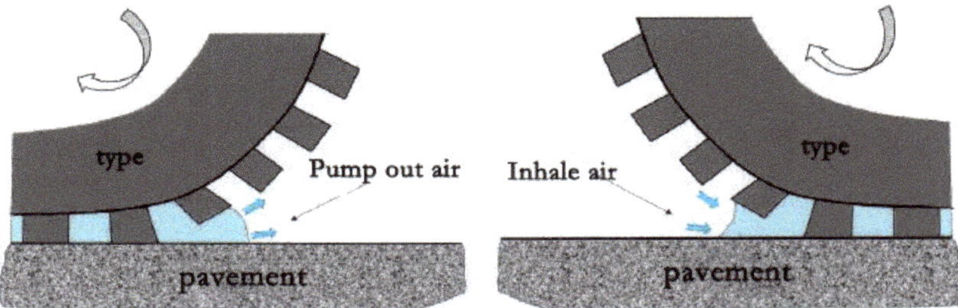

Figure 2. Air pumping effect.

1.2. Noise Evaluation Based on a Weighted Sound Pressure Level

The sound volume can be characterized by the amplitude, frequency and phase of the sound pressure, which is a function of space and time. In order to describe it more accurately, the sound pressure level is usually used to express the acoustic value. The mathematical expression is:

$$\overline{L_p} = 10\lg\left[\frac{1}{N}\sum_{i=1}^{N} 10^{0.1(L_{pi}-K_{li})}\right] - K_2 - K_3 \tag{1}$$

In the formula, $\overline{L_p}$ is the sound pressure level of the test object, dB (A); N is number of measuring points per unit time; L_{pi} is the sound pressure level of point i, dB (A); K_{li} is the background noise correction value of point I; K_2 is the correction value for the use environment; K_3 is the correction value of temperature and air pressure.

Human ears are different in sensitivity to noise of different frequencies. Therefore, in order to make the objective measurement of pavement noise consistent with human hearing, a certain frequency weighting network correction is required for pavement noise, among which the A-weighted sound pressure level frequency response is closest to the human hearing characteristics. The measurement method is simple and reasonable and has become the most widely used evaluation parameter for noise measurement. According to the international standard IEC 61672A, the A-weighting network curve function is:

$$\mathrm{A}(f) = 20\lg\left[\frac{f_4^2 f^4}{(f^2+f_1^2)(f^2+f_2^2)^{\frac{1}{2}}(f^2+f_3^2)^{\frac{1}{2}}(f^2+f_4^2)}\right] - A_{1000} \tag{2}$$

In the formula, A_{1000} is the sound pressure level of 1 kHz; f represents the calculation frequency; f_1 = 20.6 Hz, f_2 = 107.7 Hz, f_3 = 737.9 Hz and f_4 = 12194 Hz.

Moreover, the sound transmission has significant time and frequency domain characteristics; that is, pavement noise is essentially composed of sounds of different frequencies.

In order to get the whole picture of the characteristics of tire pavement noise on different roads, it is necessary to convert the time-domain waveform of pavement noise into a spectrogram through the Fast Fourier Transform (abbreviated as FFT) algorithm. The converted frequency domain oscillogram shows the frequency composition of the noise and the corresponding signal energy. The structure is analyzed to determine the specific noise source for special noise control.

The real-time processing of sound signals is realized to ensure the speed of calculation when there are a lot of data, mainly using the FFT algorithm. FFT expression is as follows:

$$X(f) = \int_{-\infty}^{\infty} x(t)e^{-2\pi ft}dt \tag{3}$$

In the formula, $x(t)$ is a continuous time-domain signal of pavement noise; f is the frequency that needs to be analyzed; $X(f)$ is the frequency-domain signal obtained after FFT of $x(t)$. The algorithm converts the pavement noise collected in the indoor test from the time-domain signal, which is difficult to process, into the frequency-domain signal (including frequency, amplitude and phase), which is easy to analyze.

In the program, the analog signal is converted into a digital signal through Analog-to-Digital Converter (ADC) sampling for the sake of the FFT transformation. The MATLAB function $fft(x)$ is utilized to calculate the amplitude of FFT, which is

$$X = abs(fft(x)) \tag{4}$$

In the formula, x is input digital signal sequence; X is the relative amplitude of the corresponding frequency of x.

In order to calculate the continuous digital signal of pavement noise, the discretization is needed. In a set of equally spaced samples, discrete points are used to approximately substitute the signal of finite length. The expression is as follows:

$$X[k] = \sum_{n=0}^{N-1} x[n]e^{-j2\pi kn/N} \tag{5}$$

In the formula, $X[k]$ is the relative amplitude of N frequency points; $x[n]$ is sampling signal; N is the length of input sequence, that is, the number of sampling points. It can be inferred from Equation (5) that the frequency resolution can be improved by increasing the sampling point or sampling time.

The FFT algorithm can automatically decompose the spectrogram into several discrete frequency regions through the A-weighting filter and then apply the A-weighting to every FFT frequency region, as shown in Figure 3, which simplifies the execution process and increases the frequency resolution.

At first, we should determine the A-weighting filter coefficient α_A(f) under the frequency of each FFT sample $X[k]$, as shown in Figure 3. The A-weighting FFT sample $X_A[k]$ is given by the following formula:

$$X_A[k] = a_A(f_k)X[k] \tag{6}$$

For the sake of the determination of the A sound level, it is necessary to require the integration of total signal energy. The Parseval's relation is used to estimate the signal energy in the frequency domain. Parseval's theorem indicates that the sum (or integral) of the function squares is equal to the sum of the squares of the Fourier transform (or integral), as shown in Equation (7). In addition, since the input signal is the real-value, the samples have complex conjugate symmetry, and the spectrogram is symmetric about half of the frequency samples, as shown in Equation (8).

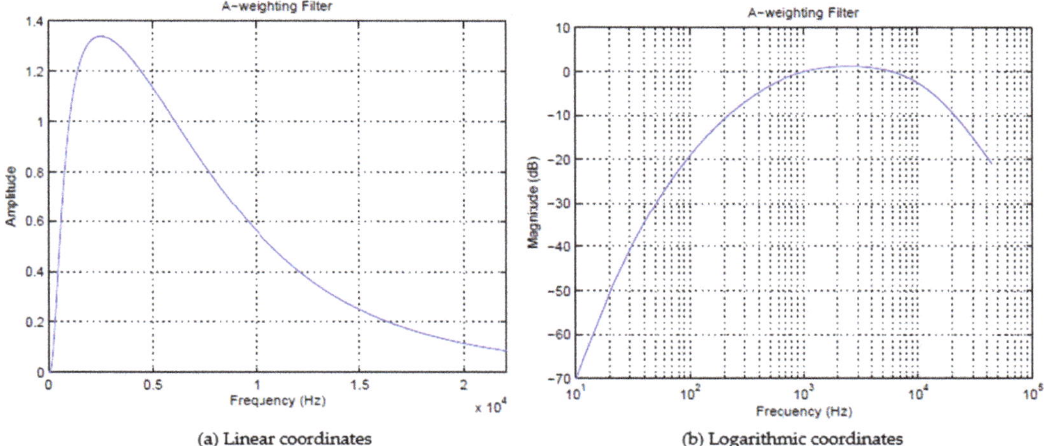

(a) Linear coordinates (b) Logarithmic coordinates

Figure 3. A-weighting filter.

$$\varepsilon_x = \sum_{n=0}^{N-1} [x[n]]^2 = \frac{1}{N} \sum_{k=0}^{N-1} [x[k]]^2 \tag{7}$$

$$X\left[\frac{N}{2} + k\right] = X\left[\frac{N}{2} - k\right] \tag{8}$$

Only the signal energy of half of the frequency samples needs to be estimated by utilizing the symmetrical feature of the spectrogram so as to reduce the calculation amount. In order to obtain the final numerical output in dBA form, a reference signal level is also needed. Assuming a suitable reference signal level is $\widetilde{\varepsilon}_{ref}$, there is:

$$dBA = 10\lg\left(\frac{\widetilde{\varepsilon}_x}{\widetilde{\varepsilon}_{ref}}\right) = 10\lg(\widetilde{\varepsilon}_x) - 10\lg\left(\widetilde{\varepsilon}_{ref}\right) \tag{9}$$

$10\lg\left(\widetilde{\varepsilon}_{ref}\right)$ is a fixed constant. The signal level expressed in dBA is as follows:

$$N = 10\lg\left(\widetilde{\varepsilon}_{ref}\right) + C \tag{10}$$

In the formula, N is the sound pressure level signal level in dBA; C is the calibration constant, which can be determined through laboratory tests.

2. Experiment

2.1. Material Design

The indoor test mainly uses four types of mixtures: OGFC-13, SMA-13, AC-13 and MS-III at the micro-surface [33]. Both coarse aggregate and machine-made sand use diabase gravel, and the key indicators of aggregate are shown in Table 1. The mineral powder uses limestone ground filler. The aggregate and mineral powder were both supplied by Furong Quarry, Heyuan, Guangdong, China. OGFC-13 uses high-viscosity composite-modified asphalt, SMA-13 and AC-13 use SBS-modified asphalt, and MS-III at the micro-surface uses modified emulsified asphalt. All kinds of asphalt binders were supplied by Guangzhou Xinyue Transportation Technology Co. Ltd., Guangdong, China. The specific indicators are shown in Tables 2–6.

Table 1. Technical indicators of coarse aggregate.

Test Item	Technical Requirement		Test Result	Monomial Assessment
	Unit	Design Requirement		
Stone crushing value	%	≤15	7.9	Eligibility
Los Angeles abrasion loss	%	≤22	9.4	Eligibility
Apparent relative density	—	≥2.60	2.888	Eligibility
Adhesion to modified asphalt	Level	≥5	5	Eligibility
Polish value	—	≥42	45	Eligibility

Table 2. The technical index of fine aggregate.

Sample Specifications			0–3 mm	
Test Item	Technical Requirement		Test Result	Monomial Assessment
	Unit	Design Requirement		
Apparent relative density	—	≥2.50	2.909	Eligibility
Robustness (>0.3 mm part)	%	≤12	2.5	—
Sand equivalent	%	≥65	73	Eligibility
Angularity (flow time)	s	≥30	39.6	Eligibility

Table 3. The technical index of high-viscosity modified asphalt.

Test Item		Technical Requirement	Test Result	Monomial Assessment
Penetration 25 °C, 100 g, 5 s, 0.1 mm		40	49	Eligibility
Ductility 5 °C, 5 cm/min, cm		≥50	70	Eligibility
Softening point (°C)		≥80	>90	Eligibility
Flash point (°C)		≥260	337	Eligibility
Viscosity 25 °C (N*m)		≥25	28	Eligibility
Tenacity 25 °C (N*m)		≥15	16	Eligibility
60 °C dynamic viscosity (Pa.S)		>250,000	>580,000	Eligibility
Rolling thin film oven test (RTFOT)	Mass change (%)	±1.0	−0.054	Eligibility
Residue (163 °C, 85 min)	Penetration ratio (%)	≥65	79.5	Eligibility

Table 4. The technical index of SBS-modified asphalt.

Test Item		Technical Requirement	Test Result	Monomial Assessment
Penetration 25 °C, 100 g, 5 s, 0.1 mm		40–60	54	Eligibility
Penetration index PI		≥+0.0	+0.17	Eligibility
Ductility 5 °C, 5 cm/min, cm		≥20	34	Eligibility
Softening point (°C)		≥75	86.5	Eligibility
Flash point (°C)		≥230	340	Eligibility
Solubility (%)		≥99	99.8	Eligibility
Storage stability *: 163 °C, 48 h, poor softening point °C		≤2.0	1.2	Eligibility
Elastic recovery 25 °C, %		≥90	96	Eligibility
Kinematic viscosity (Pa·s)	135 °C	≤3	2.38	Eligibility
	165 °C	No requirement	0.62	
Rolling thin film oven test (RTFOT)	Mass change (%)	±1.0	−0.016	Eligibility
Residue	Ductility 5 °C, 5 cm/min, cm	≥20	21	Eligibility
(163 °C, 85 min)	Penetration ratio (%)	≥65	81.6	Eligibility

Table 5. The technical index of modified emulsified asphalt.

Test Item		Unit	Technical Requirement	Test Result	Monomial Assessment
Residue on sieve (1.18 mm) sieve, no more than		%	≤0.05	0.03	Eligibility
Particle charge		—	positive ion (+)	positive ion	Eligibility
Viscosity (Asphalt Standard Viscometer $C_{25,3}$)		s	8–25	18	Eligibility
Evaporation residue	Residual content	%	≥63	65.3	Eligibility
	Penetration (25 °C)	0.1 mm	40–150	79	Eligibility
	Softening point	°C	≥55	66.0	Eligibility
	Ductility (5 °C)	cm	≥25	51	Eligibility
	Solubility	%	≥97.5	99.7	Eligibility
	Elastic recovery (10 °C)	%	≥60	71	Eligibility
Storage stability (1 d)		%	≤1	0.5	Eligibility

Table 6. The mixture design index.

Sieve Mesh	16	13.2	9.5	4.75	2.36	1.18	0.6	0.3	0.15	0.075	Asphalt-Aggregate Ratio (%)	Porosity (%)
OGFC-13	100	95	55.9	11.8	11.4	11	8.4	7	5.5	4.2	5.2	21.6
AC-13	100	90	68	38	24	15	10	7	5	4	4.3	3.8
SMA-13	100	89	63	25	19	15	14	13	12	10	5.9	4.1
MS-III	100	100	100	90	70	50	34	25	18	15	7.2	3.8

2.2. Experiment Procedure

(1) Test condition

To ensure the indoor temperature is 25 °C and the temperature is constant during the test; the tire load is 250 kN; the tire ground pressure is set at 0.7 MPa; the sampling time is 120 s; the microphone is fixed and connected to the computer.

(2) The noise measurement process is shown in Figure 4 and the noise acquisition process is shown in Figure 5.

(3) Background noise measurement and modification

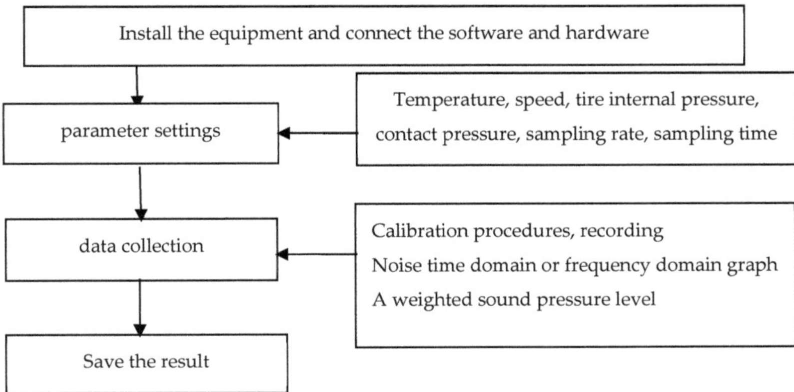

Figure 4. Indoor noise test procedure.

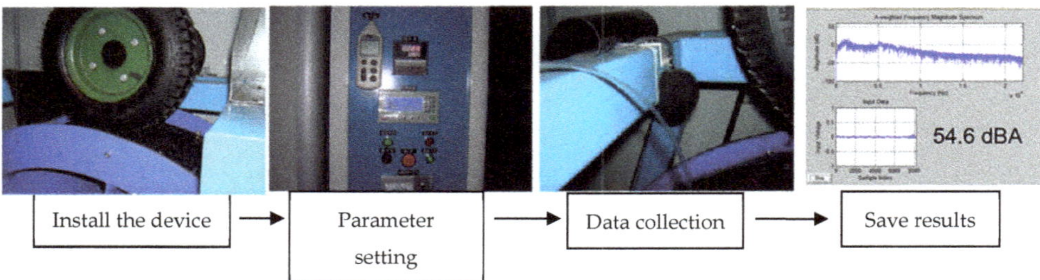

Figure 5. The noise acquisition process.

Before the test, make the driving wheel (that is, the tire) idling without contact with the surface of the test piece. The measured noise at the same position is the background noise, and the value measured in the test is 38.1 dB(A).

In order to ensure the accuracy of pavement noise measurement, it is necessary to determine the background noise of the test system environment [34]. When the difference between the measurement noise and the background noise is more than 10 dB(A), the background noise can be ignored; when the difference is between 6 and 10 dB(A), the measurement noise should be corrected, and the measurement result should be subtracted from the correction value in Table 7; when the difference is less than 6 dB(A), the measurement is invalid.

Table 7. Background noise correction value.

Difference Between the Measurement Noise and the Background Noise/dB(A)	6–8	9–10	>10
Correction value	1	0.5	0

3. Result and Discussion

3.1. *Impact of Driving Speed on Noise*

Taking the case of macroporous asphalt pavement, the noise changes of the OGFC pavement are explored at different tire-rolling speeds to adjust the running speed (in terms of the time of one revolution) and record the noise sound pressure level of the OGFC pavement at different running speeds, as shown in Figure 6.

The OGFC–asphalt mixture tire or pavement noise sound pressure level increases with the increment of driving speed, especially at a speed above 1500 ms/r, when this effect is more significant.

A semi-logarithmic spectrum graph is drawn, as shown in Figure 7. It can be seen that OGFC pavement noise is a typical broadband noise, distributed in the frequency range of 200–5000 Hz. The peak value of the spectrum curve at different driving speeds appears in the frequency range of 600–700 Hz. Compared with the background noise, the noise peak value moves to high frequency, which shows that the background environment noise is mainly affected by the driving wheel rotation system, primarily at low frequencies. When the tires interact with the road surface, the pavement noise becomes prominent, and the vibration intensity is greater in the higher frequency range. From the analysis of the frequency spectrum structure, the noise measurement system has a good ability to distinguish pavement noise.

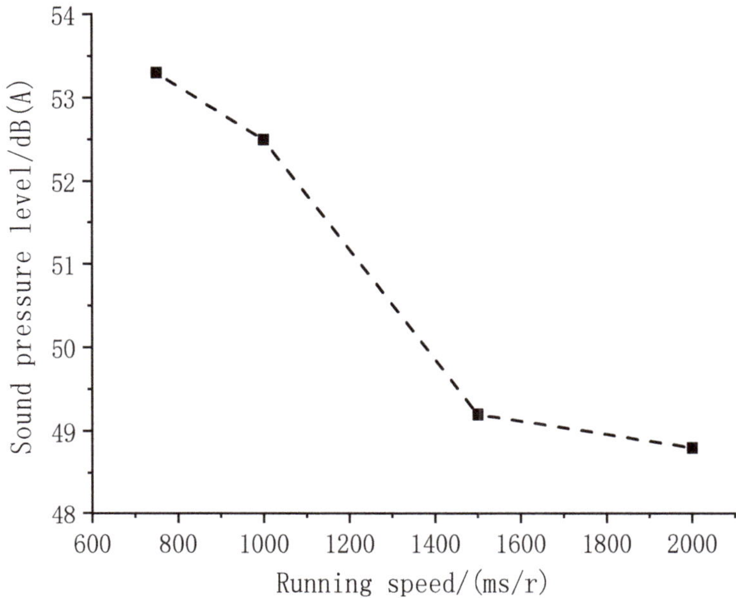

Figure 6. The pavement noise sound pressure level at different driving speeds.

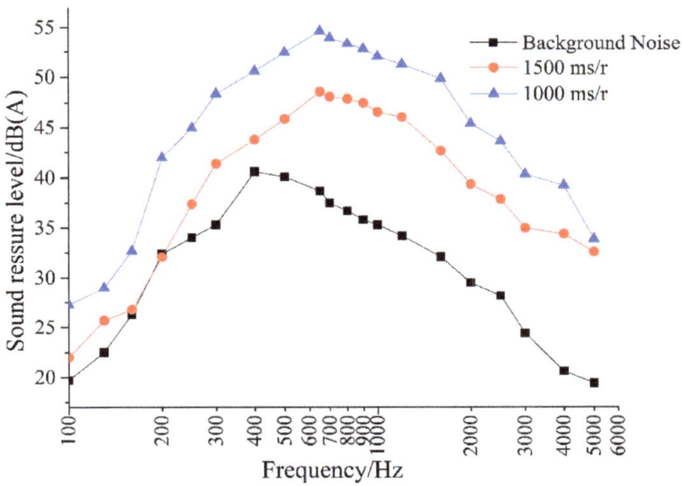

Figure 7. A pavement noise spectrogram at different driving speeds.

3.2. *Impact of Wet and Dry Conditions on Noise*

This test uses artificial watering to create wet pavement. Water was evenly sprayed on the surface of the test piece to simulate rainfall. The running speed was 1500 ms/r, and the noise test results are shown in Table 8.

Table 8. A comparison of OGFC pavement noise in dry and wet conditions.

Pavement State	Sound Pressure Level/dB(A)	Noise Reduction Level/dB(A)
Dry	49.2	0.6
Wet	49.8	

The test result shows that asphalt pavement is noisier in a wet state than in a dry state. At the same speed, OGFC asphalt pavement with high porosity has a more significant noise reduction effect in a wet state than in a dry state. The frequency spectrum characteristics of OGFC pavement noise under dry and wet conditions are further analyzed in Figure 8.

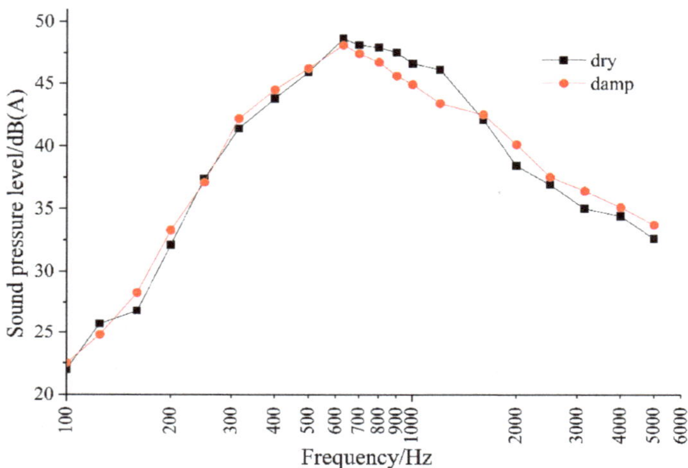

Figure 8. OGFC pavement noise spectrogram under dry or wet conditions.

The low frequency (600 Hz) of OGFC pavement noise under dry or wet conditions does not change significantly. The noise on dry pavement in the mid-frequency (600–1200 Hz) range is slightly greater than that on wet pavement. The increase in sound pressure of wet pavement noise compared to dry pavement noise is mainly reflected in the high-frequency range (≧2000 Hz). The reason is that the water film produced by the OGFC pavement in a wet state reduces the adhesion between the tire and the pavement, causing the noise in the mid-frequency range to decrease, while the amount of water and wheel speed mainly cause the increase in the high-frequency range noise.

3.3. Impact of Mixture Type on Pavement Noise

In order to explore the impact of mixture type on pavement noise, four representative pavements, namely porous drainage asphalt pavement (OGFC), ordinary asphalt pavement (AC), asphalt mastic pavement (SMA) and micro-surface pavement (MS-III), are used as the research objects. The measurement test condition is a dry state. The running speed is 1500 ms/r, and the test results are shown in Table 9.

Table 9. Measurement and analysis results of different pavement noises.

Gradation Type	Background Noise	OGFC-13	MS-III	AC-13	SMA-13
Porosity /%	—	21.6	3.8	3.8	4.1
Structural depth/mm	—	1.94	0.83	0.42	1.14
Sound pressure level /dB(A)	38.1	49.2	57.3	52.9	51.8

It can be known from Table 5 that the porosity and structural depth of OGFC are much larger than those of SMA, AC and micro-surface. The corresponding A-weighted sound pressure level shows that OGFC has the best noise reduction effect, SMA has good noise reduction performance. Meanwhile, AC is poorer, and the micro-surface has the largest noise. The noise mechanism of different pavement structures is not the same, which can be reflected in the composition of the noise. The noise sound pressure level is only a part of the composition. In order to find the main frequency bands that affect the pavement noise, the spectrum analysis of the four pavement noises is performed, as shown in Figure 9.

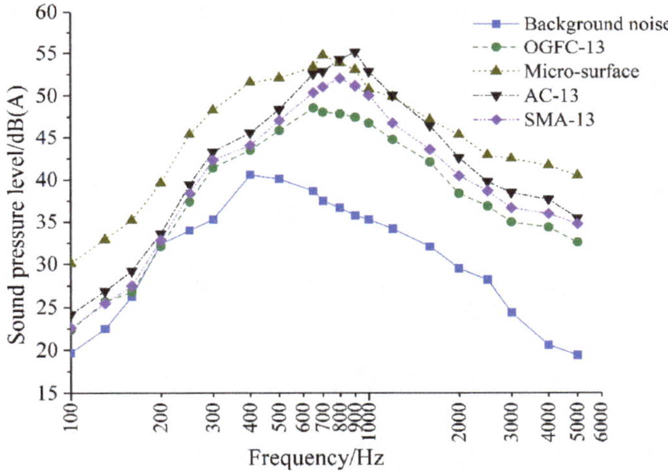

Figure 9. Noise frequency diagram of different pavement types.

OGFC, AC, SMA and micro-surface pavements have significant tire or pavement noise characteristics, whose pavement noise spectra are all broad and continuous. The peaks of AC, SMA, micro-surface and OGFC move to low frequencies in sequence. The peak of pavement noise is located in the most sensitive frequency range of the human ears, so this article mainly studies the frequency range of 500–2000 Hz. Comparing the spectrum curve, it is found that the pavement noise difference in the frequency range of 700–1600 Hz is significant, and this frequency domain is the main area where the air pump effect occurs.

There are many pores in OGFC, and the pores are connected to each other and communicate with the outside through the surface. When sound waves occur on the surface of the material, they are reflected or penetrated into the interior and propagate forward, causing the air in the pores to move and create friction with the irregular pore inner walls. The viscous effect and heat conduction effect convert sound energy into heat energy and consume it, which greatly reduces the pumping noise of 700–1600 Hz and effectively reduces pavement noise. The peak frequency is lower than other pavements (600–700 Hz). In previous studies, it was found that the macroporous structure of the pavement changes the propagation characteristics of tire noise, including the fact that the slits in the surface structure directly reduce aerodynamic noise, while the connected pores inside the pavement structure also absorb part of the tire noise [35]. In Europe, there are test results showing that vehicle noise can even be reduced by up to 10 dB(A) with porous pavement structures. European test results show that vehicle noise can even be reduced by up to 10 dB(A) with macroporous pavement structure [36].

Although the porosity of SMA is low, its surface has more coarse aggregates and rich surface texture. With the decrease in the wavelength of the texture structure and the increase in the amplitude, the Helmholtz resonance phenomenon occurs with the reflection in the internal cavity, which provides a free channel for air movement in the contact area and effectively weakens the pumping noise. Compared with AC and micro-surface, SMA pavement noise is also reduced in low-frequency bands, indicating that its dynamic modulus and internal damping help to attenuate the low-frequency noise of tire vibration. The sound pressure level in the full frequency range of the spectrum curve at the micro-surface is far greater than that of other pavements such as AC, especially the low-frequency noise increase rate, which is more obvious. It is because the micro-surface is not compacted, the surface structure is uneven, the top surface of large-size aggregates is uneven, the superposition of common-size aggregates and the fine aggregate enrichment area is recessed, resulting in a significant increase in tire impact noise and vibration noise. Narayanan et al. conducted noise tests on different pavements by orienting the design with different porosity, pore size and morphology of the surface structure (depth, width, shape) and using TPTA (Tire-Pavement Test Apparatus). They initially analyzed that the noise level is related to the depth of the road surface structure, but the influence of the depth of the road surface structure and the porosity inside the pavement remains to be studied [37].

3.4. Analysis on the Law of Pavement Noise Attenuation

3.4.1. Impact of Tire Action Times on Noise Decay

It can be known from engineering practice that OGFC has a good pavement noise reduction effect. Nevertheless, with the extension of operating time, the squeeze of the vehicle and the blockage of dust and other debris will result in a smaller porosity and a small increase in the noise level of the OGFC. However, lacking long-term monitoring data of OGFC noise, it is impossible to grasp the law of OGFC noise attenuation.

The indoor accelerated loading test system can simulate the impact of long-term traffic load well by increasing the number of tire actions on the OGFC pavement, thereby obtaining the long-term noise characteristics of the OGFC pavement. After the preset number of actions (7×104 times), the long-term noise measurement results of OGFC-13, SMA-13, AC-13 and MS-3 at the micro-surface are shown in Figure 10.

It can be seen from Figure 10 that the noise on the four pavements increases with the number of tire actions, among which the tire noise at the micro-surface has always been at the highest level, mainly related to the roughness and flatness of the surface texture. The noise level of densely graded SMA-13 is slightly better than that of AC-13 pavement, which is mainly affected by the damping coefficient of SMA pavement and the larger macroscopic structure depth. The noise attenuation laws of the two densely graded pavements are also similar. As for the large-pore OGFC pavement, its initial noise reduction performance is relatively good, which is at the lowest noise level; when the number of driving operations reaches more than 40,000, the tire noise increases rapidly, which is mainly due to the falling off of aggregate and mortar on the pavement and the decrease in interconnected pores in the mixture. The attenuated noise level in the later stage is closer to the densely graded noise, indicating that the noise reduction function of the macroporous pavement is basically lost at this time. Through the road noise tests at different ages, it was found that the tire noise increased with the aggravation of road wearing [38]. It is consistent with our main finding of the four pavement noise variation patterns under the actions of different wheel-tires conducted by the laboratory accelerated loading system developed in this paper. This also proves that the device can simulate the action of wheel tires on the pavement well.

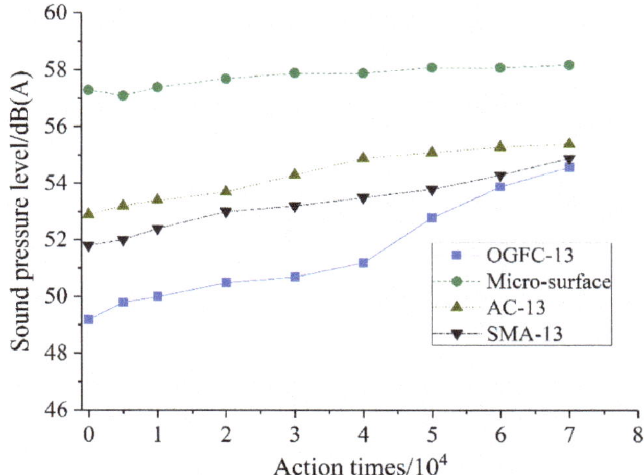

Figure 10. Pavement long-term noise characteristics.

According to the frequency spectrum analysis based on the measured noise data, the changes in the noise characteristics of OGFC are observed as the number of tire actions increases, as shown in Figure 11. The noise level of the newly built OGFC pavement is greater than that of the OGFC pavement after the action of tires in the range of 100–600 Hz, while the higher-frequency domain above 700 Hz is the opposite, which is because the effect of the traffic load may cause the single stone protruding on the OGFC pavement to fall off and become smoother, the impact on the tires is reduced, and the low-frequency vibration and noise generated are also reduced accordingly. Furthermore, with the compaction of the tires, the connected porosity of the OGFC pavement decreases, and the effect of reducing the pumping noise becomes smaller accordingly. The internal and external pressure between the tire and the road cannot be kept constant, resulting in increased noise in the high-frequency range of the OGFC pavement, and the peak of the spectrum has to move to the high-frequency direction.

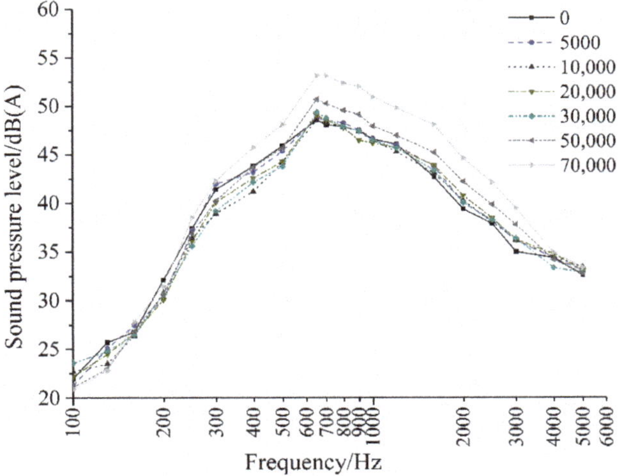

Figure 11. Changes in the OGFC pavement noise characteristics.

3.4.2. Impact of Structural Depth on Noise Decay

After OGFC pavement has passed tire action times (7×10^4 times), the structural depth change is shown in Figure 12. OGFC pavement is compacted under the action of tires, and the structural depth decreases rapidly, and then the decrease rate becomes slower and tends to be stable.

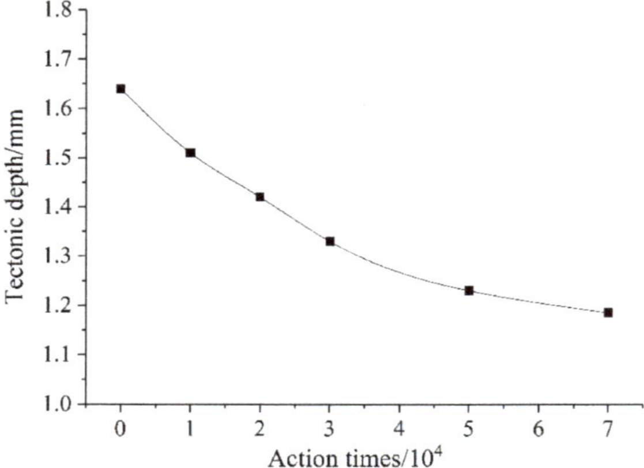

Figure 12. OGFC structural depth change curve.

The correlation between the structural depth attenuation process and noise of constructing a large-pore pavement is shown in Figure 13. OGFC structural depth and noise show a good secondary correlation. As the number of tire actions increases, the structure depth decreases, OGFC pavement noise tends to increase, and the correlation is good. The main reason is that the initial OGFC pavement has a large structural depth, rich surface texture, and dense through-holes formed on the surface and inside of the road, which has good sound absorption performance. When the depth of the structure becomes lower under the load of the tire, the connected pores are correspondingly reduced, and the air in the pores is squeezed when the tire interacts with the road surface. When the interaction between the tire and the road surface ends, the pores suck in a large amount of air due to the imbalance of internal and external pressure, and the pumping noise increases.

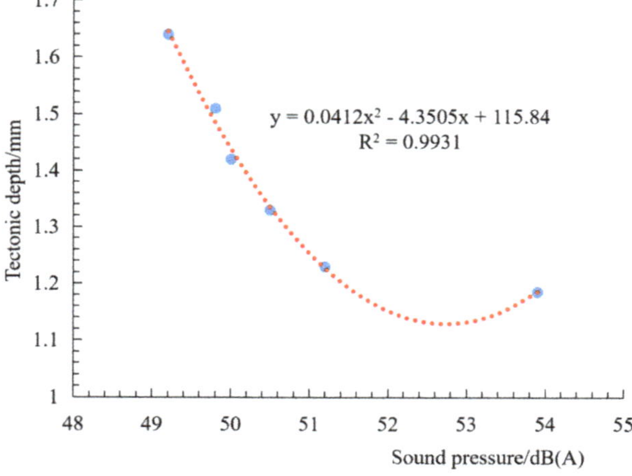

Figure 13. The relationship between structural depth and noise.

4. Conclusions

(1) OGFC tire or pavement noise increases with the increase in driving speed, and the noise structure is different at different speeds. OGFC pavement noise at low speeds increases the noise level in the high-frequency range due to tire or road sticky action. Regarding the noise at higher speeds, in addition to the effect of sticky action, there is also pumping noise and vibration noise from tire patterns, which violently beat the pavement.

(2) The noise reduction performance of OGFC comes from its porosity up to 20%, which effectively reduces air pump noise. SMA has a rich surface texture and high internal damping, and its noise reduction performance is inferior only to OGFC pavement. The higher noise at the micro-surface is due to the unevenness and nonuniformity of the pavement macroscopic structure. The noise reduction effect of different asphalt pavements is OGFC-13> SMA-13> AC-13> MS-III.

(3) With the decrease in the structural depth, the OGFC pavement noise has an increasing tendency, and the OGFC pavement structural depth has a good quadratic parabolic relationship with the noise sound pressure level. In the later stage of the driving action, the noise reduction function of OGFC tends to disappear due to the speed of the connected pores. Meanwhile, the pavement noise characteristics are similar to those of densely graded asphalt pavement.

Future studies will be focused on the slipperiness characteristics and the in situ validation of the macroporous asphalt pavement based on the weighted sound pressure level sensor.

Author Contributions: Data curation, F.L.; funding acquisition, Z.H.; investigation, F.L. and F.G.; methodology, F.L.; writing—original draft, F.L. and F.G; writing—review and editing, Z.H. All authors have read and agreed to the published version of the manuscript.

Funding: The authors would like to acknowledge the China Postdoctoral Science Foundation project (2020M672639).

Institutional Review Board Statement: Not applicable.

Informed Consent Statement: Informed consent was obtained from all subjects involved in the study.

Data Availability Statement: The data presented in this study are available on request from the corresponding author.

Conflicts of Interest: The authors declare no conflict of interest.

References

1. Wolfgang, B. Traffic noise and risk of myocardial infarction. *Epidemiology* **2005**, *16*, 33–40.
2. Yang, W.; He, J.; He, C.; Cai, M. Evaluation of urban traffic noise pollution based on noise maps. *Transp. Res. Part D Transp. Environ.* **2020**, *87*, 102516. [CrossRef]
3. Iglesias-Merchan, C.; Laborda-Somolinos, R.; González-Ávila, S. Elena-Rosselló, R. Spatio-temporal changes of road traffic noise pollution at ecoregional scale. *Environ. Pollut.* **2021**, *286*, 117291. [CrossRef] [PubMed]
4. Li, T.; Burdisso, R.; Sandu, C. Literature review of models on tire-pavement interaction noise. *J. Sound Vib.* **2018**, *420*, 357–445. [CrossRef]
5. Lu, X.; Kang, J.; Zhu, P.; Cai, J.; Guo, F.; Zhang, Y. Influence of urban road characteristics on traffic noise. *Transp. Res. Part D Transp. Environ.* **2019**, *75*, 136–155. [CrossRef]
6. Umeda, S.; Kawasaki, Y.; Kuwahara, M.; Iihoshi, A. Risk evaluation of traffic standstills on winter roads using a state space model. *Transp. Res. Part C Emerg. Technol.* **2021**, *125*, 103005. [CrossRef]
7. Yu, H.Y.; Deng, G.S.; Wang, D.Y.; Zhang, Z.Y.; Oeser, M. Warm asphalt rubber: A sustainable way for waste tire rubber recycling. *J. Cent. South Univ.* **2020**, *27*, 3477–3498. [CrossRef]
8. Bachok, K.S.R.; Hamsa, A.A.K.; Mohamed, M.Z.b.; Ibrahim, M. A theoretical overview of road hump effects on traffic noise in improving residential well-being. *Transp. Res. Procedia* **2017**, *25*, 3383–3397. [CrossRef]
9. Chen, L.; Cong, L.; Dong, Y.; Yang, G.; Tang, B.; Wang, X.; Gong, H. Investigation of influential factors of tire/pavement noise: A multilevel Bayesian analysis of full-scale track testing data. *Constr. Build. Mater.* **2021**, *270*, 121484. [CrossRef]
10. Mohammadi, S.; Ohadi, A. A novel approach to design quiet tires, based on multi-objective minimization of generated noise. *Appl. Acoust.* **2021**, *175*, 107825. [CrossRef]

11. Vázquez, V.F.; Hidalgo, M.E.; García-Hoz, A.M.; Camara, A.; Terán, F.; Ruiz-Teran, A.M.; Paje, S.E. Tire/road noise, texture, and vertical accelerations: Surface assessment of an urban road. *Appl. Acoust.* **2020**, *160*, 107153. [CrossRef]
12. Ganji, M.R.; Golroo, A.; Sheikhzadeh, H.; Ghelmani, A.; Bidgoli, M.A. Dense-graded asphalt pavement macrotexture measurement using tire/road noise monitoring. *Autom. Constr.* **2019**, *106*, 102887. [CrossRef]
13. Jin, J.; Gao, Y.; Wu, Y.; Li, R.; Liu, R.; Wei, H.; Qian, G.; Zheng, J. Performance evaluation of surface-organic grafting on the palygorskite nanofiber for the modification of asphalt. *Constr. Build. Mater.* **2021**, *268*, 121072. [CrossRef]
14. Jin, J.; Gao, Y.; Wu, Y.; Liu, S.; Liu, R.; Wei, H.; Qian, G.; Zheng, J. Rheological and adhesion properties of nano-organic palygorskite and linear SBS on the composite modified asphalt. *Powder Technol.* **2021**, *377*, 212–221. [CrossRef]
15. Yu, H.; Zhu, Z.; Leng, Z.; Wu, C.; Zhang, Z.; Wang, D.; Oeser, M. Effect of mixing sequence on asphalt mixtures containing waste tire rubber and warm mix surfactants. *J. Clean. Prod.* **2020**, *246*, 119008. [CrossRef]
16. Yu, H.; Leng, Z.; Zhou, Z.; Shih, K.; Xiao, F.; Gao, Z. Optimization of preparation procedure of liquid warm mix additive modified asphalt rubber. *J. Clean. Prod.* **2017**, *141*, 336–345. [CrossRef]
17. Li, D.; Leng, Z.; Zou, F.; Yu, H. Effects of rubber absorption on the aging resistance of hot and warm asphalt rubber binders prepared with waste tire rubber. *J. Clean. Prod.* **2021**, *303*, 127082. [CrossRef]
18. Jin, J.; Liu, S.; Gao, Y.; Liu, R.; Huang, W.; Wang, L.; Xiao, T.; Lin, F.; Xu, L.; Zheng, J. Fabrication of cooling asphalt pavement by novel material and its thermodynamics model. *Constr. Build. Mater.* **2021**, *272*, 121930. [CrossRef]
19. Yu, H.; Leng, Z.; Dong, Z.; Tan, Z.; Guo, F.; Yan, J. Workability and mechanical property characterization of asphalt rubber mixtures modified with various warm mix asphalt additives. *Constr. Build. Mater.* **2018**, *175*, 392–401. [CrossRef]
20. Yu, H.; Deng, G.; Zhang, Z.; Zhu, M.; Gong, M.; Oeser, M. Workability of rubberized asphalt from a perspective of particle effect. *Transp. Res. Part D Transp. Environ.* **2021**, *91*, 102712. [CrossRef]
21. Zhang, J.; Huang, W.; Hao, G.; Yan, C.; Lv, Q.; Cai, Q. Evaluation of open-grade friction course (OGFC) mixtures with high content SBS polymer modified asphalt. *Constr. Build. Mater.* **2021**, *270*, 121374. [CrossRef]
22. Gołebiewski, R.; Makarewicz, R.; Nowak, M.; Preis, A. Traffic noise reduction due to the porous road surface. *Appl. Acoust.* **2003**, *64*, 481–494. [CrossRef]
23. Paje, S.E.; Bueno, M.; Terán, F.; Viñuela, U. Monitoring road surfaces by close proximity noise of the tire/road interaction. *J. Acoust. Soc. Am.* **2007**, *122*, 2636–2641. [CrossRef]
24. Dong, S.; Han, S.; Zhang, Q.; Han, X.; Zhang, Z.; Yao, T. Three-dimensional evaluation method for asphalt pavement texture characteristics. *Constr. Build. Mater.* **2021**, *287*, 122966. [CrossRef]
25. Sandberg, U. Abatement of traffic, vehicle, and tire/road noise-The global perspective. *Noise Control. Eng. J.* **2001**, *49*, 170–181. [CrossRef]
26. Chen, D.; Ling, C.; Wang, T.; Su, Q.; Ye, A. Prediction of tire-pavement noise of porous asphalt mixture based on mixture surface texture level and distributions. *Constr. Build. Mater.* **2018**, *173*, 801–810. [CrossRef]
27. Vaitkus, A.; Andriejauskas, T.; Vorobjovas, V.; Jagniatinskis, A.; Fiks, B.; Zofka, E. Asphalt wearing course optimization for road traffic noise reduction. *Constr. Build. Mater.* **2017**, *152*, 345–356. [CrossRef]
28. Lee, S.-H.; Eum, K.-Y.; Le, T.H.M.; Park, D.-W. Evaluation on mechanical behavior of asphalt concrete trackbed with slab panel using full-scale static and dynamic load test. *Constr. Build. Mater.* **2021**, *276*, 122207. [CrossRef]
29. Deng, Q.; Zhan, Y.; Liu, C.; Qiu, Y.; Zhang, A. Multiscale power spectrum analysis of 3D surface texture for prediction of asphalt pavement friction. *Constr. Build. Mater.* **2021**, *293*, 123506. [CrossRef]
30. Zhang, H.; Liu, Z.; Meng, X. Noise reduction characteristics of asphalt pavement based on indoor simulation tests. *Constr. Build. Mater.* **2019**, *215*, 285–297. [CrossRef]
31. Mak, K.L.; Hung, W.T.; Lee, S.H. Exploring the impacts of road surface texture on tyre/road noise–A case study in Hong Kong. *Transp. Res. Part D Transp. Environ.* **2012**, *17*, 104–107. [CrossRef]
32. Eisenblaetter, J.; Walsh, S.J.; Krylov, V.V. Air-related mechanisms of noise generation by solid rubber tyres with cavities. *Appl. Acoust.* **2010**, *71*, 854–860. [CrossRef]
33. Ding, S.; Wang, K.C.P.; Yang, E.; Zhan, Y. Influence of effective texture depth on pavement friction based on 3D texture area. *Constr. Build. Mater.* **2021**, *287*, 123002. [CrossRef]
34. Li, T. A state-of-the-art review of measurement techniques on tire–pavement interaction noise. *Measurement* **2018**, *128*, 325–351. [CrossRef]
35. Liao, G.; Sakhaeifar, M.; Heitzman, M.; West, R.; Waller, B.; Wang, S.; Ding, Y. The effects of pavement surface characteristics on tire/pavement noise. *Appl. Acoust.* **2014**, *76*, 14–23. [CrossRef]
36. Ling, S.; Yu, F.; Sun, D.; Sun, G.; Xu, L. A comprehensive review of tire-pavement noise: Generation mechanism, measurement methods, and quiet asphalt pavement. *J. Clean. Prod.* **2021**, *287*, 125056. [CrossRef]
37. Neithalath, N.; Garcia, R.; Weiss, J.; Olek, J. Tire-Pavement Interaction Noise: Recent Research on Concrete Pavement Surface Type and Texture. In Proceedings of the 8th International Conference on Concrete Pavements, Colorado Springs, CO, USA, 14–18 August 2005; Volume 2, pp. 523–540.
38. Irali, F.; Kivi, A.; Tighe, S.L.; Sangiorgi, C. Tire/Pavement Noise and Wearing Course Surface Characteristics of Experimental Canadian Road Pavement Sections. *Can. J. Civ. Eng.* **2015**, *42*, 818–825. [CrossRef]

 materials

MDPI

Article

Development of Water Retentive and Thermal Resistant Cement Concrete and Cooling Effects Evaluation

Xiaowei Wang [1,*], Xinyu Hu [2], Xiaoping Ji [2], Bo Chen [3] and Hongqing Chen [4]

1 School of Civil Engineering, Xi'an University of Architecture & Technology, Xi'an 710055, China
2 Key Laboratory of Special Area Highway Engineering of Ministry of Education, Chang'an University, Xi'an 710064, China; xinyu_hu2021@163.com (X.H.); jixp@163.com (X.J.)
3 No. 1 Municipal Administration Institute, Xianyang Planning & Design Institute, Xianyang 712021, China; song_quan@163.com
4 CCCC-SHEC Third Highway Engineering Co., Ltd., Xi'an 710016, China; suyunfei030@163.com
* Correspondence: xwwang@xauat.edu.cn

Abstract: The high pavement temperature plays an important role in the development of urban heat island (UHI) in summer. The objective of this study was to develop water retentive and thermal resistant cement concrete (WTCC) to enhance the pavement cooling effects. The WTCC was prepared by combining a water retentive material and a high aluminum refractory aggregate (RA) with porous cement concrete (PCC). Water retention capacity test, fluidity test, and compressive strength test were used to determine the composition ratio of the water retentive material. Mechanical performance and cooling effects of WTCC were evaluated by compressive and flexural strength tests and temperature monitoring test. The mass ratios of fly ash, silica fume, cement, and water in the water retentive material were determined as 65:35:15:63.9. The compressive strength and the flexural strength of WTCC after 28 days curing were 30.4 MPa and 4.6 MPa, respectively. Compared with stone mastic asphalt (SMA) mixture, PCC, and water retentive cement concrete (WCC), surface temperature of WTCC decreased by 11.4 °C, 5.5 °C, and 4.1 °C, respectively, and the internal temperatures of WTCC decreased by 10.3 °C, 6.1 °C, and 4.6 °C, respectively. The water retentive material has benefits of strength improvements and temperature reduction for WTCC. Based on the results, WTCC proved to have superior cooling effects and the potential to efficiently mitigate the UHI effects and be used in medium traffic roads.

Keywords: urban heat island effects; water retentive material; water retentive and thermal resistant cement concrete; pavement temperature reduction

Citation: Wang, X.; Hu, X.; Ji, X.; Chen, B.; Chen, H. Development of Water Retentive and Thermal Resistant Cement Concrete and Cooling Effects Evaluation. *Materials* **2021**, *14*, 6141. https://doi.org/10.3390/ma14206141

Academic Editors: Markus Oeser, Michael Wistuba, Pengfei Liu and Di Wang

Received: 28 September 2021
Accepted: 14 October 2021
Published: 16 October 2021

1. Introduction

With global warming and fast urbanization, temperatures in metropolitan areas are significantly greater than the surrounding suburban areas in summer, and this phenomenon is named urban heat island (UHI) [1,2]. Studies demonstrated that pavement contributes highly to the development of UHI because the pavement temperature is very high in summer, and the pavement area percentage in urban areas may be up to 40% [3–5]. Especially, asphalt pavements have a more significant influence on the UHI effects because the heat absorptivity and the temperature of asphalt pavement are higher than other pavements [6]. The high temperature not only causes pavement distress [7] but also aggravates the UHI effects [1,8]. Therefore, developing new pavement technologies to reduce the pavement temperature has a great significance for pavement durability and sustainability of the environment.

There are mainly four ways to reduce the pavement temperature: (a) increasing the albedo of pavement surface; (b) decreasing the thermal conductivity of pavement materials; (c) utilizing water retentive pavements; and (d) using phase change materials. Increasing the albedo of pavement, known as reflective pavement, has been used for a long time to reduce the pavement temperature [1,9]. In the eastern San Francisco Bay Area of California, the albedo and the temperature of different asphalt pavements were measured. Results indicated that the temperature and the albedos correlated well with each other. The albedo of chip-seal was about 70% of the aggregates, resulting in a surface temperature reduction of about 9.0 K compared with the traditional asphalt pavements [10]. Furthermore, coloring additives together with proper aggregates were used to increase the albedo of the pavements, and surface temperature decreased about 12 K compared to weathered asphalt [11]. In general, reflective pavements are mainly carried out using the following strategies: (a) use of white high reflective paints on the surface of the pavement [11,12]; (b) use of infrared reflective colored paints on the surface of the pavement [3,13]; (c) use of heat reflecting paints to cover aggregates [9]; and (d) use of color changing paints on the surface of the pavement [14]. The durability of the reflective pavements is the main concern in the practices.

Apart from the reflective pavements, permeable pavements and thermal resistant pavements are other pavement technologies available to reduce the pavement temperature through decreasing the thermal conductivity of pavement materials. The permeable pavements are paved by porous cement concrete or porous asphalt mixtures, which have a high air void (AV) content and lower thermal conductivity than traditional dense-graded asphalt mixtures [15,16]. Studies reported that the insulation properties of permeable pavements contributed to a temperature reduction [17,18]. Alternative aggregates with a lower thermal conductivity incorporated into asphalt mixtures have been applied to reduce pavement temperature. The thermal conductivity of ceramic, floating beads, and refractory gravel is lower than ordinary aggregates. Pavement temperature reduction was observed when the ordinary aggregates were substituted by these lower thermal conductivity aggregates [2,6]. In addition, water retentive pavements were used to decrease the pavement temperature by dissipating the heat through water evaporation [8]. Water absorption capability and water retention capacity of sintered ceramic pervious brick (CB), pervious concrete brick (PB), and open-graded pervious concrete (PC) were investigated. Results revealed that CB and PB could reduce the surface temperature by up to 20 °C and 12 °C with cooling periods of 16 h and 12 h, respectively [19]. To extend the evaporative cooling period of water retentive pavements, water retentive materials that use highly absorptive filler, such as sepiolite, fly ash, or blast furnace slag, were added into the pore structures of pervious pavements [5,20,21]. Super absorbent polymer (SAP) was added into porous asphalt mixtures, showing a significant cooling effect [22]. Recently, phase change materials incorporated with a lightweight aggregate were used in asphalt mixtures to reduce pavement surface temperature [23,24]. Surface temperature decreased about 4 °C under a test temperature range between 30 °C and 60 °C [24]. The phase change materials also were used in a cement mortar to accumulate heat [25].

In this study, a water retentive material was designed and combined with a thermal resistant aggregate and porous cement concrete (PCC) to develop a water retentive and thermal resistant cement concrete (WTCC). With lower thermal conductivity PCC and water evaporation of a water retentive material, the WTCC may have the potential to enhance the pavement cooling effects. The mechanical performance and the cooling effects of WTCC were also evaluated and compared with stone mastic asphalt mixtures (SMA), PCC, and water retentive cement concrete (WCC). The schematic development of WTCC and the test program is shown in Figure 1.

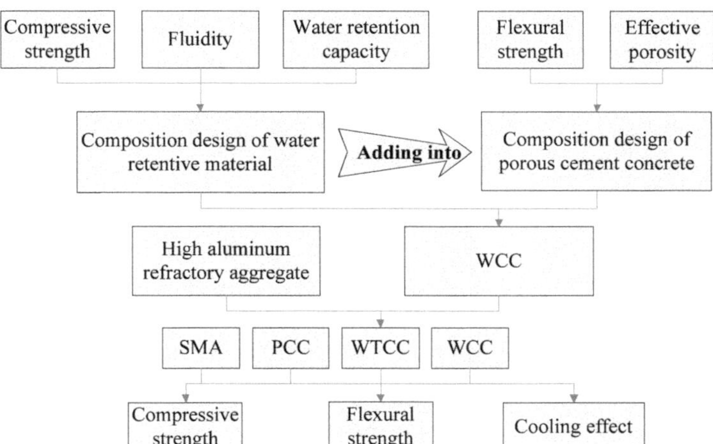

Figure 1. Schematic development of WTCC and test program.

2. Materials and Sample Preparation

The main raw materials for preparing WTCC included thermal resistant aggregate, cement, silica fume, fly ash, and water. In addition, SMA, PCC, and WCC were also prepared and compared with WTCC. A summary of the main composition of these mixtures is shown in Table 1. A water retentive material was designed for preparing the WCC and the WTCC.

Table 1. Brief of the composition of different mixtures.

Mixture Type	Binder	Aggregate	Water Retentive Material
SMA	SBS modified asphalt	Serpentinite	No
PCC	Cement	Serpentinite	No
WCC	Cement	Serpentinite	Yes
WTCC	Cement	High aluminum refractory aggregate	Yes

Styrene-butadiene-styrene (SBS) modified asphalt was used for SMA, while cement graded 42.5 was used for PCC, WCC, and WTCC. The properties of SBS modified asphalt and cement are listed in Tables 2 and 3, respectively. Two types of coarse aggregates—serpentinite aggregate (SA) and thermal resistant aggregate of high aluminum refractory aggregate (RA)—were used, and the properties of the coarse aggregates are summarized in Table 4. SA with different sizes was used for SMA, PCC, and WCC, while RA with a uniform size of 5–10 mm was used for WTCC. RA was produced by Xi'an Hua Rong Refractories Co., Ltd. The thermal conductivity of RA was 0.5 $W\cdot(m\cdot K)^{-1}$, which is significantly lower than the ordinary aggregates of SA.

Table 2. Properties of SBS modified asphalt.

Properties	Value
Penetration (25 °C, 100 g, 5 s) (0.1 mm)	66.2
Ductility (5 cm/min, 5 °C) (cm)	36.5
Kinematic viscosity (135 °C) (Pa·s)	1.965
Softening point (°C)	74.9
Elastic recovery (%)	99.0

Table 3. Properties of cement.

Properties		Value
Specific surface area (Blaine method) ($m^2 \cdot kg^{-1}$)		362
Normal consistency (%)		28.0
Setting time (min)	Initial setting time	230
	Final setting time	295
Flexural strength (MPa)	3 days	4.2
	28 days	7.9
Compressive strength (MPa)	3 days	20.1
	28 days	44.5

Table 4. Properties of coarse aggregates.

Properties	SA	RA	Limits
Crushed value (%)	6.8	26.8	≤30
Los Angeles abrasion loss (%)	5.6	28.4	≤35
Apparent specific gravity ($g \cdot cm^{-3}$)	2.964	2.830	≥2.45
Bulk specific gravity ($g \cdot cm^{-3}$)	2.896	2.732	-
Water absorption (%)	0.71	5.89	≤3.0
Thermal conductivity ($W \cdot (m \cdot K)^{-1}$)	2.2	0.5	-

2.1. Water Retentive Material

Water retentive material was designed to improve the absorption and the retention capacities of WCC and WTCC. The water retentive material was composed of fly ash, cement, and silica fume. Fly ash is a porous material with strong water absorption and retention capacity. Silica fume is a non-crystalline polymorph spherical particle with smooth surface, large specific surface area, and high activity. In the water retentive material, silica fume mainly acted as a lubricant to improve the fluidity, and cement provided the strength. The chemical compositions of fly ash and silica fume are shown in Table 5.

Table 5. Chemical compositions of fly ash and silica fume.

Elements	Fly Ash	Silica Fume
SiO_2 (%)	38.29	85.6
Al_2O_3 (%)	22.83	0.81
Fe_2O_3 (%)	20.15	0.9
CaO (%)	1.18	0.3
MgO (%)	7.76	0.7
Ignition Loss (%)	15.2	1.0
Na_2O (%)	/	1.3
Specific surface area (m^2/g)	1.7	20.8

To prepare the water retentive material, first, cement, fly ash, and silica fume were weighted according to the composition ratio. Next, all the materials were mixed together with water, and the mixture was stirred until uniform consistency slurry was formed. The water retentive material should have good workability in order to easily infiltrate the pore structure of PCC as well as good water retention capacity to store water. The composition ratio of the water retentive material was determined based on water retention capacity (W_R), fluidity (F_L), and compressive strength and is discussed in the following section.

2.2. Stone Mastic Asphalt (SMA)

SMA with a nominal maximum aggregate size (NMAS) of 13.2 mm (SMA-13) was investigated and used as the surface layer of SMA pavements. SMA-13 was designed based on the Marshall mixture design method [26]. The asphalt content was 6.1%, and the AV content was 3.5%. The gradation of SMA-13 is shown in Figure 2. A rolling compaction machine was used to prepare the specimen with a length of 300 mm, a width of 300 mm, and a height of 50 mm. Another dense graded asphalt mixture with a NMAS of 19 mm (AC-20) was compacted with the same dimensions. AC-20 was the bottom layer of SMA pavement. The SMA-13 specimen was bonded with the AC-20 specimen using emulsified asphalt to simulate the SMA pavement structure. Thus, the total height of the SMA specimen was 100 mm.

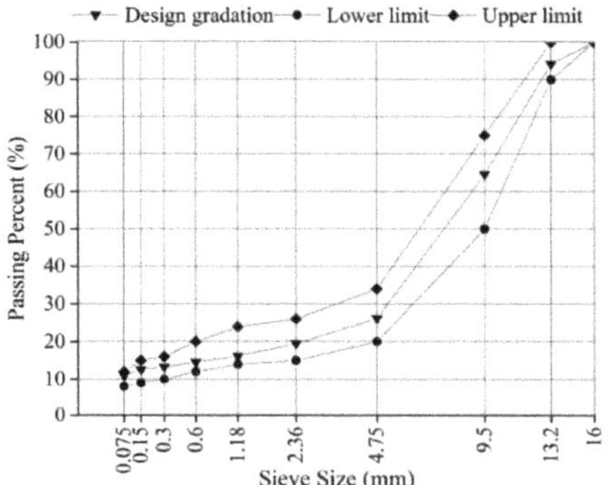

Figure 2. Gradation of SMA-13.

2.3. Porous Cement Concrete (PCC)

PCC is the matrix for preparing WCC and WTCC and providing strength for them. Two types of PCC were prepared in this study. One was PCC prepared by SA (PCC-SA), and the other was PCC prepared by RA (PCC-RA). WCC was prepared by pouring the water retentive material into PCC-SA, and WTCC was prepared by pouring the water retentive material into PCC-RA. PCC-SA and PCC-RA were designed based on target valid interconnected AV content (AV_{valid}), workability, and flexural strength. The target AV_{valid} of the PCC was 18–22%, and the AV_{valid} was calculated based on the Equation (1). The specimens used in the AV_{valid} measurement were cubic specimens with lengths, widths, and heights of 150 mm after 7 days of curing. The compositions of the PCC-SA and the PCC-RA are shown in Table 6.

$$AV_{valid} = (1 - \frac{m_{water} - m_{dry}}{V \cdot \rho_w}) \times 100\% \quad (1)$$

where m_{water} is saturated specimen mass in water, g; m_{dry} is mass of specimen in air, g; V is the volume of specimens, cm^3; and ρ_w is density of water, g/cm^3.

Table 6. Compositions of PCC-SA and PCC-RA.

Type	Materials Mass in One Cubic Meter/kg				AV_{valid} (%)
	Cement	Silica Fume	Aggregate	Water	
PCC-SA	266	30	1427	107	19.5
PCC-RA	266	30	1395	121	21.4

3. Test Methods

Water retention capacity test, fluidity test, and compressive strength test were used to evaluate W_R, F_L, and compressive strength of the water retentive material. In addition, the compressive strength test, the flexural strength test, and the temperature monitoring test were utilized to evaluate the mechanical performance and the cooling effects of WTCC.

3.1. Water Retention Capacity Test

The water retentive material should have good W_R to store as much as water as possible. The greater W_R is, the longer the evaporation time lasts. W_R was determined by the following steps:

1. Water retentive material was prepared according to the composition ratio. A cylindrical specimen with a height of 5 cm and diameter of 5 cm was fabricated.
2. The specimens were cured at a constant temperature of 20 ± 2 °C and a humidity of 95% for seven days.
3. Specimens were dried at a constant temperature of 105 ± 2 °C for 24 h, and the original mass of the specimens (m_0) was measured.
4. Specimens were immersed into water until the mass was constant. The mass of the saturated specimen was measured as m_1. The W_R is calculated by the Equation (2).

$$W_R = \frac{m_1 - m_0}{V} \times 100\% \quad (2)$$

where V is the volume of specimens, cm^3.

3.2. Fluidity Test

The F_L was determined via flow cone method according to JTG 3420-2020 [27]. A volume of 1725 mL ± 5 mL of water retentive material was poured into a flow cone by opening the outlet and allowing the water retentive material to free flow out from the flow cone. The flowing time was recorded as the F_L. The greater F_L was, the better the fluidity was.

3.3. Compressive Strength and Flexural Strength Test

For the water retentive material, compressive strength test was performed on cylindrical specimens with a height of 50 mm and diameter of 50 mm. The specimens were cured at a temperature of 20 ± 2°C and a humidity of 95% for seven days. A loading rate of 5 mm/min was applied until the specimens failed. The compressive strength (R_c) of the water retentive material was calculated based on Equation (3).

$$R_c = \frac{P}{A} \quad (3)$$

where P is the peak force, kN; and A is the area of upper surface of specimen, mm^2.

For PCC, WTCC, and WCC, both compressive strength and flexural strength were evaluated. Prismatic specimens with a height of 150 mm, a width of 150 mm, and a length of 550 mm were prepared for the measurement of flexural strength. Four point bending test was used to measure the flexural strength. Cubic specimens with height, width, and length of 150 mm were prepared for the compressive strength test [27]. The compressive strength was calculated based on the Equation (3), and the flexural strength (F_f) was calculated based on the Equation (4). The compressive strength and the flexural strength of the specimens after 7 days, 28 days, and 90 days curing were evaluated.

$$F_f = \frac{PL}{bh^2} \quad (4)$$

where L is the length between two fixing supports, mm; and b and h are width and height of specimen, mm.

3.4. Temperature Monitoring Test

Temperature monitoring test was designed to evaluate the cooling effects of SMA, PCC, WCC, and WTCC. The specimens of PCC, WCC, and WTCC used in the temperature monitoring test had a length of 300 mm, a width of 300 mm, and a height of 100 mm. Vibration compaction was used to prepare the specimen of PCC, WCC, and WTCC. Before compaction, one temperature sensor was embedded at the depth of 50 mm, and another temperature sensor was embedded at the surface of the center of each specimen.

In actual pavement, the heat only can be transferred from the surface of the pavement. To simulate the field condition, a testing box was designed to avoid the heat transferring from the bottom and the side of the specimen. As shown in Figure 3, the testing box was a wood box filled with insulating foam and tested mixtures. The thickness of the insulating foam was 5 cm. The bottom and the side of the specimens were wrapped by the same mixtures with the tested specimens.

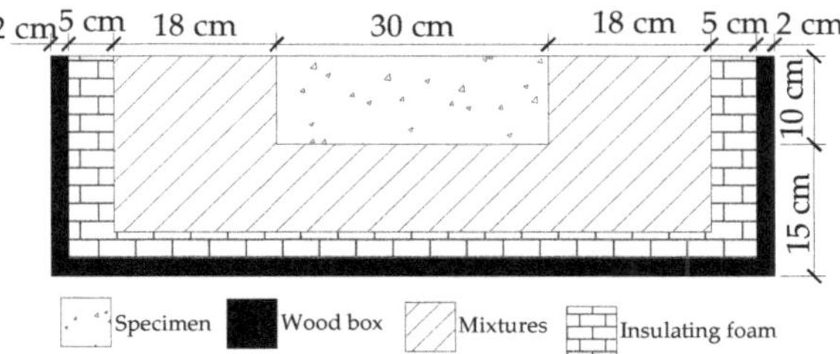

Figure 3. Schemes of testing box.

A multi-channel automatic temperature recorder, produced by Hang Zhou Ze Da Instrument Co., Ltd., (Hangzhou, China) was used in the temperature monitoring test. The measuring accuracy was 0.1 °C, and the temperature was recorded every 10 min. Before testing, a mass of 500 mL water was sprinkled on the surface of the specimens to simulate the rainfall. Temperature sensors were connected to the multi-channel automatic temperature recorder. To evaluate the field cooling effects, one day with maximum and minimum air temperatures of 35 °C and 19 °C in summer was selected to perform the temperature monitoring test.

4. Results and Discussion

4.1. Design of Water Retentive Material

The composition ratio of fly ash, cement, silica fume, and water in the water retentive material was determined based on W_R, F_L, and compressive strength. The cement not only provided strength for the water retentive material but also prevented fly ash and silica fume from losing the water retentive material. However, excessive cement may clog the air voids and had a negative effect on the absorption capacity. Based on common sense and practices, the maximum percentage of cement is recommended as 15% of the total mass of fly ash and silica fume [28]. In addition, W_R greater than 0.25 g/cm^3, F_L between 15 and 20 s, and compressive strength greater than 2.0 MPa are generally required to ensure workability and strength for the water retentive material [28].

The mass ratio of fly ash to silica fume (F/S) was firstly determined. Three F/Ss of 65/35, 60/40, and 55/45 were investigated by the water retention capacity test. Figure 4a shows the W_R under different F/S. W_R decreased with the decrement of F/S. Results indicate that increasing the percentage of fly ash can increase the W_R of the water retentive material. When the F/S was greater than 60/40, the W_R could satisfy the requirement of being greater than 0.25 g/cm^3. Secondly, the mass of water to the total mass of fly ash, silica fume, and cement (W/FSC) was determined based on the F_L. The fluidities of three W/FSCs of 1.8/1, 2.0/1, and 2.2/1 were evaluated. Figure 4b presents the F_L of different W/FSCs under the F/Ss of 65/35 and 60/40. The F_L decreased with the increasing W/FSC. Under the same W/FSC, F_L under F/S of 60/40 was greater than F/S of 65/35, demonstrating that silica fume had a positive influence on the fluidity of the water retentive material. Under the F/S of 65/35 with W/FSC of 1.8/1 and F/S of 60/40 with W/FSC of 2.0/1, F_L satisfies the requirement of between 15 and 20 s. Compared with W/FSC of 2.0/1, the water retentive material under W/FSC of 1.8/1 with less water had better durability because water had a negative influence on the durability. Thus, the composition of the water retentive material was determined as F/S of 65/35, cement mass of 15% of the total mass of fly ash and silica fume, and W/FSC of 1.8/1. In other words, the mass ratio of fly ash, silica fume, cement, and water was 65:35:15:63.9. Finally, the compressive strength test was performed, and the compressive strength was 2.7 MPa, which is greater than 2.0 MPa. The design procedure of water retentive material is summarized in Figure 5.

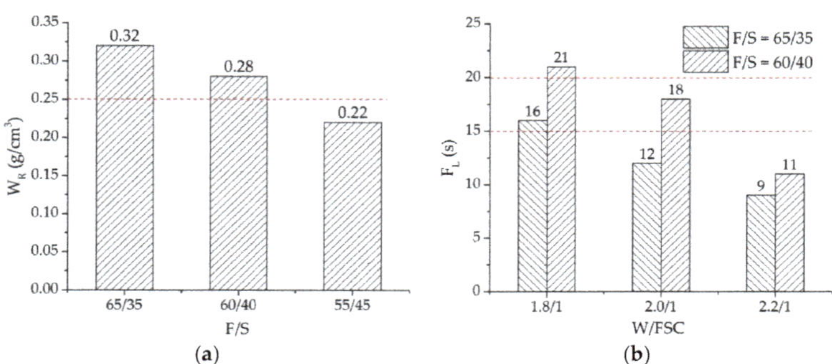

Figure 4. W_R and F_L: (**a**) W_R under different F/S; (**b**) F_L under different W/FSC.

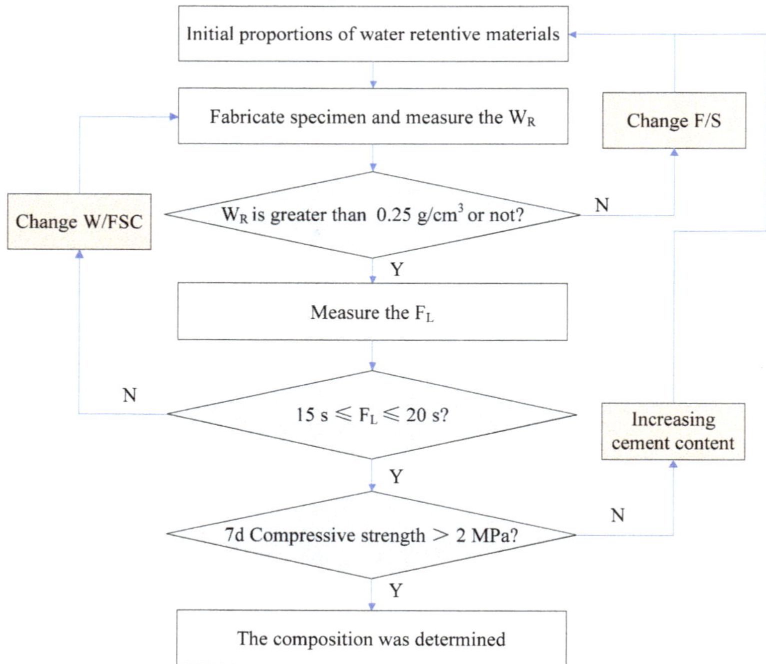

Figure 5. Design procedure of water retentive material.

4.2. *Preparation of WTCC*

The WTCC was prepared by pouring water retentive material into PCC-RA, while WCC was prepared by pouring water retentive material into PCC-SA after 7 days of curing. The flexural strengths of PCC-SA and PCC-RA after 28 days of curing were 4.8 MPa and 4.3 MPa, respectively. The flexural strength of RCC-RA was smaller than RCC-SA because the strength of RA was smaller than SA. Based on the specification of design of highway cement concrete pavement of China, the minimum flexural strength of 4.5 MPa is specified for medium traffic roads, and the minimum flexural strength of 4.0 MPa is specified for light traffic roads [29]. Therefore, PCC-SA could be used for the medium traffic roads, while PCC-RA only could be used for the light traffic roads.

The preparation of WCC and WTCC was based on the following steps: (1) Dust was removed from the surface of the specimens, and some water was sprayed on the surface of the specimens. (2) We placed the specimens on a vibration table and poured the water retentive material on the surface of specimen. The surface of the specimen was covered by a geotextile. (3) We turned on the vibration table until the water retentive material could not infiltrate the specimens. (4) The water retentive material that remained on the surface was removed, and the specimens were cured under the temperature of 20 ± 2 °C and the humidity of 95% for 28 days.

4.3. Mechanical Performance of WTCC

Figure 6a,b show the compressive strength and flexural strength of different pavements after 7 days, 28 days, and 90 days curing, respectively. The compressive strength and flexural strength of WTCC at 28 days is 30.4 MPa and 4.6 MPa, respectively, which satisfies the requirement of medium traffic roads. Even though the WTCC used in the square, bicycle lane, and sidewalk is more appropriate, the WTCC has a potential used in the medium traffic roads based on the mechanical performance.

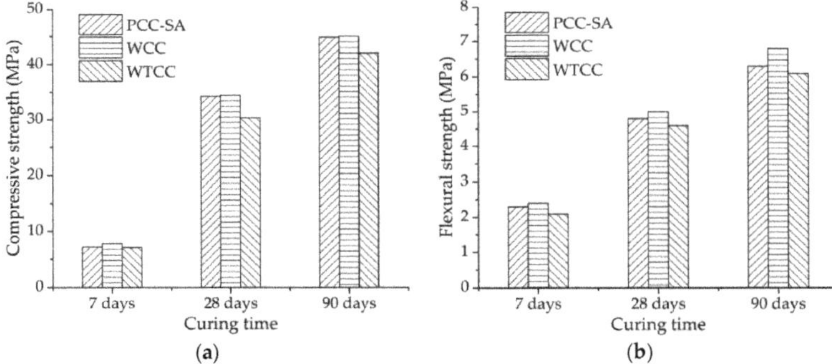

Figure 6. Compressive and flexural strength of different pavements: (**a**) compressive strength; (**b**) flexural strength.

The compressive strength and the flexural strength increased with the curing time. The flexural strength of WCC was greater than PCC-SA, and the flexural strength of WTCC was greater than PCC-RA, indicating that the water retentive material functionally improved the flexural strength. After the water retentive material was added into the PCC-SA and the PCC-RA, the pore structures were filled with water retentive materials, leading to AV content decrement. The strength increased with the decrement of AV content, and the self-strength of the water retentive material could also improve the strength of the specimen. In addition, compared with PCC and WTCC, both compressive strength and flexural strength of WCC were the largest. It also demonstrates that the water retentive material and the higher strength of SA had contributions to the strength of the specimen.

4.4. Cooling Effects of WTCC

Figure 7 shows the surface temperature of four specimens under the field air temperature. As shown in Figure 7a, the temperature changes showed a consistency with the air temperature. At the start of the test, the temperature of cement concrete pavement was lower than SMA due to a lower heat absorptivity of cement concrete than asphalt mixtures.

The highest surface temperature appeared at about 15:00, and the surface temperature of SMA was the highest. The highest surface temperatures of SMA, PCC, WCC, and WTCC were 62.9 °C, 57.0 °C, 55.6 °C, and 51.5 °C, respectively. Compared with SMA, PCC, and WCC, the highest surface temperatures of WTCC decreased by 11.4 °C, 5.5 °C, and 4.1 °C, respectively. Figure 7b shows the average surface temperature of the four mixtures. Compared with SMA, PCC, and WCC, the average surface temperatures of WTCC decreased by 2.9 °C, 4.0 °C, and 6.1 °C, respectively. Figure 8a shows the temperature distribution at the depth of 50 mm under the field air temperature. The temperature at the depth of 50 mm of SMA was also the highest. Compared with SMA, PCC, and WCC, the highest temperatures at the depth of 50 mm of WTCC decreased by 10.3 °C, 6.1 °C, and 4.6 °C, respectively. Figure 8b shows the average of the temperature at the depth of 50 mm. Compared with SMA, PCC, and WCC, the average surface temperatures of WTCC decreased by 3.5 °C, 4.0 °C, and 6.4 °C, respectively. A significant temperature reduction was found in WTCC.

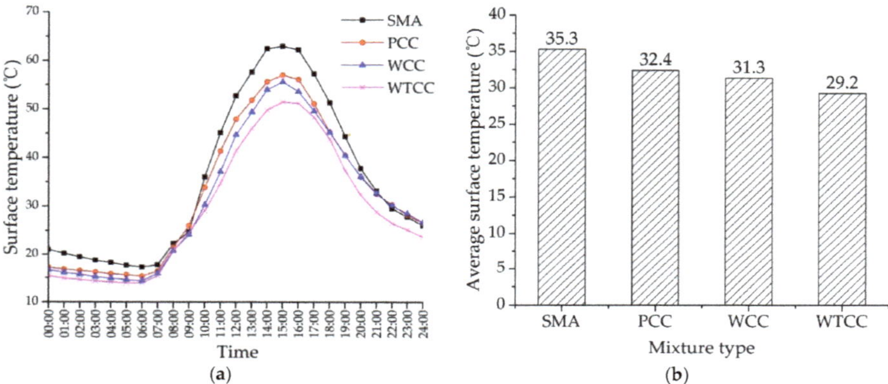

Figure 7. Surface temperature under field air temperature: (**a**) temperature variation curve; (**b**) average surface temperature.

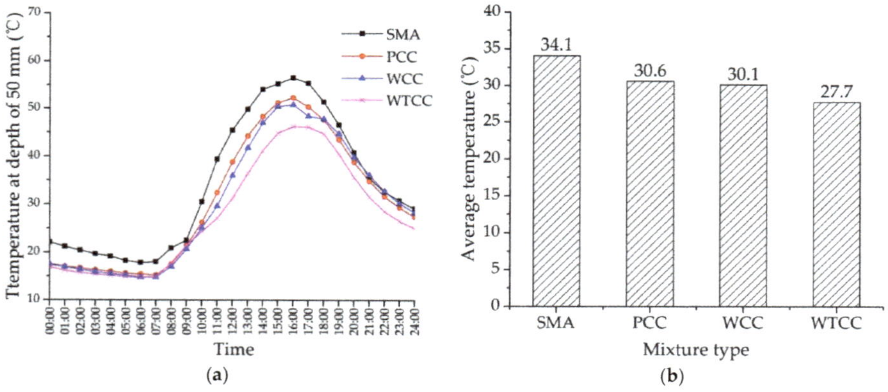

Figure 8. Temperature at the depth of 50 mm under the field air temperature: (**a**) temperature variation curve; (**b**) average temperature.

The temperature of WCC was slightly smaller than PCC due to the heat dissipation through water evaporation of the water retentive material. However, as shown in Figures 7 and 8, the temperature of WCC approached the same as the temperature of PCC after 18:00. Water in the water retentive material was consumed during continuous evaporation. When the water was exhausted, temperature reduction through water evaporation failed. Therefore, it is suggested to supplement water for pavement at 12:00 to prolong and enhance the cooling effects. This work can be done by incorporating it in the cleaning work of pavements. Because WTCC uses thermal resistant aggregates, the temperature of WTCC was still the lowest, even when the water in the water retentive material was exhausted.

Based on the results, WTCC presents a great cooling effect via the combination of porous cement concrete, a thermal resistant aggregate, and water retentive material. WTCC has potential to efficiently mitigate the UHI effects and be used in medium traffic roads.

5. Conclusions

In this study, water retentive material and WTCC were developed to alleviate the UHI effects. The mechanical performance and the cooling effects of WTCC were evaluated and compared with SMA, PCC, and WCC. Several conclusions could be obtained from the study.

(1) Based on water retention capacity, fluidity, and compressive strength, the composition ratio of the water retentive material was determined. The mass ratios of fly ash, silica fume, cement, and water were determined as 65:35:15:63.9.
(2) WTCC was prepared by pouring the water retentive material into PCC incorporating a thermal resistant aggregate. Even though the WTCCs used in the square, the bicycle lane, and the sidewalk were more appropriate, the WTCC has potential to be used in medium traffic roads based on the mechanical performance.
(3) The water retentive material has benefits of improving the strength of pavements and temperature reduction. The freeze–thaw resistance of WTCC will be further evaluated in future tests.
(4) Compared with SMA, PCC, and WCC, surface temperatures of WTCC decreased by 11.4 °C, 5.5 °C, and 4.1 °C, respectively, and internal temperatures decreased by 10.3 °C, 6.1 °C, and 4.6 °C, respectively. Results demonstrate that WTCC has superior cooling effects due to the lower thermal conductivity of porous cement concrete, the thermal resistant aggregate, and the water retentive material. WTCC can be used to efficiently to mitigate UHI effects.

Author Contributions: Conceptualization, X.W. and X.J.; methodology, X.H.; validation, X.J.; formal analysis, X.H. and B.C.; investigation, X.H.; resources, H.C. and B.C.; data curation, X.W.; writing—original draft preparation, X.W.; writing—review and editing, X.H.; visualization, X.H.; supervision, X.J.; funding acquisition, X.W. All authors have read and agreed to the published version of the manuscript.

Funding: This research was funded by National Natural Science Foundation of China (52008333), Natural Science Basic Research Program of Shaanxi (2020JQ-680), China Postdoctoral Science Foundation (2020M673354), and Scientific Research Program Funded by Shaanxi Provincial Education Department (20JK0711).

Institutional Review Board Statement: Not applicable.

Informed Consent Statement: Not applicable.

Data Availability Statement: The data presented in this study are available on request from the correspondence author.

Acknowledgments: The authors would like to express sincere thanks to the National Natural Science Foundation of China, Natural Science Basic Research Program of Shaanxi, China Postdoctoral Science Foundation, and Shaanxi Provincial Education Department.

Conflicts of Interest: The authors declare no conflict of interest.

References

1. Santamouris, M. Using cool pavements as a mitigation strategy to fight urban heat island—A review of the actual developments. *Renew. Sustain. Energy Rev.* **2013**, *26*, 224–240. [CrossRef]

2. Jiang, Y.J.; Ye, Y.Q.; Xue, J.S.; Chen, Z.J. Thermal Resistant Stone Mastic Asphalt Surface and Its Antirutting Performance. *J. Mater. Civ. Eng.* **2018**, *30*. [CrossRef]
3. Cheela, V.R.S.; John, M.; Biswas, W.; Sarker, P. Combating Urban Heat Island Effect-A Review of Reflective Pavements and Tree Shading Strategies. *Buildings* **2021**, *11*, 93. [CrossRef]
4. Sen, S.; Roesler, J. Thermal and optical characterization of asphalt field cores for microscale urban heat island analysis. *Constr. Build. Mater.* **2019**, *217*, 600–611. [CrossRef]
5. Nwakaire, C.M.; Onn, C.C.; Yap, S.P.; Yuen, C.W.; Onodagu, P.D. Urban Heat Island Studies with emphasis on urban pavements: A review. *Sustain. Cities Soc.* **2020**, *63*. [CrossRef]
6. Wang, J.R.; Zhang, Z.Q.; Guo, D.T.; Xu, C.; Zhang, K. Study on Cooling Effect and Pavement Performance of Thermal-Resistant Asphalt Mixture. *Adv. Mater. Sci. Eng.* **2018**, *2018*. [CrossRef]
7. Wang, X.; Gu, X.; Hu, X.; Zhang, Q.; Dong, Q. Three-stage Evolution of Air Voids and Deformation of Porous-Asphalt Mixtures in High-temperature Permanent Deformation. *J. Mater. Civ. Eng.* **2020**, *32*, 04020233. [CrossRef]
8. Santamouris, M.; Synnefa, A.; Karlessi, T. Using advanced cool materials in the urban built environment to mitigate heat islands and improve thermal comfort conditions. *Sol. Energy* **2011**, *85*, 3085–3102. [CrossRef]
9. Zheng, M.L.; Tian, Y.J.; He, L.T. Analysis on Environmental Thermal Effect of Functionally Graded Nanocomposite Heat Reflective Coatings for Asphalt Pavement. *Coatings* **2019**, *9*, 178. [CrossRef]
10. Pomerantz, M.; Pon, B.; Akbari, H.; Chang, S. *The Effects of Pavement Temperatures on Air Temperatures in Large Cities*; LBNL-43442; Berkeley Lab: Berkeley, CA, USA, 2000.
11. Santamouris, M.; Gaitani, N.; Spanou, A.; Saliari, M.; Giannopoulou, K.; Vasilakopoulou, K.; Kardomateas, T. Using cool paving materials to improve microclimate of urban areas–Design realization and results of the flisvos project. *Build. Environ.* **2012**, *53*, 128–136. [CrossRef]
12. Xie, N.; Li, H.; Abdelhady, A.; Harvey, J. Laboratorial investigation on optical and thermal properties of cool pavement nano-coatings for urban heat island mitigation. *Build. Environ.* **2019**, *147*, 231–240. [CrossRef]
13. Sha, A.M.; Liu, Z.Z.; Tang, K.; Li, P.Y. Solar heating reflective coating layer (SHRCL) to cool the asphalt pavement surface. *Build. Environ.* **2017**, *139*, 355–364. [CrossRef]
14. Cao, X.J.; Tang, B.M.; Zhu, H.Z.; Zhang, A.M.; Chen, S.M. Cooling Principle Analyses and Performance Evaluation of Heat-Reflective Coating for Asphalt Pavement. *J. Mater. Civ. Eng.* **2011**, *23*, 1067–1075. [CrossRef]
15. Hassn, A.; Chiarelli, A.; Dawson, A.; Garcia, A. Thermal properties of asphalt pavements under dry and wet conditions. *Mater. Des.* **2016**, *91*, 432–439. [CrossRef]
16. Wang, X.W.; Ren, J.X.; Gu, X.Y.; Li, N.; Tian, Z.Y.; Chen, H.Q. Investigation of the adhesive and cohesive properties of asphalt, mastic, and mortar in porous asphalt mixtures. *Constr. Build. Mater.* **2021**, *276*. [CrossRef]
17. Stempihar, J.J.; Pourshams-Manzouri, T.; Kaloush, K.E.; Rodezno, M.C. Porous Asphalt Pavement Temperature Effects for Urban Heat Island Analysis. *Transp. Res. Rec.* **2012**, *2293*, 123–130. [CrossRef]
18. Hassn, A.; Aboufoul, M.; Wu, Y.; Dawson, A.; Garcia, A. Effect of air voids content on thermal properties of asphalt mixtures. *Constr. Build. Mater.* **2016**, *115*, 327–335. [CrossRef]
19. Wang, J.S.; Meng, Q.L.; Zhang, L.; Zhang, Y.; He, B.J.; Zheng, S.L.; Santamouris, M. Impacts of the water absorption capability on the evaporative cooling effect of pervious paving materials. *Build. Environ.* **2019**, *151*, 187–197. [CrossRef]
20. Jiang, W.; Sha, A.M.; Xiao, J.J.; Wang, Z.J.; Apeagyei, A. Experimental study on materials composition design and mixture performance of water-retentive asphalt concrete. *Constr. Build. Mater.* **2016**, *111*, 128–138. [CrossRef]
21. Yamagata, H.; Nasu, M.; Yoshizawa, M.; Miyamoto, A.; Minamiyama, M. Heat island mitigation using water retentive pavement sprinkled with reclaimed wastewater. *Water Sci. Technol.* **2008**, *57*, 763–771. [CrossRef]
22. Geng, J.G.; Chen, M.Y.; Shang, T.; Li, X.; Kim, Y.R.; Kuang, D.L. The Performance of Super Absorbent Polymer (SAP) Water-Retaining Asphalt Mixture. *Materials* **2019**, *12*, 1964. [CrossRef]
23. Ryms, M.; Lewandowski, W.M.; Klugmann-Radziemska, E.; Denda, H.; Wcislo, P. The use of lightweight aggregate saturated with PCM as a temperature stabilizing material for road surfaces. *Appl. Therm. Eng.* **2015**, *81*, 313–324. [CrossRef]
24. Ryms, M.; Denda, H.; Jaskula, P. Thermal stabilization and permanent deformation resistance of LWA/PCM-modified asphalt road surfaces. *Constr. Build. Mater.* **2017**, *142*, 328–341. [CrossRef]
25. Ryms, M.; Januszewicz, K.; Haustein, E.; Kazimierski, P. Thermal properties of a cement composite containing phase change materials (PCMs) with post-pyrolytic char obtained from spent tyres as a carrier. *Energy* **2022**, *239*, 121936. [CrossRef]
26. Research Institute of Highway Ministry of Transport of China. *Technical Specifications for Construction of Highway Aspahlt Pavements*; JTG F40-2004; Research Institute of Highway Ministry of Transport of China: Beijing, China, 2004.
27. Research Institute of Highway Ministry of Transport of China. *Testing Methods of Cement and Concrete for Highway Engineering*; JTG 3420-2020; Research Institute of Highway Ministry of Transport of China: Beijing, China, 2020.
28. Hu, L.; Sha, A.M. Research on water-holding and cooling concrete design method and its performance. *J. Funct. Mater.* **2012**, *43*, 1348–1351.
29. Ministry of Transport of the People's Republic of China. *Specifications for Design of Highway Cement Concrete Pavement*; JTG D40-2011; Ministry of Transport of the People's Republic of China: Beijing, China, 2011.

MDPI

Article

Numerical Investigation of the Temperature Field Effect on the Mechanical Responses of Conventional and Cool Pavements

Pengfei Liu [1,*], Xiangrui Kong [1], Cong Du [1], Chaohe Wang [1], Di Wang [2] and Markus Oeser [1,3]

[1] Institute of Highway Engineering, RWTH Aachen University, Mies-van-der-Rohe-Str. 1, 52074 Aachen, Germany
[2] Department of Civil Engineering, Aalto University, Rakentajanaukio 4, 02150 Espoo, Finland
[3] Federal Highway Research Institute (BASt), Brüderstr. 53, 51427 Bergisch Gladbach, Germany
* Correspondence: liu@isac.rwth-aachen.de; Tel.: +49-241-80-22780

Abstract: Conventional asphalt pavement has a deep surface color and large thermal inertia, which leads to the continuous absorption of solar thermal radiation and the sharp rise of surface temperature. This can easily lead to the permanent deformation of pavement, as well as aggravate the urban heat island (UHI) effect. Cool pavement with a reflective coating plays an important role in reducing pavement temperature and alleviating the UHI effect. It is of great significance to study the influence of temperature on the mechanical response of different types of pavement under vehicle loading. Therefore, this study examined the heat exchange theory between pavement and the external environment and utilized the representative climate data of a 24 h period in the summer. Two kinds of three-dimensional finite element models were established for the analysis of temperature distribution and the mechanical responses of conventional pavement and cool pavement. The results show that in this environmental condition, conventional pavement temperatures can exceed 50 °C under high temperatures in summer, which allows for the permanent deformation of pavement and further affects the service life of asphalt pavement. The temperature difference in a conventional pavement surface between 6 h (24.7 °C) and 22 h (30.2 °C) is much less than that between 22 h (30.2 °C) and 13 h (50.1 °C) in the summer. However, the difference in the vertical displacements of the pavement surface between 6 h and 22 h is much larger than that between 22 h and 13 h. One reason is that the difference in temperature distribution between the morning and night leads to changes in pavement structure stiffness, resulting in significant differences in vertical displacement. Cool pavement has a significant cooling effect, which can reduce the surface temperature of a road by more than 15 °C and reduce the vertical displacement of the pavement by approximately 11.3%, which improves the rutting resistance of the pavement. However, the use of cool pavement will not change the horizontal strain at the bottom of the asphalt base and will not improve the fatigue resistance of asphalt pavement. This research will lay the foundation for further clarifying the difference in the mechanical properties between the two types of pavements in the management and maintenance stage.

Keywords: asphalt pavement; cool pavement; temperature distribution; numerical simulation; mechanical response

Citation: Liu, P.; Kong, X.; Du, C.; Wang, C.; Wang, D.; Oeser, M. Numerical Investigation of the Temperature Field Effect on the Mechanical Responses of Conventional and Cool Pavements. *Materials* **2022**, *15*, 6813. https://doi.org/10.3390/ma15196813

Academic Editor: Simon Hesp

Received: 6 September 2022
Accepted: 27 September 2022
Published: 30 September 2022

1. Introduction

As a typical temperature-sensitive material, the modulus of an asphalt mixture changes with its temperature. As a result, the load-carrying capacities and pavement performances of asphalt pavements are significantly influenced by temperature. Various common types of damage to asphalt pavements, such as low-temperature cracking, high-temperature rutting, and fatigue damage, are also directly or indirectly related to the temperature distribution within the pavement [1–4].

Compared with green vegetation, asphalt pavement has different characteristics such as a darker surface color and greater thermal inertia. Asphalt pavement constantly absorbs solar heat radiation, causing the surface temperature to rise sharply, which not only accelerates its own high-temperature rutting damage, but also exacerbates the urban heat island (UHI) effect. The use of cool pavement to mitigate the UHI effect is a popular direction in recent research. The phrase "cool pavements" was recently defined by the United States Environmental Protection Agency (USEPA) as "cool pavements include a range of established and emerging technologies that communities are exploring as part of their heat island reduction efforts. The term currently refers to paving materials that reflect more solar energy, enhance water evaporation, or have been otherwise modified to remain cooler than conventional pavements." [5]. The cool pavement cooling principle mainly includes heat reflection, water evaporation, and road heat storage [6].

The reflective coating is a pre-treatment solution that prevents the first surface layer from absorbing heat and reflects a significant amount of energy back into the environment, thus reducing the downward conduction of heat sources in the pavement structure [7]. Zheng et al. [8] developed different reflective coating materials and studied their effects on human comfort through field tests. The results show that the use of reflective coatings can effectively improve human comfort and alleviate the UHI effect. Evaporative pavements are often designed as water-locked structures, with water absorbing heat to transition from a liquid to a gas state. This process requires the absorption of heat from the surrounding environment, which cools the pavement. Parking lots, walkways, highway shoulders, and city streets are all examples of evaporative pavements [9]. Another use of permeable pavement is for storing rainwater after high-intensity rainfall events [10]. In absorbing the heat extracted by embedded asphalt solar collectors, collected heat pavement can be used as a sustainable energy source.

Research on the performance of asphalt pavement under a multi-physical field has been rapidly developing in recent years [11]. Wang et al. [12] studied the thermodynamics and mechanical properties of porous asphalt under high temperature and high-strength rainfall. The resistance of porous asphalt to rutting under this multi-physical field condition was evaluated. The results show that the rutting resistance of a porous asphalt mixture under rainfall conditions is lower than that under dry conditions. The rutting resistance of a porous asphalt mixture is more sensitive to temperature than to rainfall conditions. Ma et al. [13] used measured weather data to simulate temperature transfer during asphalt pavement construction. The results show that the initial temperature and layer thickness affect the overall temperature field during compaction, and wind speed and temperature mainly affect the upper-temperature field of hot mix asphalt (HMA). Zhao et al. [14] developed a three-dimensional (3D) finite element (FE) model based on transient heat transfer. The results show that the influence of temperature decreases with the increase in depth. Therefore, future subgrade temperature stress is less of a concern. Temperature differences should be considered in pavement design in regions with large temperature differences.

In summary, a comparison of the influence of temperature on the mechanical properties of conventional pavement and cool pavement with reflective coating is seldom seen in the literature. The main objective of this paper is to study the effects of temperature on the two types of asphalt pavements using the FE method, mainly in terms of the variation of their mechanical parameters. To this end, the heat transfer model for asphalt pavement is first introduced. The development of FE models for conventional and cool pavements is then described. The results derived from the FE simulations with respect to the heat transfer within and the mechanical responses of the two types of asphalt pavements are analyzed. Finally, the conclusions and outlook are provided.

2. Heat Transfer Model in Asphalt Pavement

For asphalt pavement structures, it is generally assumed that the horizontal temperature gradient is zero. Therefore, the side boundary conditions can be ignored, and the main boundaries considered are the surface boundary and the bottom boundary. The pavement surface is the main boundary of the pavement heat transfer, and the heat exchange between the pavement and the external environment is carried out mainly in three ways: solar radiation, thermal convection, and pavement surface radiation [4], as shown in Figure 1.

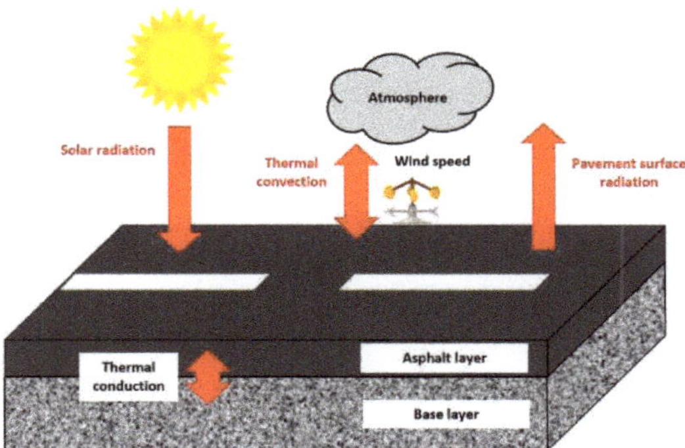

Figure 1. Schematic diagram of heat transfer modes for asphalt pavement.

For solar radiation, the radiation flux incident on the surface or medium passes through three paths, namely, transmission, reflection, and absorption. According to the definition of energy conservation, the relationship between the transmissivity α, reflectivity ρ, and absorptivity τ of the radiation flux is as shown in Equation (1) [15]:

$$\alpha + \rho + \tau = 1 \tag{1}$$

Asphalt pavement is a black opaque solid. For the opaque surface, the transmittance α is 0, and then $\rho + \tau = 1$. Therefore, the lower the absorption rate is, the higher the reflectivity of the opaque surface is. The higher the reflectivity, the lower the heat loss.

According to the research of Liao et al., solar radiation can be calculated as follows [16]:

$$q_s(t) = \begin{cases} 0 & 0 \le t < 12 - \frac{c}{2} \\ q_0 \cos\, m\beta(t-12) & 12 - \frac{c}{2} \le t \le 12 + \frac{c}{2} \\ 0 & 12 + \frac{c}{2} \le t < 24 \end{cases} \tag{2}$$

$$q_0 = 0.131 m Q_d \tag{3}$$

$$m = 12/c \tag{4}$$

where $q_s(t)$ is the function of heat flux of solar radiation with time (mJ/(h·mm^2)), q_0 is the maximum value of q_s, Q_d is the total heat flux of solar radiation per day (mJ/(mm^2)), c is the effective duration of sunshine (h), β is a parameter (rad, $2\pi/24 = 0.2618$), and t is the time (h).

The q_s values obtained by Equations (2)–(4) are not smooth and continuous, and there will be jumping points when calculating the temperature field. In accordance with the relevant principle of the Fourier series, the cosine function was expanded into the corresponding Fourier series form. When k reaches 30, it can meet the requirements of engineering accuracy [16].

$$q_s(t) = \frac{a_0}{2} + \sum_{k=1}^{\infty} a_k \cos\frac{k\pi(t-12)}{12} \quad a_0 = \frac{2q_0}{m\pi} \tag{5}$$

$$a_k = \begin{cases} \frac{q_0}{\pi}\left[\frac{1}{m+k}\sin\,(m+k)\frac{\pi}{2m} + \frac{\pi}{2m}\right] & k = m \\ \frac{q_0}{\pi}\left[\frac{1}{m+k}\sin\,(m+k)\frac{\pi}{2m} + \frac{1}{m-k}\sin\,(m-k)\frac{\pi}{2m}\right] & k \ne m \end{cases} \tag{6}$$

$$Q_s = \tau \times q_s(t) \tag{7}$$

where Q_s is the effective solar radiation absorbed by the pavement mJ/(h·m^2), and τ is the absorptivity of solar radiation [16].

When affected by solar radiation, the air temperature will also show corresponding changes. The air temperature and wind speed are important factors affecting the heat exchange (thermal convection) between the road surface and the atmosphere. It is not accurate to use a single sine function to simulate the temperature change process during the day, and so a linear combination of two sine functions was used to simulate the temperature change process [16], as shown in Equation (8):

$$T_a = \overline{T_a} + T_m[0.96\sin\omega(t - t_0) + 0.14\sin\omega(t - t_0)],\ \omega = 2\pi/24 \tag{8}$$

where $\overline{T_a}$ is the average of the highest and lowest temperatures of the day, which can be expressed as [16]:

$$\overline{T_a} = \frac{1}{2}\left(T_a^{max} + T_a^{min}\right) \tag{9}$$

where T_m is the magnitude of the temperature change during a day, i.e., half of the difference between the highest and lowest temperature values, which can be shown as [16]:

$$T_m = \frac{1}{2}\left(T_a^{max} - T_a^{min}\right) \tag{10}$$

The exchange coefficient of the heat exchange, h_c (mJ/(h·mm^2 °C)), between the road surface and the atmosphere is mainly affected by the wind speed v_w, m/s, and the relationship between them is linear [16], as shown in Equation (11):

$$h_c = 13.32v_w + 33.84 \tag{11}$$

The pavement surface radiation can be represented by Equation (12) [16]:

$$Q_{ps} = \varepsilon\sigma\left[(T_{ps} - T_Z)^4 - (T_a - T_Z)^4\right] \tag{12}$$

where Q_{ps} is the effective road surface radiation (mJ/(h·mm^2)), ε is the road surface radiation emissivity, $\sigma = 2.04 \times 10^{-7}$ mJ/(h·mm^2·K^4) is a Stefan-Boltzmann constant, T_{ps} and T_a are the pavement surface temperature and air temperature, respectively, and T_z is the absolute zero temperature.

Due to the influence of the atmospheric environment, the temperature fluctuation of the pavement surface is large, while the temperature fluctuation of the deeper part of the subgrade is small compared to the atmospheric environment fluctuation, which can be considered as a constant. Some researchers have considered the bottom boundary as an adiabatic boundary. However, it has been found that whether the bottom boundary condition is set to a constant temperature or is adiabatic, its effect on the temperature field of the upper layer of the asphalt pavement is negligible.

3. Development of FE Model of Asphalt Pavement

Two asphalt pavement models were developed on the general-purpose FE software ABAQUS (2017, Dassault Systèmes SE, Vélizy-Villacoublay, France): one was thermal analysis, which was designed to simulate the heat transfer of asphalt pavement under the effects of solar radiation and atmospheric temperature, and the other was mechanical analysis, which had the same geometry as the first model, with the temperature field of the first model imported into this model. A time-dependent half-sine wave load was applied to observe and compare the stress and displacement distributions when considering the case of temperature variation effects and changes.

3.1. The FE Heat Transfer Model

In this study, two FE pavement heat transfer models without (conventional pavement) and with (cool pavement or reflective pavement) reflective coatings were developed. Both models had the same pavement structure, which was divided into six layers and designed mainly according to the German design standard RStO [17]. In this standard, the width of a single-lane highway should be 3750 mm; however, a single-lane highway can be divided into two parts by the middle axis, and the left and right parts are essentially the same in structure and material properties. Therefore, the model could be simplified as an axisymmetric model, and the width of the model was set to 1875 mm. In addition, since the length of the asphalt pavement does not affect the whole temperature conduction process, the model length was selected as 2000 mm to improve the calculation efficiency of the model. To summarize, the size of the asphalt pavement model was 2000 mm × 1875 mm × 1750 mm, as shown in Figure 2. The interaction relationship between the first three layers of asphalt surface was fully connected. The vertical displacement in the last four layers of the asphalt pavement structure was set to continuous, while sliding could occur in a horizontal direction in the three interfaces.

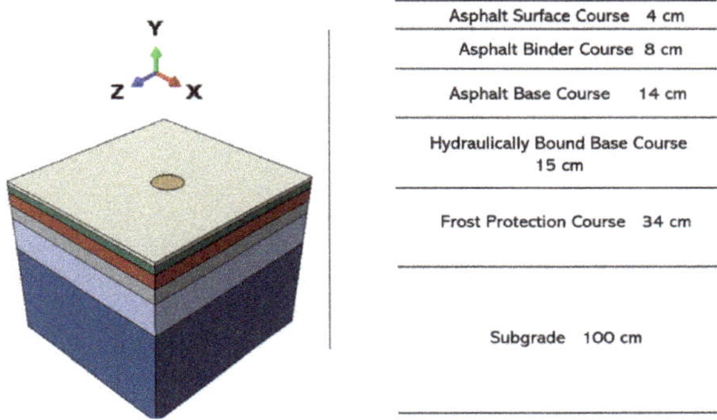

Figure 2. Asphalt pavement model in ABAQUS.

To simulate the cooling effect of the cool pavement, Perfect Cool, a dark pavement coating with a high reflectivity (recently developed by NIPPO Corporation Co. Ltd (Tokyo, Japan)) [18] was selected. Perfect Cool aims to lower the temperature of pavement during the day by enhancing its reflectivity and minimizing the amount of heat absorbed. To decrease heat transmission, Perfect Cool combines dark, low-reflective colored pigments with high infrared-heat-reflecting pigments and small hollow ceramic particles [18]. For the FE pavement heat transfer model with a reflective coating, one additional layer with a thickness of 0.6 mm was created on top of the asphalt surface layer. The thermal parameters used by Perfect Cool are shown in Table 1.

Table 1. Thermal properties of the cool pavement on a global scale.

Albedo	19%
Pavement radiation emissivity	0.828
Thermal conductivity	907.2 mJ/(mm·h·K)

The two modes of heat transfer—steady and transient—were set in the simulation step. The steady-state heat transfer state was set to a very small time, and the transient state simulated the temperature change process of a 24 h period.

The solar radiation acts as a surface heat flux load on the pavement models in the heat transfer model. The action surface was chosen as the upper surface of the asphalt pavement. The solar radiation fluxes defined by Equations (5)–(7) were written by the subroutine

DFLUX in ABAQUS [19]. As established by the analytical Equations (8) and (11), the air temperature and wind speed will influence the simulated thermal convection. In this study, hot summer weather was considered. The specific data are shown in Table 2.

Table 2. Representative summer weather data used in the FE model [20].

Wind Speed	Sunshine Time	Air Temperature			
		Time	°C	Time	°C
2.6 m/s	10.7 h	1:00	25.8	13:00	34.4
		2:00	24.7	14:00	35.3
		3:00	23.7	15:00	35.6
		4:00	23.1	16:00	35.3
		5:00	22.8	17:00	34.7
		6:00	23.1	18:00	33.7
		7:00	24	19:00	32.6
		8:00	25.4	20:00	31.5
		9:00	27.2	21:00	30.3
		10:00	29.2	22:00	29.2
		11:00	31.2	23:00	28.1
		12:00	33	0:00	26.9

The heat exchange coefficient and air temperature were defined using the subroutine FILM in ABAQUS, according to Equations (8) and (11). In this study, the bottom boundary of the pavement structure was set to be insulated.

The reliability of this FE heat transfer model has been validated by comparing the results from this model to the results from [16], using the same parameters.

3.2. The FE Mechanical Pavement Model

Because the reflective coating is very thin compared to the structural layers of the pavement, its influence on the mechanical response of the pavement can be ignored. Therefore, in the FE mechanical pavement modeling, only the pavement structure with six structural layers was used. The relevant mechanical parameters of the model mainly refer to RStO [17] and RDO Asphalt 09 [21], and the damping coefficient refers to the data used in the existing reference [16]. Table 3 shows the parameters of each layer of the pavement model in ABAQUS.

Table 3. Thickness and material properties of the pavement structure.

Layer	Poisson's Ratio	Density [ton/mm^3]	Damping Factor
Asphalt surface course	0.35	2.38×10^{-9}	0.9
Asphalt binder course	0.35	2.49×10^{-9}	0.9
Asphalt base course	0.35	2.30×10^{-9}	0.9
Hydraulically bound base course	0.25	2.40×10^{-9}	0.8
Frost protection course	0.5	2.40×10^{-9}	0.4
Subgrade	0.5	2.40×10^{-9}	0.4

The temperature-dependent material properties were applied for the three asphalt layers. Due to the limited testing results, the viscoelastic parameters of the asphalt surface course and asphalt base course were derived from laboratory tests. The asphalt surface course and the asphalt base course are related to the rutting and fatigue cracking, respectively. The respective Prony series data used in ABAQUS are shown in Tables 4 and 5. For the asphalt binder course, the temperature-dependent Young's modulus referred to RDO Asphalt 09 [21].

Table 4. Prony series data of the asphalt surface course.

Item	Relaxation Time (s)	The mth Maxwell Spring Modulus (MPa)	Item	Relaxation Time (s)	The mth Maxwell Spring Modulus (MPa)
1	1.00×10^{-7}	1.25	7	0.1	1141.19
2	1.00×10^{-6}	3.90	8	1	2214.31
3	1.0×10^{-5}	12.29	9	10	1624.26
4	1.0×10^{-4}	38.33	10	100	425.02
5	0.001	121.51	11	1000	39.42
6	0.01	371.69	12	10,000	1.00
The parallel spring modulus (MPa)					107.04
Tr (°C)					−5
C1					10.16
C2					46.64

Table 5. Prony series data of the asphalt base course.

Item	Relaxation Time (s)	The mth Maxwell Spring Modulus (MPa)	Item	Relaxation Time (s)	The mth Maxwell Spring Modulus (MPa)
1	1.00×10^{-7}	1.73	7	0.1	13,237.69
2	1.00×10^{-6}	9.67	8	1	7135.05
3	1.0×10^{-5}	42.93	9	10	2066.84
4	1.0×10^{-4}	248.43	10	100	678.16
5	0.001	1148.68	11	1000	187.77
6	0.01	6775.57	12	10,000	65.36
The parallel spring modulus (MPa)					22.29
Tr (°C)					10
C1					254,154,208
C2					1,637,000,000

The completed heat transfer model results (temperature distribution) at specific times were imported to the mechanical model for the mechanical response of the pavement. All nodes at the bottom of the model were restricted from moving in all degrees of freedom. The four sides of the FE model were restrained from making any perpendicular movements to the side of the model. The interaction between the asphalt layers was fully coupled. The interfaces between the asphalt base course, hydraulically bound base course, frost protection course, and subgrade were assumed to be partially bound, which means that the vertical displacements of the adjacent layers were consistent, while the horizontal displacements could be different. When studying the mechanical response of the asphalt pavement model, to simplify the computational time and effect, the conventional standard load of 0.7 MPa was used. The load was applied to a circular area with a diameter of 300 mm [17]. In order to simulate the process of a tire passing over the road (45 km/h), a time-dependent half-sine wave load was created, i.e., the load was applied gradually and reached the maximum value at 0.012 s, and then the load was gradually removed until it reached zero at 0.024 s. The completed pavement model with structural boundary conditions is shown in Figure 3. The reliability of this mechanical FE model has been validated in previous investigations [22–25].

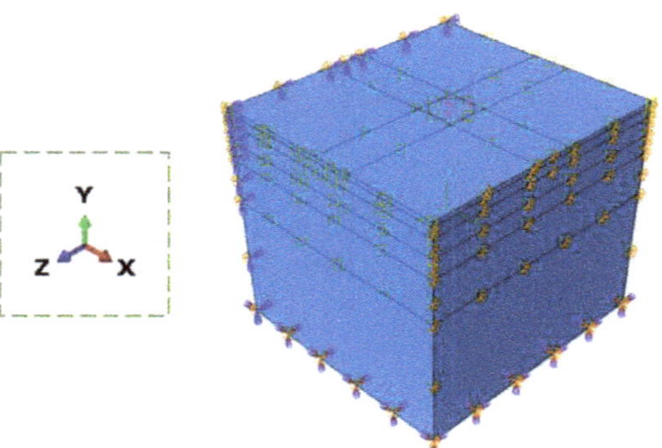

Figure 3. The pavement model with structural boundary conditions.

4. Result and Analysis

4.1. Results of the Responses of Conventional Pavement

4.1.1. Results of the Thermal Response of Conventional Pavement

The analysis of the temperature changes at different depths in the model during the summer shows that the pavement heats up in summer under the effect of solar radiation. The temperature changes at different road depths within a 24 h day are shown in Figure 4.

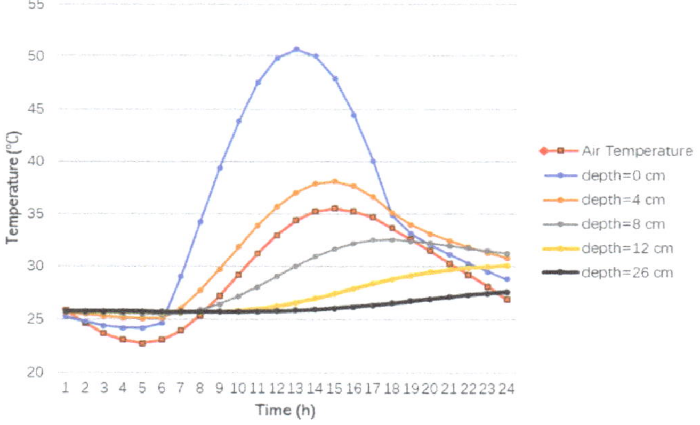

Figure 4. Hourly air temperature and temperature of the asphalt layers at different depths in the summer.

It can be seen from Figure 4 that the maximum temperature of the road during the day reaches 50.7 °C, which is 15.6 °C higher than the air temperature at the same time. The road surface temperature is consistent with the air temperature trend, reaching the highest value at 13 h. However, with the increase in depth and the consumption of heat between the asphalt layers, the temperature variation is increasingly stable, which is approximately in the range of 25–30 °C.

The temperature curves of the asphalt layers at different times are shown in Figure 5. It can be seen that the temperature gradient varies with time and depth. With the increase in depth, the temperature is more stable with time.

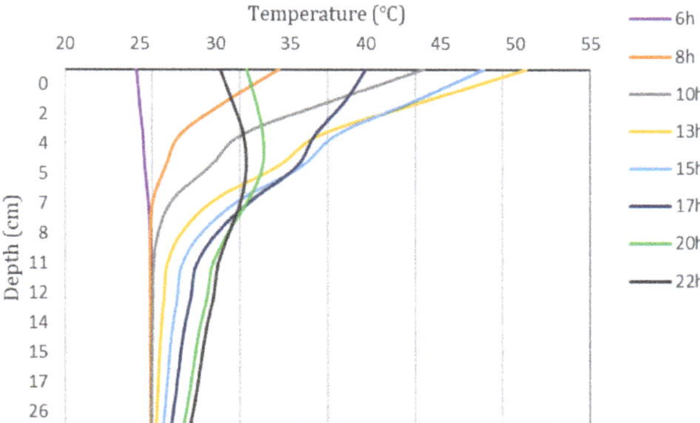

Figure 5. Temperature profiles of the asphalt layers along pavement depth at different moments in the summer.

The temperature difference between the top (y = 0 cm) and the bottom (y = 4 cm) of the asphalt surface course is largely positive during the day due to the absorption of solar radiation. Conversely, at night, the top temperature of the asphalt surface is higher than the air temperature, and the radiation from the pavement surface dominates the heat exchange activity. Therefore, the top temperature of the asphalt surface decreases significantly, and the temperature difference between the top and bottom of the asphalt surface is largely negative.

Solar radiation is applied to the surface of the road as heat flux. Heat flux refers to the heat energy passing through a unit area per unit time, and it is a directional vector. The variation of heat flux with time in the asphalt layers is shown in Figure 6. It reaches its highest at 13 h, and the variation in heat flux is consistent with the variation in temperature.

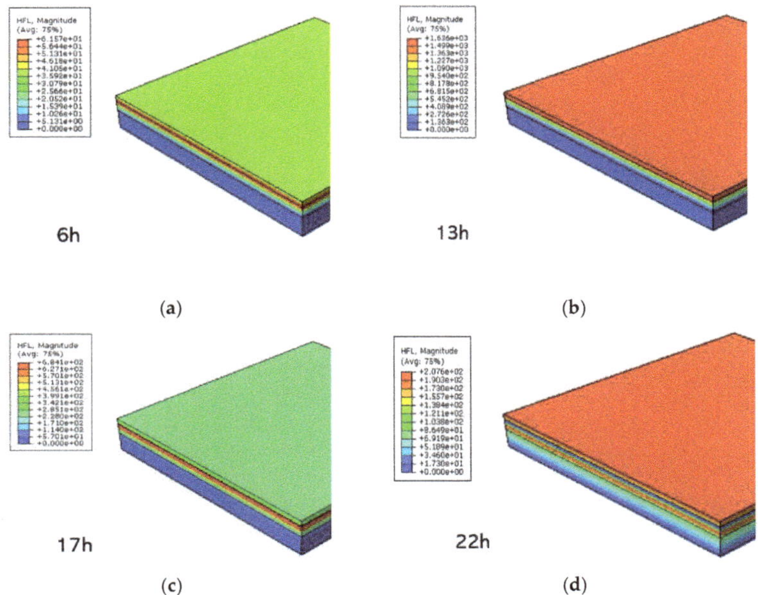

Figure 6. Cloud image of heat flux at different times in the summer: (**a**) 6 h, (**b**) 13 h, (**c**) 17 h, and (**d**) 22 h.

4.1.2. Results of the Mechanical Response of Conventional Pavement

For the heat transfer simulation in summer temperatures, the pavement surface temperature at 24.7 °C and 50.1 °C corresponded to 6 h and 13 h, respectively. Since the temperature distribution between the asphalt layers at night is different from that at daytime, 22 h was selected as the research object again, and the surface temperature of the asphalt pavement was 30.2 °C. The vertical displacement distribution of the road at the different temperatures and different times is shown in Figure 7.

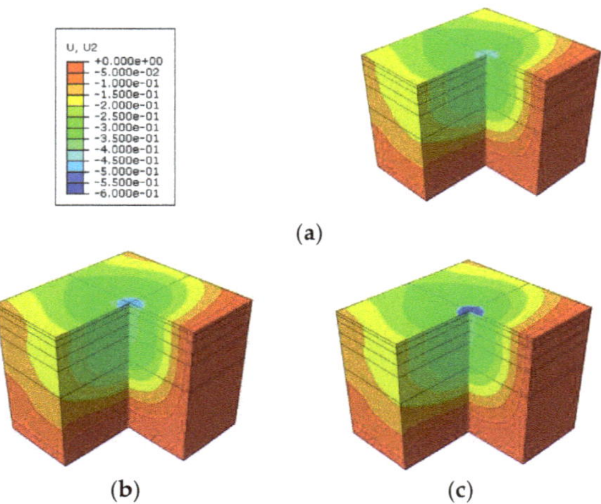

Figure 7. Distribution of the vertical displacement at different moments: (**a**) 6 h, 24.7 °C; (**b**) 22 h, 30.2 °C; and (**c**) 13 h, 50.1 °C.

It can be seen from Figure 7 that under the influence of different temperatures, the vertical displacement of the pavement under the same load is obviously different. Figure 8 shows the time-dependent variation of the vertical displacement of the center point in the loading area of the pavement surface. The vertical displacement increases with increasing load at the beginning and reaches a peak at 0.012 s. Immediately afterward, the vertical displacement again gradually decreases to 0 as the load is gradually removed. The change of vertical displacement with time is consistent with the change of the applied load. It can be seen that although the surface temperature difference between 6 h (24.7 °C) and 22 h (30.2 °C) is less than 6 °C, the difference in the vertical displacement is much larger than that between 22 h (30.2 °C) and 13 h (50.1 °C). The main reason is that the temperature distribution along the pavement depth is different in the morning than it is at night. At night, the maximum temperature of the road structure appears in the bottom of the asphalt surface course, while the maximum temperature of the daytime structure always appears on the surface of the pavement, which can be seen in Figure 5. This difference in temperature distribution leads to changes in structure stiffness, resulting in significant differences in vertical displacement. It is worth mentioning that the computational mechanical responses were derived from the pavement structures (conventional pavement and cool pavement), with full connection between the asphalt layers. For the influence of the different interlayer bonding conditions on the mechanical responses of the asphalt pavement, many previous studies from both experimental and numerical aspects have been carried out, and it is not the research focus of this study.

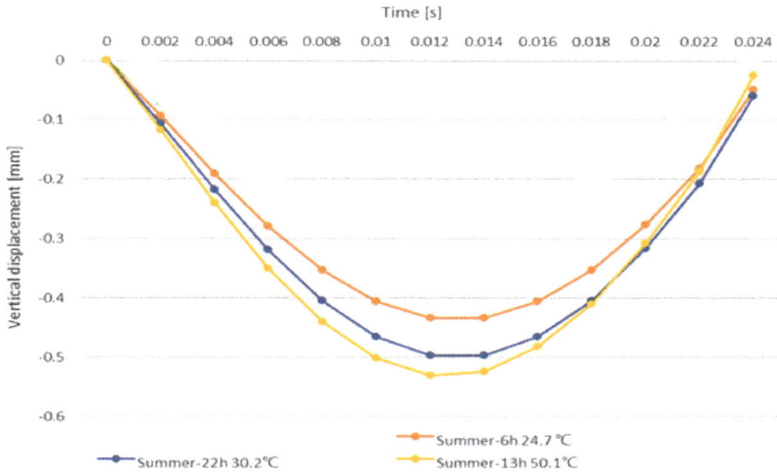

Figure 8. Vertical displacement of the center point of the asphalt surface course at different moments.

4.2. Results of the Responses of Cool Pavement

4.2.1. Results of the Thermal Response of Cool Pavement

Figure 9 shows the temperature trend of the cool pavement at different depths over time. The temperature trend of the cool pavement at different depths is approximately the same as that of the conventional pavement under the same conditions. The cool pavement has a significant cooling effect such that the temperature of the asphalt pavement surface is approximately the same as the air temperature.

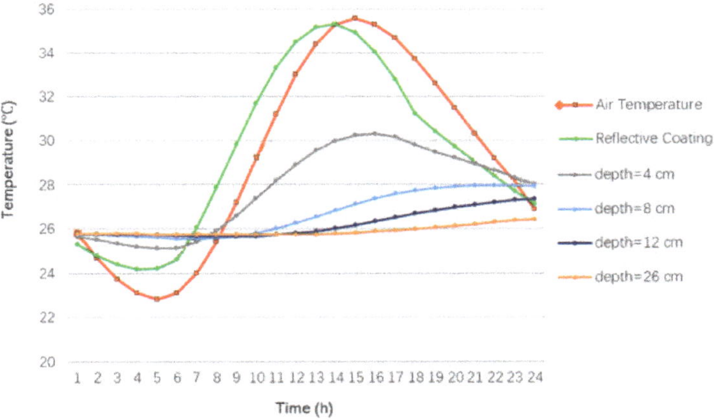

Figure 9. Hourly air temperature and temperature of the asphalt layers of the cool pavement at different depths in the summer.

The temperature difference between the conventional pavement and cool pavement at different depths is shown in Figure 10.

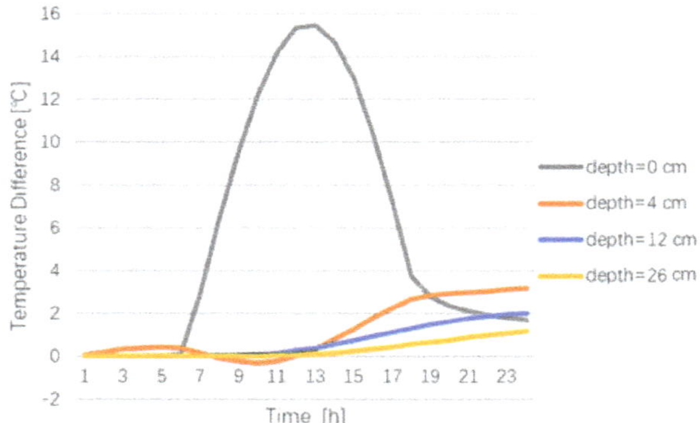

Figure 10. Temperature difference between the conventional pavement and cool pavement at different moments.

As can be seen from Figure 10, the cool pavement coating mainly decreases the temperature at the top of the asphalt pavement (y = 0 cm), which can be cooled by up to 15.5 °C. With the increase in pavement depth, the temperature reduction become less. In the range of 12–26 cm in the asphalt base course, the cooling effect brought by the cool pavement is not obvious in the daytime, but it can bring about a 2 °C cooling effect at night.

4.2.2. Results of the Mechanical Response of Cool Pavement

According to the above research results, the use of a cool pavement coating can effectively reduce the temperature of the pavement surface. At the highest temperature at 13 h in the summer, the pavement temperature can be reduced by as much as 15.5 °C, and the cooling effect is the best at this time. Therefore, the influence of the mechanical parameters caused by the application of cool pavement at this time was studied.

The comparison of the vertical displacement of the conventional pavement and the cool pavement at 13 h in the summer is shown in Figure 11.

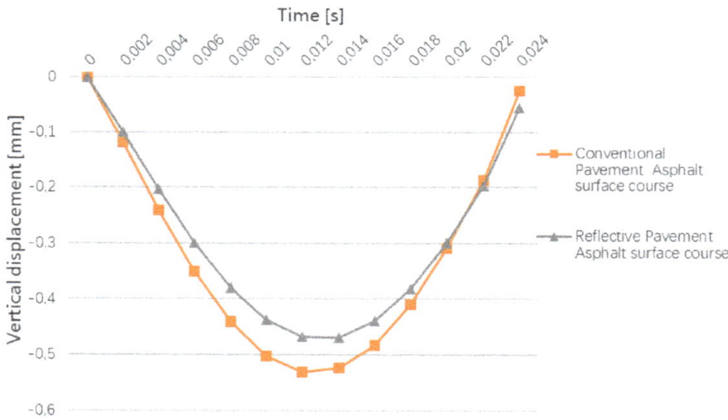

Figure 11. Comparison of the vertical displacement of the conventional pavement and cool pavement at 13 h in the summer.

As shown in Figure 11, on the asphalt surface course, the cooling effect is obvious, and the change of vertical displacement is brought by the temperature decrease. At the center of the load, the vertical displacement of the road surface is decreased by 0.06 mm (11.3% of the displacement in conventional pavement), which reduces the permanent deformation (rutting) of the asphalt pavement under the cumulative load over time.

Figure 12 shows the horizontal strains for the conventional and cool pavements in the summer under a uniform contact pressure. The application of the cool pavement does not affect the horizontal strain and there is no difference between the maximum and minimum values. The likelihood of fatigue damage on the asphalt base course is approximately the same for both pavements. However, the cool pavement has a better resistance to rutting because there is still a large difference in the road surface temperature.

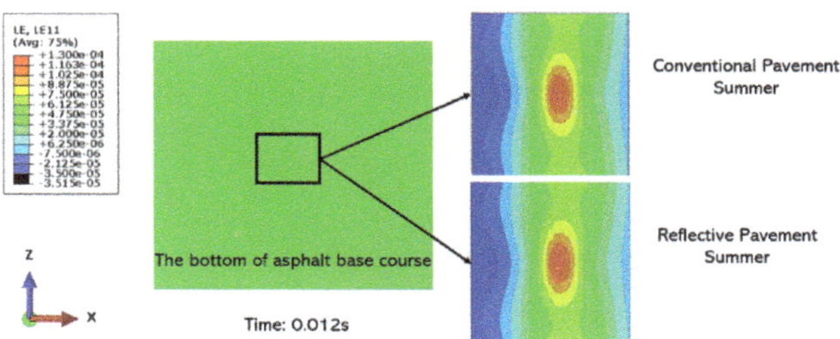

Figure 12. Distribution of logarithmic horizontal strain of the conventional and cool pavements.

5. Conclusions and Outlook

In this study, the 3D models of asphalt pavements were established through the FE method. The stable temperature field of the pavement structure was calculated by applying different temperature boundary conditions to the pavement surface and applying the mechanical model to analyze the mechanical response of conventional pavement and cool pavement in a real environment. The simulation of the pavement structure temperature distribution under summer weather conditions shows that the pavement surface temperature can exceed 50 °C, which can easily lead to permanent deformation of the pavement surface and affect the service life of asphalt pavement. The cool pavement has a significant cooling effect, which makes the surface temperature of asphalt pavement and the atmospheric temperature nearly flat, and the temperature of each depth has a certain degree of reduction. The cool pavement mainly reduces the temperature of the top of the asphalt pavement. Particularly, cool pavement reduces the surface temperature of pavement by more than 15 °C, and it reduces the vertical displacement of pavement by approximately 11.3%, which proves that cool pavement can effectively improve anti-rutting performance. However, the use of cool pavement does not change the horizontal strain at the bottom of the asphalt base course, and so it does not increase the fatigue resistance of asphalt pavement.

In summary, the mechanical response of asphalt pavement was studied considering the characteristics of asphalt material changing with temperature. In future research, more environmental conditions should be considered, such as variable wind speed and humidity. The different interlayer bonding conditions between the asphalt layers should be investigated by experimental tests and numerical simulations. Additionally, wheel load should be calculated by establishing a vehicle dynamics model.

Author Contributions: Conceptualization, P.L.; methodology, P.L. and C.D.; software, X.K. and C.D.; validation, C.D. and P.L.; formal analysis, X.K.; investigation, X.K., C.D. and P.L.; resources, P.L. and M.O.; data curation, X.K.; writing—original draft preparation, X.K. and C.W.; writing—review and editing, P.L. and D.W.; visualization, X.K.; supervision, C.D., P.L. and M.O.; project administration, P.L. and M.O.; funding acquisition, P.L. and M.O. All authors have read and agreed to the published version of the manuscript.

Funding: This paper was funded by the Deutsche Forschungsgemeinschaft (DFG, German Research Foundation)—OE 514/15-1 (Project ID 459436571) and SFB/TRR 339 (Project-ID 453596084). The authors are solely responsible for the content.

Institutional Review Board Statement: Not applicable.

Informed Consent Statement: Not applicable.

Data Availability Statement: The data are available from the first author and can be shared upon reasonable request.

Conflicts of Interest: The authors declare no conflict of interest.

References

1. Xu, J.; Fan, Z.; Lin, J.; Liu, P.; Wang, D.; Oeser, M. Study on the effects of reversible aging on the low temperature performance of asphalt binders. *Constr. Build. Mater.* **2021**, *295*, 123604. [CrossRef]
2. Hu, J.; Liu, P.; Wang, D.; Oeser, M.; Canon Falla, G. Investigation on interface stripping damage at high temperature using microstructural analysis. *Int. J. Pavement Eng.* **2019**, *20*, 544–556. [CrossRef]
3. Otto, F.; Liu, P.; Zhang, Z.; Wang, D.; Oeser, M. Influence of temperature on the cracking behavior of asphalt base courses with structural weaknesses. *J. Transp. Sci. and Technol.* **2018**, *7*, 208–216. [CrossRef]
4. Sun, Y.; Du, C.; Gong, H.; Li, Y.; Chen, J. Effect of temperature field on damage initiation in asphalt pavement: A microstructure-based multiscale finite element method. *Mech. Mater.* **2020**, *144*, 103367. [CrossRef]
5. Santamouris, M. Using cool pavements as a mitigation strategy to fight urban heat island - A review of the actual developments. *Renew. Sustain. Energy Rev.* **2013**, *26*, 224–240. [CrossRef]
6. Sankar Cheela, V.R.; John, M.; Biswas, W.; Sarker, P. Combating Urban Heat Island Effect - A Review of Reflective Pavements and Tree Shading Strategies. *Build* **2021**, *11*, 93. [CrossRef]
7. Sha, A.; Liu, Z.; Tang, K.; Li, P. Solar heating reflective coating layer (SHRCL) to cool the asphalt pavement surface pavement surface. *Constr. Build. Mater.* **2017**, *139*, 355–364. [CrossRef]
8. Zheng, M.; Tian, Y.; He, L. Analysis on Environmental Thermal Effect of Functionally Graded Nanocomposite Heat Reflective Coatings for Asphalt Pavement. *Coatings* **2019**, *9*, 178. [CrossRef]
9. Qin, Y. A review on the development of cool pavements to mitigate urban heat island effect. *Renew. Sustain. Energy Rev.* **2015**, *52*, 445–459.
10. Hill, K.; Beecham, S. The Effect of Particle Size on Sediment Accumulation in Permeable Pavements. *Water* **2018**, *10*, 403. [CrossRef]
11. Office, J.E.; Chen, J.; Dan, H.; Ding, Y.; Gao, Y.; Guo, M.; Guo, S.; Han, B.; Hong, B.; Hou, Y.; et al. New innovations in pavement materials and engineering: A review on pavement engineering research 2021. *J. Traffic Transp. Eng. (Engl. Ed.)* **2021**, *8*, 815–999.
12. Wang, X.; Gu, X.; Ni, F.; Deng, H.; Dong, X. Rutting resistance of porous asphalt mixture under coupled conditions of high temperature and rainfall. *Constr. Build. Mater.* **2018**, *174*, 293–301. [CrossRef]
13. Ma, X.; Zhou, P.; Jiang, J.; Hu, X. High-temperature failure of porous asphalt mixture under wheel loading based on 2D air void structure analysis. *Constr. Build. Mater.* **2020**, *252*, 119051. [CrossRef]
14. Zhao, X.; Shen, A.; Ma, B. Temperature response of asphalt pavement to low temperatures and large temperature differences. *Int. J. Pavement Eng.* **2018**, *21*, 49–62. [CrossRef]
15. Palmer, J.M. The measurement of transmission, absorption, emission, and reflection. In *Handbook of Optics*, 2nd ed.; Optical Society of America: Washington, DC, USA, 1995.
16. Liao, G.; Huang, X. *Application of ABAQUS Finite Element Software in Road Engineering*; Southeast University Press: Nanjing, China, 2014.
17. FGSV. *Guidelines for the Standardization of the Upper Structure of Traffic Areas*; Research Society for Road and Transportation: Cologne, Germany, 2012.
18. Wan, W.C.; Hien, W.N.; Ping, T.P.; Zhi, A.; Aloysius, W. A study on the effectiveness of heat mitigating pavement coatings in Singapore. *J. Heat Isl. Inst. Int.* **2012**, *7*, 238–247.
19. ABAQUS. *ABAQUS Version 6.10 HTML—ABAQUS/CAE User's Manual*; Dassault Systems: Paris, France, 2020.
20. NASA POWER Data Access Viewer. Available online: https://power.larc.nasa.gov/data-access-viewer/ (accessed on 1 October 2021).
21. FGSV. *Guidelines for the Computational Dimension of the Upper Structure of Road with Asphalt Surface Course*; Research Society for Road and Transportation: Cologne, Germany, 2009.
22. Liu, P.; Xing, Q.; Wang, D.; Oeser, M. Application of dynamic analysis in semi-analytical finite element method. *Materials* **2017**, *10*, 1010. [CrossRef] [PubMed]
23. Liu, P.; Wang, D.; Oeser, M. Application of semi-analytical finite element method to analyze asphalt pavement response under heavy traffic loads. *J. Traffic Transp. Eng. (Engl. Ed.)* **2017**, *4*, 206–214. [CrossRef]
24. Liu, P.; Wang, D.; Hu, J.; Oeser, M. SAFEM-Software with graphical user interface for fast and accurate finite element analysis of asphalt pavements. *J. Test. Eval.* **2016**, *45*, 1301–1315. [CrossRef]
25. Liu, P.; Wang, D.; Otto, F.; Oeser, M. Application of semi-analytical finite element method to analyze the bearing capacity of asphalt pavements under moving loads. *Front. Struct. Civ. Eng.* **2018**, *12*, 215–221. [CrossRef]

materials

MDPI

Article

Optimizing the Texturing Parameters of Concrete Pavement by Balancing Skid-Resistance Performance and Driving Stability

Jiangmiao Yu [1], Binhui Zhang [1], Peiqi Long [2], Bo Chen [1,*] and Feng Guo [3]

[1] School of Civil Engineering and Transportation, South China University of Technology, Guangzhou 510006, China; yujm@scut.edu.cn (J.Y.); ctbhzhang@scut.edu.cn (B.Z.)
[2] Foshan Branch, Suzhou Planning & Design Research Institute Co., Ltd., Foshan 528300, China; longpeiqi163@163.com
[3] Department of Civil and Environmental Engineering, University of South Carolina, Columbia, SC 29201, USA; fengg@email.sc.edu
* Correspondence: chenb@scut.edu.cn; Tel.:+86-132-8868-2471

Abstract: Curved texturing is an effective technique to improve the skid-resistance performance of concrete pavements, which relies on the suitable combination of the groove parameters. This study aims to optimize these parameters with the consideration of skid-resistance performance and driving stability. A pressure film was adopted to obtain the contact stress distribution at the tire–pavement interface. The evaluated indicator of the stress concentration coefficient was established, and the calculation method for the stationary steering resistance torque was optimized based on actual tire–pavement contact characteristics. Test samples with various groove parameters were prepared use self-design molds to evaluate the influence degree of each groove parameter at different levels on the skid-resistance performance through orthogonal and abrasion resistance tests. The results showed that the groove depth and groove spacing had the most significant influence on the stress concentration coefficient and stationary steering resistance torque, respectively, with the groove depth having the most significant influence on the texture depth. Moreover, the driving stability and durability of the skid-resistance performance could be balanced by optimizing the width of the groove group. After analyzing and comprehensively comparing the influences of various parameters, it was found the parameter combination with width, depth, spacing, and the groove group width, respectively, in 8 mm, 3 mm, 15 mm, and 50 mm can balance the skid-resistance performance and driving stability. The actual engineering results showed that the R^2 of the fitting between the stress concentration coefficient and SFC (measured at 60 km/h) was 0.871, which proved the effectiveness of the evaluation index proposed in this paper.

Keywords: skid-resistance performance; orthogonal test; abrasion test; groove parameters

Citation: Yu, J.; Zhang, B.; Long, P.; Chen, B.; Guo, F. Optimizing the Texturing Parameters of Concrete Pavement by Balancing Skid-Resistance Performance and Driving Stability. *Materials* **2021**, *14*, 6137. https://doi.org/10.3390/ma14206137

Academic Editor: Francesco Canestrari

Received: 17 August 2021
Accepted: 10 October 2021
Published: 15 October 2021

1. Introduction

The anti-skid performance of concrete pavements has been the focus of research. Moreover, a good skid-resistance performance can help to effectively reduce the slip accident rate. Conventional techniques of improving the anti-skid performance include dragging, grinding, and grooving [1–6]. Grooving is the most commonly used in engineering construction, and the design of the groove dimension and shape is key [7–10]. Fwa and Ong established a simulation model and concluded that the skid-resistance effect is significant when using a rectangular groove with width, depth, and spacing in the ranges of 2–10, 1–10, and 5–25 mm, respectively [11]. By adopting a continuous friction tester to measure the sideway-force coefficient (SFC) on various pavement surfaces at the same vehicle speed, Zhang recommended a rectangular groove with a large center spacing of 25 cm [12]. With the use of diamond saw blades to cut rectangular grooves on cured concrete pavements, the grooving technology can help improve the macrotexture of the pavement; however, the cutting surface is relatively flat, which reduces the effective tire–pavement contact area and has no significant effect on improving the microtexture [13–17]. Related research combined diamond grinding and grooving technology to improve the microtexture and macrotexture of the surface; however, this method has many drawbacks such as poor maneuverability and high costs [18,19]. In 2012, a new texturing technology was introduced in China to improve the skid-resistance performance of concrete pavements by carving dense longitudinal corrugated grooves on a cured pavement. Additionally, the average SFC of the curved textured pavement reached 68, which is higher than that of rectangular grooved pavements (approximately in the range of 15–30%). A fuller contact provides a greater lateral force, resulting in a more pronounced wheel shimmy [20,21]. A survey on a completed textured pavement found that some vehicles have poor driving stability at speeds close to 100 km/h. The shimmy is associated with complex nonlinear dynamics of the influencing factors, and the mechanical parameters are difficult to measure when a vehicle is in motion [22–24]. Related studies have shown that the tire–pavement friction is a factor influencing wheel shimmy [25–27].

The skid resistance force has been considered a critical indicator of the concrete pavement performance by the Portland Cement Association (PCA) and American Association of State Highway and Transportation Officials (AASHTO) [28–32]. Related studies have proven that the skid-resistance is influenced by many factors, including the pavement surface texture, tire type, tire–pavement friction, or other factors such as the speed, water film depth, and temperature, among which the tire–pavement friction is a primary factor [33–36]. Skid resistance is generally evaluated directly by the friction coefficient. The conventional measurement methods of the pavement skid resistance include the sand patch test, outflow meter, British Pendulum tester (BPT) and dynamic friction tester (DFT) testing, which mainly focus on the characterization of pavement macrotexture by using the volumetric index under specific conditions (certain load, test speed, friction mode, water-film thickness) [37–39]. Some novel measurement methods including but not limited to laser scanning, mechanical stylus, and image processing have also been developed to describe the surface texture characterization more accurately by capturing and showing the 2-D curve or 3-D surface texture and morphology within the measured region [4,40]. Many experimental results have shown that the groove parameters significantly influence the tire–pavement contact surface and that a suitable combination of parameters can help effectively improve the skid-resistance performance [41–43]. Although the relationship between the contact stress distribution and the skid-resistance performance has been studied, the actual characteristics at the tire–pavement contact have been largely simplified. The actual tire–pavement contact stress cannot be obtained using the conventional evaluation method of the skid resistance because of technical limitations.

Additionally, there is currently no clear standard for texturing technology. The selection of the groove parameters of the textured pavement is mainly based on engineering experience, and the lack of long-term monitoring data of related projects makes it difficult to verify the reliability of the empirical parameters. Therefore, it is necessary to study the contact stress distribution characteristics and develop a more accurate measurement method to evaluate the skid-resistance performance.

Based on tire–pavement friction, the stress concentration indicators were proposed, and the calculation method for the stationary steering resistance torque was optimized. In addition, an L9 (3^4) orthogonal table was determined on the basis of the selected parameters and levels, and a range analysis was conducted using these evaluation indices. Next, the effect of the groove group width (GGW) was studied, and an optimal combination of the parameters was determined through a comprehensive analysis of orthogonal and abrasion resistance test results. Finally, the skid-resistance performance under a set of optimized parameters was verified by actual project cases.

2. Objectives and Scope of this Study

The objectives and scope of this study are as follows:

(1) An evaluation method for the contact mechanics was established to describe the skid resistance and influence of the cement pavement on the driving stability, which should be verified based on engineering test results.

(2) A set of small texturing equipment that can help prepare specimens with different texture parameters in the laboratory was developed. Pressure-sensitive films were used to obtain the contact stress distribution between the tire and different pavements. Texture parameters that can provide a balance between driving stability and skid resistance performance are recommended by conducting orthogonal and kneading tests.

3. Methodology

3.1. Tire–Pavement Friction

Persson concluded that the tire–pavement contact is incomplete [44]. The embedding of the convex texture makes the tire deform, resulting in a stress concentration at the contact interface. Moore explained the friction phenomenon between tire and pavement surface. The adhesion (F_a), hysteresis (F_h), and ploughing (F_p) components of the frictional forces in elastomers are shown in Figure 1 and Equation (1).

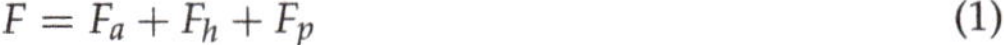

$$F = F_a + F_h + F_p \tag{1}$$

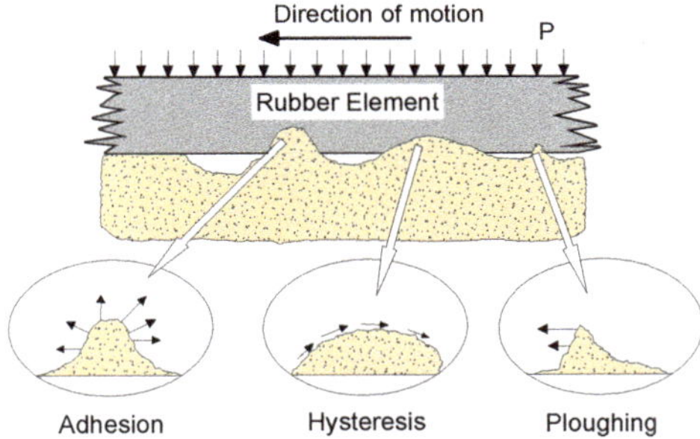

Figure 1. Tire–pavement friction.

An adhesive friction is mainly produced between the rubber and microtexture, which comprises 90% of the friction at low speeds [45,46]. The stress on the asperities is proportional to the adhesion friction, and a greater adhesion provides a better skid resistance performance. The adhesion friction function is expressed in Equation (2).

$$F_a = K_1 K_2 \sigma_{\mathrm{m}} \frac{N}{H} \tan\delta \tag{2}$$

where K_1 and K_2 are constants, σ_m is the maximum normal stress at the top of the asperity, N is the normal load, H is the rubber hardness, and $tan\delta$ is the tangent modulus of rubber.

When vehicles pass across the asperities of a rough surface pavement, the hysteresis component reflects the energy lost during this process, as the rubber is alternately compressed and decompressed [47,48]. The hysteresis friction function is expressed in Equation (3).

$$F_h = c \sum_{i=1}^{n} (E_{ci} - E_{ei}) \tag{3}$$

where c is a constant, $E_{ci} - E_{ei}$ is the energy loss of the tire rubber on a single asperity surface.

The asperities on the pavement surface will have a microcutting effect on the tire, and the ploughing force is shown in Equation (4).

$$F_p = K_3 htg(\delta)T_{\max} = K_3 K_4 \left| \frac{P}{tg\frac{\theta}{2}} \right|^{\frac{1}{2}} tg(\delta)T_{\max} \quad (4)$$

where K_3 and K_4 are constants, N is the normal stress acting on the asperity; θ is the apex angle of the simplified asperity profile; h is the depth to which the asperity penetrates the rubber; $tg\delta$ is the rubber tangent modulus; T_{max} is the maximum tangential stress that breaks the rubber molecular chain.

3.2. Pressure Film Testing

The pressure film (Figure 2) can accurately measure the contact area and the pressure distribution (the minimum effective measurement is 0.125 mm^2). In this study, double-slice pressure films were chosen, including an A-film with a color generation agent and a C-film with a color developer. Under the application of pressure, the pressure level was described in terms of the color density. Due to the limited range, a complete tire–pavement contact stress cannot be obtained by adopting a single-range pressure film. Therefore, various specification films were adopted, including LLLW (Ultra Super Low Pressure) (0.2–0.6 MPa), LLW (Super Low Pressure) (0.5–2.5 MPa), and LW (Low Pressure) (2.5–10 MPa).

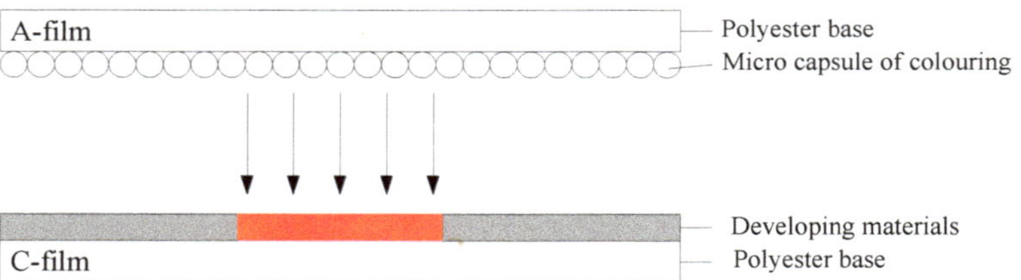

Figure 2. Working principle of a pressure film.

The stress distribution information of the contact interface was stored in a 2D matrix after processing, as shown in Equation (5).

$$F(X,Y) = \begin{bmatrix} f(0,0) & f(0,1) & \dots & f(0,n) \\ f(1,0) & \dots & \dots & \dots \\ \dots & \dots & \dots & \dots \\ f(m,0) & \dots & \dots & f(m,n) \end{bmatrix} \quad (5)$$

where $F(X, Y)$ is the overall normal stress acting on the contact area, and $f(m, n)$ is the mean contact stress at the measurement point.

The use of the pressure film is as follows:

(1) The film is placed between the tire and the road and statically loaded for more than two mins (Figure 3a).
(2) The temperature and humidity of the test site are recorded, and the correct model of the pressure and color density is determined.
(3) After the color reaction is complete, the test film is calibrated and scanned and identified in the FPD-8010E (Version 1.1, 2007, FIJIFILM Corporation; Tokyo, Japan) dedicated software (Figure 3b,c).
(4) The test results corresponding to the different specifications of the pressure film are analyzed in MATLAB(Version 9.1, 2016, MathWorks company; Natick, MA, USA) and a numerical quantification and statistical analysis is performed (Figure 3d).

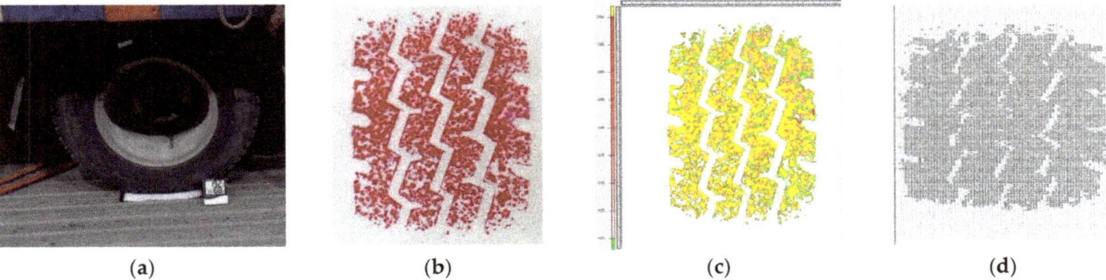

(a) (b) (c) (d)

Figure 3. Schematics of pressure film operation process: (**a**) loading; (**b**) tire mark; (**c**) scanning image; (**d**) numerical statistics of contact stress.

3.3. Stress Concentration Effect

To visually show the difference in the stress distribution, the contact pressure distributions of a pavement with no grooves, a rectangular groove pavement (groove width: 4 mm, groove depth: 4 mm, spacing: 25 mm), a curved groove pavement (groove width: 8 mm, groove depth: 1 mm, spacing: 8 mm), and an asphalt pavement (AC-16) were obtained, as shown at Figure 4, and the contact stress was divided into several parts in steps of 0.1 MPa. The proportion of the stress distribution area of each part in the total effective contact area is counted, and the cumulative proportion is shown in Figure 5. The contact stress of concrete pavements with no grooves, rectangular grooves, curved grooves, and asphalt pavement without grooves are mainly concentrated in 0~3 MPa, 0~7 Mpa, 0~8 MPa, and 0~10 MPa, respectively. According to the results of Figure 5, the smooth concrete slab is flatter and more fully in contact with vehicle tire, resulting in a more even contact stress distribution of tire; the values of contact stress are mainly less than 1.8 MPa. The contact between the AC-16 asphalt pavement and vehicle tire produced a larger stress concentration phenomenon, with the values of contact stress mainly more than 6 MPa; this phenomenon is consistent with the findings of the literature [40]. As the roughness of the pavement construction increases, the embedded effect on the tire becomes more significant, while the top of the construction pierces the tread rubber to the greatest depth, thus generating a significant stress concentration phenomenon. It is clear that the degree of dispersion of the stress distribution is related to the surface texture characteristics.

The Weibull function (Equation (6)) is widely used in material science to characterize the uniformity of the material strength distribution. c is called the Weibull modulus. The higher the c value, the lower the dispersion degree of the material and the better the uniformity.

$$F(x) = \begin{cases} 1 - \exp\left\{-\left(\frac{x-a}{b}\right)^c\right\}, & x \geq a \\ 0, & x < a \end{cases} (b, c > 0) \quad (6)$$

where a is the position parameter, b is the scale parameter, and c is the shape parameter of the Weibull distribution (also known as the Weibull modulus).

Based on the significance of the Weibull modulus in material science, the value of the Weibull modulus was adopted to characterize the degree of dispersion of the contact stress. Previous studies have shown that the coarse particles and texture depth of a pavement structure directly affect the degree of stress dispersion [21]. Table 1 shows that the smooth concrete pavement has the highest Weibull modulus value, followed by the rectangular grooved pavement, curved textured pavement, and asphalt pavement, in that order. In addition, the non-uniformity of the textured pavement structure is better than that of the grooved pavement, indicating that the macro- and micro-textures of the textured pavement are more complex and that the stress distribution is more discrete.

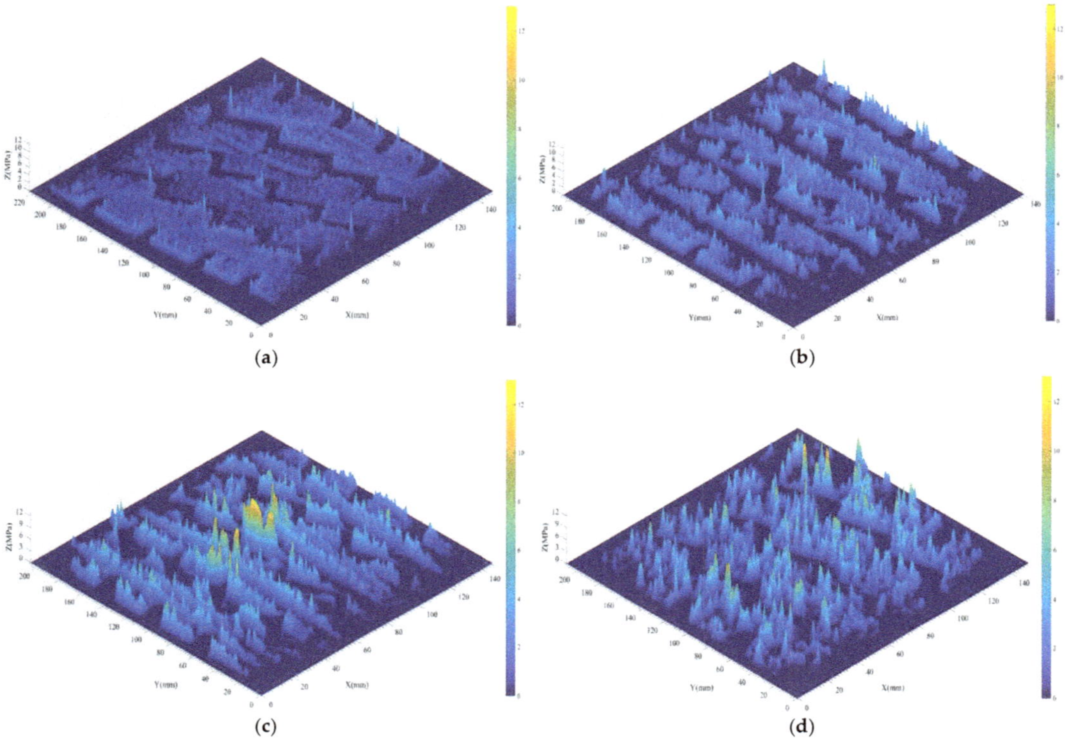

(**a**) (**b**)

(**c**) (**d**)

Figure 4. Tire contact stress of different pavements: (**a**) concrete pavements with no grooves, (**b**) rectangular grooves, (**c**) curved grooves, (**d**) asphalt pavement without grooves.

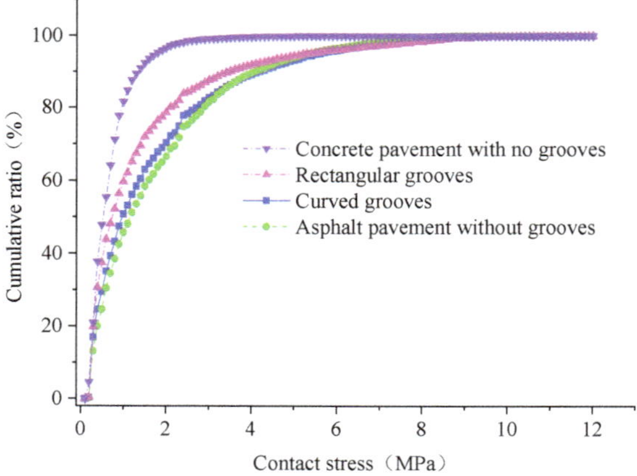

Figure 5. Contact stress distribution on different pavements.

Table 1. Test results of stress distribution on different contact interfaces.

Contact Interface	*a*	*b*	*c*	R^2
Asphalt pavement without grooves	0.2068	0.9199	0.6558	0.999
Curved grooves	0.1987	1.4898	0.7052	0.998
Rectangular grooves	0.1645	1.2691	0.8465	0.998
Concrete pavements with no grooves	0.1809	0.4947	0.9754	0.999

The stress concentration phenomenon is due to tire–pavement friction. Although the Weibull modulus can be used to explain the dispersion of the stress distribution on the contact surface, it cannot describe the degree of stress concentration effect.

To quantify the degree of stress concentration, a stress concentration coefficient was adopted, which is expressed in Equation (7).

$$K_s = \frac{\iint_{A'} f(x,y)dxdy}{\iint_{A} f(x,y)dxdy} \times 100\% \tag{7}$$

where K_S is the stress distribution concentration (%), and A' is a high-stress area (mm^2). According to a previous study [46], the stress above 1.8 MPa is defined as high stress. A is the actual contact area between the tire and the pavement (mm^2), and $f(x, y)$ is the single-point contact stress measured by the pressure film.

3.4. Stationary Steering Resistance Torque

The steering resistance torque of a vehicle is maximum under a static condition, which is affected by the tire–pavement friction and vehicle steering system. The friction coefficient of the steering system is typically considered a constant; therefore, the contact friction is the most important factor [49,50]. Additionally, the mechanical parameters are difficult to measure while the vehicle is in motion; nevertheless, static indicators can indirectly reflect and help evaluate the dynamic process. The tire ground pressure is typically considered a uniform or linear parabolic load distribution, which simplifies the actual contact pressure. This study used the typical distributed stress as the actual contact stress measured by the pressure film. The shape of curved groove is similar to the longitudinal pattern of tire; therefore, the meshing effect between tire and pavement is more significant (Figure 6).

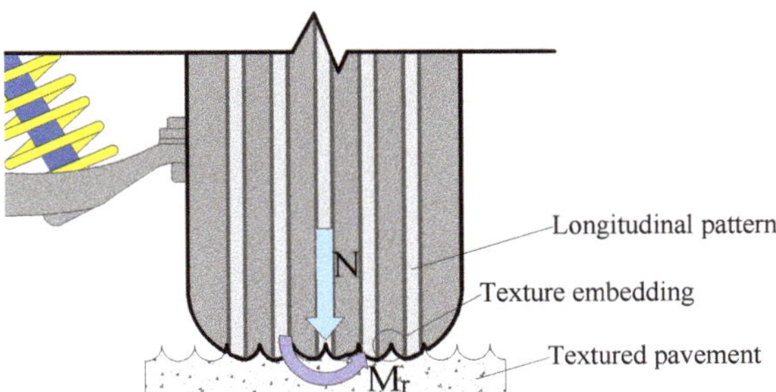

Figure 6. The meshing effect: N is vertical load; Mr is stationary steering resistance torque.

The stationary steering resistance torque was calculated by the effective contact areas and the actual contact stress distribution. The calculation of the microelement of the stationary steering resistance torque at the center of the contact interface is expressed in Equation (8).

$$dM_r = r \times dF_r = \sqrt{x^2 + y^2} \times \mu \times f(x,y)dxdy \tag{8}$$

Using the MATLAB calculation program to cyclically accumulate the friction torque of each contact unit, we finally obtain the total steering torque value M_r, as expressed in Equation (9).

$$M_r = \int dM_r = \iint \sqrt{x^2 + y^2} \times \mu \times f(x,y)dxdy \quad (9)$$

where x and y are the coordinate values of the single-point contact stress (the origin of the coordinate is the contact center point), and the road friction coefficient μ was measured using pendulum slip resistance testing equipment; $f(x, y)$ is the single-point contact stress measured by the pressure film.

By calculating the stationary steering resistance torque on the above four types of pavements, it can be seen from Figure 7 that the stationary steering resistance torque of the textured pavement is greater than that of other types of pavements, which indicates that greater opposite rotary moment of steering wheel is needed to ensure driving stability. Therefore, it is necessary to enhance the driving stability by optimizing texture parameters.

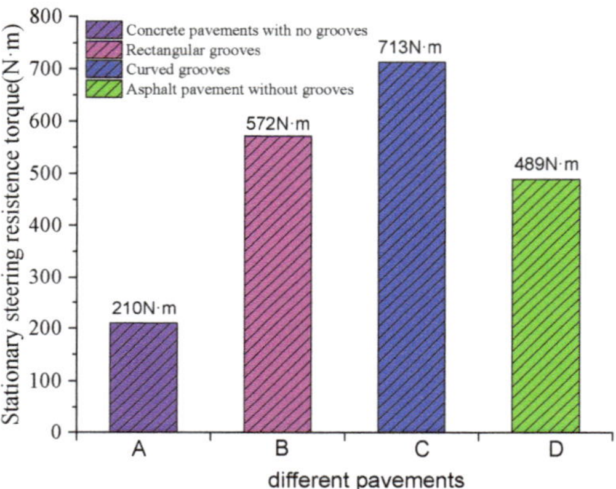

Figure 7. Comparison of stationary steering resistance torque.

4. Materials and Methods

4.1. Materials and Mixture Design

Type P.O42.5 cement (Yunfu, China), medium sand (a fineness modulus of 2.77 and a mud content of 0.8%), limestone aggregate (size ranges of 10–30, 10–20, and 5–10 mm, respectively), and a high-range water reducer (CNF-13, Yunfu, China) were used to prepare the mix. Tables 2 and 3 list the parameters of the cement and proportion of the mix, respectively.

4.2. Sample Preparation

Through the calculation of the different curved groove width and depth, a sample preparation tester was developed (Figure 8).

(1) Based on the concrete mix proportion, listed in Table 3, and the groove parameters, listed in Table 4, the samples were molded into a mold of dimensions 300 mm × 300 mm × 50 mm.
(2) Before the concrete sets and solidifies, the slurry on the specimen surface is scraped off.
(3) The rail mold is set, and the steel wire is made to adhere to the sample surface.
(4) A texturing tool is used to carve curved grooves between the two steel wires.
(5) The samples are placed in a standard curing room for seven days, and the texturing tool is used to scrape off the slurry on the groove surface to restore the rough texture.

Table 2. Cement parameters.

Initial Setting Time/min	Final Setting Time/min	Seven-Days Bending Strength/MPa	Compressive Strength/MPa
235	287	4.95	38.1

Table 3. Proportion of mix.

Material	Cement	Sand	10–30 mm	10–20 mm	5–10 mm	Water	CNF-13	Water–Cement Ratio
Amount/(kg)	378	691	706	449	128	140	7.6	0.37
Weight ratio	1	1.828	1.868	1.188	0.34	0.37	0.02	

(**a**)

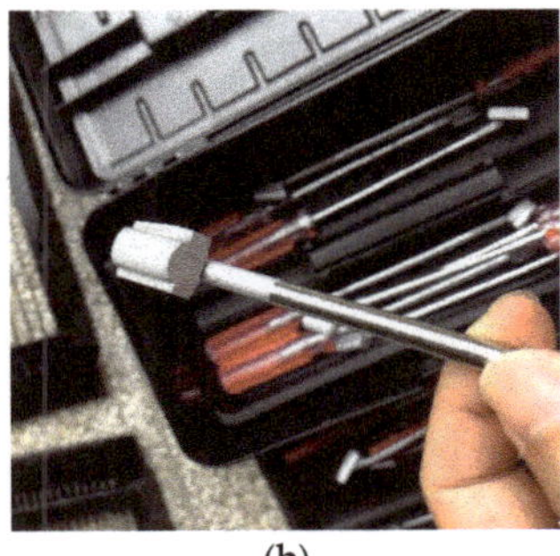

(**b**)

Figure 8. Tester for preparing samples: (**a**) rail mold; (**b**) texturing tool.

Table 4. Factors and levels of the orthogonal test.

Test No.	W/mm	D/mm	GS/mm
1	6	1	0
2	6	2	15
3	6	3	25
4	8	1	15
5	8	2	25
6	8	3	0
7	10	1	25
8	10	2	0
9	10	3	15

5. Orthogonal Designs

5.1. Orthogonal Design

Orthogonal experiments are one of the most commonly used experimental design methods, wherein an orthogonal table is used to scientifically analyze the influence laws of multiple factors. Therefore, the orthogonal test method was chosen in this study to analyze the skid-resistance and driving stability performance of a curved textured pavement with different parameters.

Among the impact factors, the curved groove width *W*, depth *D*, and spacing GS were selected; these have a direct impact on the tire–pavement contact interface. Figure 9 shows the diagram of the groove parameters. The three levels of the groove width *W* are 6, 8, and 10 mm, the three levels of the groove depth *D* are 1, 2, and 3 mm, and the three levels of the groove spacing GS are 0, 15, and 25 mm. An L9 (3^4) orthogonal table, including three factors and three levels, is shown in Table 4.

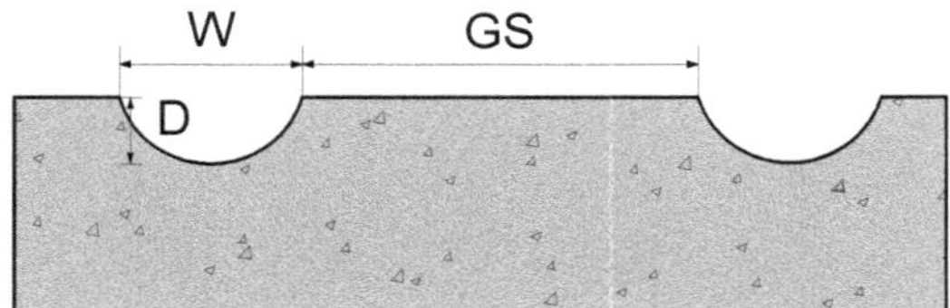

Figure 9. Curved groove parameters (W: the curved groove width, mm; D: the curved groove depth, mm; GS: the curved groove spacing, mm).

5.2. Analysis of Orthogonal Test of Results

A single wheel load of 15.8 kN and a pressure of 770 kPa were selected for the test, samples with different groove parameters were statically loaded, and then pressure films were adopted to obtain the stress distribution information. The sand paving method and pendulum tester were used to obtain the texture depth TD and lateral friction coefficient, respectively. The stress concentration coefficient Ks and the stationary steering resistance torque M_r of the samples were calculated using the method introduced above. Table 5 lists the calculation results.

Table 5. Orthogonal test results.

Test No.	Texture Depth/mm	Stress Concentration Coefficient/%	Stationary Steering Resistance Torque/N·m
1	0.72	47.64	531.89
2	0.80	52.63	555.31
3	0.86	55.33	607.80
4	0.76	51.84	540.62
5	0.64	51.90	640.19
6	1.02	60.79	591.57
7	0.61	47.43	547.07
8	0.72	53.09	554.13
9	1.08	59.32	549.32

To evaluate the influence degree of the three factors on the test results, a range analysis of the orthogonal method is necessary. In the calculation of the range analysis, the value K_{ij} and the influence degree R_j are shown in Equations (10) and (11), respectively.

$$K_{ij} = \sum_{i=1}^{n} y_{ij} \tag{10}$$

$$Rj = \max\{K_{1j}, K_{2j}, K_{3j}\} - \min\{K_{1j}, K_{2j}, K_{3j}\} \tag{11}$$

where i is the level number; j represents the impact factor; n = 3; y is the test result.

In the range analysis, the value K_{ij} when i = 1, 2, and 3 describes the effect of the factor j on the test result. Moreover, the influence degree R_j measures the degree of impact of the factor j. Using the analysis method above, the range analysis results with different K_{ij} and R_j values are shown in Table 6.

As shown in Table 6, the range analysis is implemented at different factors and levels to facilitate a comparative analysis. In terms of the texture depth TD, stress concentration coefficient, and stationary steering resistance torque, the results show that $R_D > R_{GS} > R_W$, $R_D > R_W > R_{GS}$, and $R_{GS} > R_D > R_W$, respectively.

Based on the selection criteria of the orthogonal test, the factors with the highest influence degree should be selected with an appropriate level, and the less important factors can be arbitrarily selected. Figure 10 shows the average value of each factor at each level for different test results. Figure 10 shows that Ks and Mr increase first and then decrease with the increase in the groove width, and *TD* and *Ks* increase with the increase in the groove depth.

Table 6. Range analysis of the orthogonal test results.

Results		Level Factors		
		W	D	GS
Texture depth	K_{11}	2.38	2.09	2.46
	K_{21}	2.42	2.16	2.64
	K_{31}	2.41	2.96	2.11
	Rj	0.01	0.29	0.18
Stress concentration coefficient	K_{12}	155.6	146.91	161.52
	K_{22}	164.53	157.62	163.79
	K_{32}	159.84	175.44	154.66
	Rj	2.98	9.51	3.04
Stationary steering resistance torque	K_{13}	1695	1619.58	1677.59
	K_{23}	1772.38	1749.63	1645.25
	K_{33}	1650.52	1748.69	1795.06
	Rj	40.62	43.35	49.94

For *Ks* and *TD*, to obtain a high value, the level of the groove depth *D* should be D_3 (3 mm). Considering that the influence of groove spacing on *Mr* is the highest, the level of groove spacing should be GS_2 (15 mm). At this level, *Mr* takes a small value, which is conducive to the driving stability. Additionally, from Table 6, we find that the groove width is the second most important factor influencing K_S, which is maximum when *W* is W2 (8 mm).

To determine whether the changes in the various factors and test errors significantly affect the test indicators, the analysis of variance was applied to further analyze the test data. The F distribution table was adopted to quantitatively evaluate the significance level of each factor and test the significance of each factor. The commonly used significance levels (α) are 0.01, 0.05, and 0.10, and the critical values are $F_{0.01}$ (2,2) = 99, $F_{0.05}$ (2,2) = 19, and $F_{0.10}$ (2,2) = 9, respectively. F > $F_{0.01}$ (2,2) indicates high significance, $F_{0.01}$ (2,2) > F > $F_{0.05}$ (2,2) indicates moderate significance, and F < $F_{0.05}$ (2,2) indicates no significance. Based on the results of the orthogonal experiment, the SPSS statistical analysis software was used to analyze the data.

Based on the results of the variance analysis listed in Table 7, the three groove parameters are found to have a significant impact on the stress concentration coefficient and stationary steering resistance torque. The groove depth and groove spacing significantly affect the texture depth, whereas the groove width has the least significant impact. Therefore, considering the above analysis, the levels (values) of each factor should be selected as W_2 (8 mm), D_3 (3 mm), and GS_2 (15 mm). The test under these levels can be considered an optimized test.

To further improve the skid-resistance performance of cement pavements while maintaining the driving stability, the width of the groove group was selected, and the long-term influence corresponding to different texture parameter combinations on the stress concentration coefficient and stationary steering resistance torque is further studied through a kneading test.

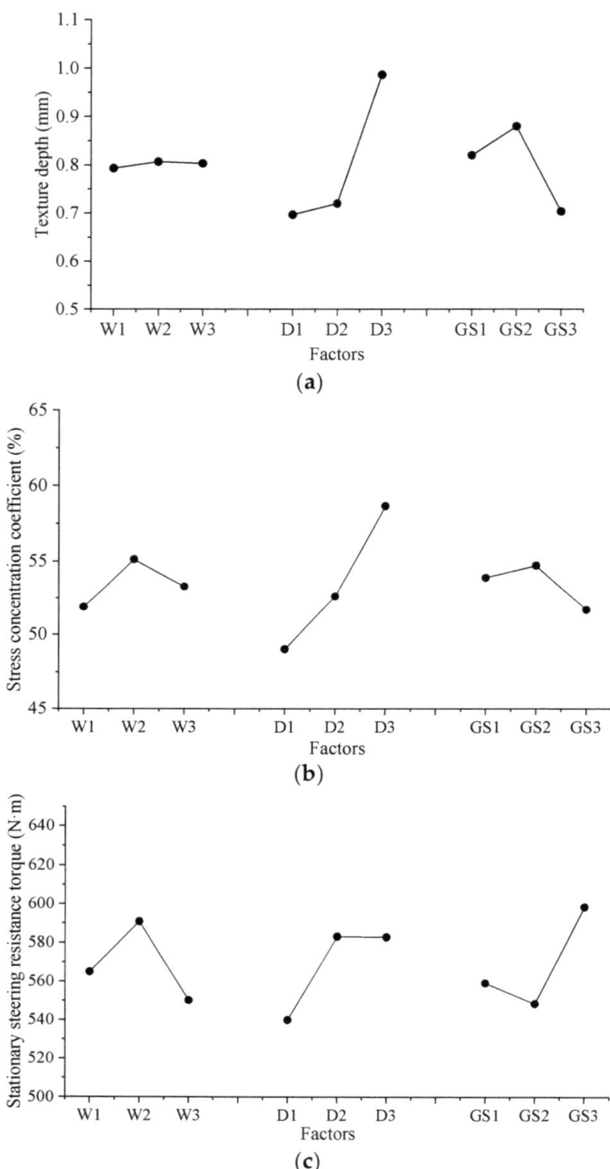

Figure 10. Average value per level for each factor: (**a**) texture depth; (**b**) stress concentration coefficient; (**c**) stationary steering resistance torque.

Table 7. Analysis of variance table.

Evaluation Index	Factors	Sum of Square between Groups	Degree of Freedom	*f* Value	Significance Degree
Texture depth	W	2.89×10^{-4}	2	0.14	not significant
	D	1.56×10^{-1}	2	77.02	moderately significant
	GS	4.84×10^{-2}	2	23.95	moderately significant
Stress concentration coefficient	W	13.30	2	19.10	moderately significant
	D	138.47	2	198.82	highly significant
	GS	15.06	2	21.63	moderately significant
Stationary steering resistance torque	W	2535.11	2	19.29	moderately significant
	D	3731.48	2	28.39	moderately significant
	GS	4143.12	2	31.53	moderately significant

6. Durability Research Based on Abrasion Test

6.1. Design of Abrasion Test

The combination of the texture parameters obtained from the orthogonal test can only reflect the initial skid-resistance performance. The samples (8 mm in width, 3 mm in depth, and 1 mm in spacing) exhibiting the best skid resistance in the forementioned tests were selected, and the groove parameters were further explored by refining the GGW to identify the differences in their skid-resistance durability.

Figure 11 shows the GGW. In accordance with the sample preparation method presented in Section 3.2, six types of cement board samples with different texture parameters were prepared for the experiment, as shown in Figure 12 and Table 8.

A self-developed abrasion tester (Figure 13) was adopted, and the abrasion time was used to characterize the running time of the tire on the actual pavement. The abrasion tester is mainly composed of a kneading wheel and a horizontal plate that can carry samples. When the tester is in operation, the plate can move laterally and repeatedly, while the kneading wheel moves vertically and repeatedly. Table 9 lists the operation parameters of the abrasion tester.

Each sample was placed on a horizontal plate of the abrasion tester and fastened by a clamp. Samples, loaded at 0.7 MPa, were taken out every 1 h to measure the stress acting at the contact interface and kneaded for 12 h. In this study, to enhance the abrasion effect, the test environment was set at normal temperature under an overflowing water condition.

6.2. Analysis of Abrasion Test Results

Different grooving parameters lead to different attenuation laws of the concrete pavement in the process of abrasion, which can be fitted using an appropriate mathematical model, and the relevant parameters of the mathematical model can reflect this difference.

According to the experimental analysis, the skid-resistance performance attenuation curve of the curved textured pavement is nonlinear and exponential. In this study, an asymptotic model was selected, as expressed in Equation (12). Figures 14 and 15 show the fitting curves. Tables 10 and 11 list the regression parameters of the fitting curve. The R^2 are greater than 0.95 in all cases, which can better characterize the attenuation law of the evaluation indicators.

$$y = Ae^{Bx} + C \tag{12}$$

where y is the value of the evaluation indicator; x is the abrasion time; A, B, and C are regression parameters, where A + C, C, A, B represent the initial value, final value, attenuation amplitude, and rate of reaching stable state of the indicators, respectively.

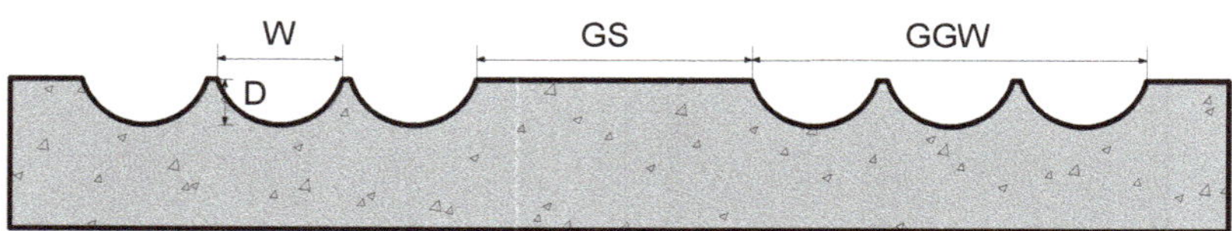

Figure 11. New curved groove parameters (W: the curved groove width, mm; D: the curved groove depth, mm; GS: the curved groove spacing, mm; GGW- the curved groove group width, mm).

Figure 12. Test specimens and groove size schematic diagram: (**a**) T-1:8-3-0-8; (**b**) T-2:8-3-15-8; (**c**) T-3:8-3-15-30; (**d**) T-4:8-3-15-50; (**e**) T-5:8-3-15-70; (**f**) rectangular groove.

Table 8. Test specimen parameters.

No.	W	D	GS	GGW
T-1	8	3	0	8
T-2	8	3	15	8
T-3	8	3	15	30
T-4	8	3	15	50
T-5	8	3	15	70
T-6	Parameters of rectangular groove: 8 mm in width; 3 mm in depth; 15 mm in spacing.			

Table 9. Operating parameters of abrasion tester.

Lateral Speed (cm/min)	Wheel Movement Frequency (times/min)	Pressure (MPa)
10	42 ± 1	0.7

Figure 13. Abrasion tester.

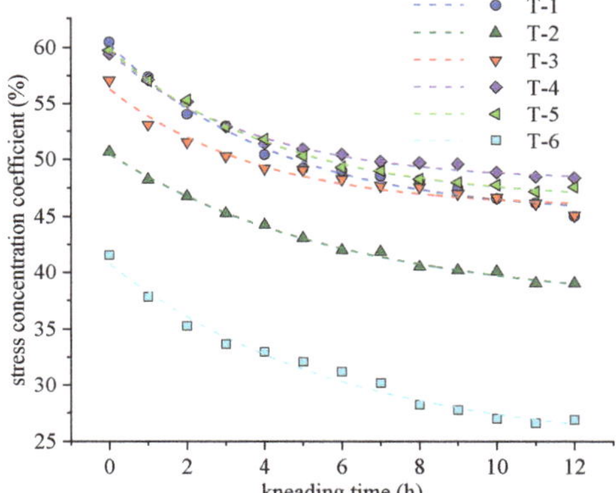

Figure 14. Fitting curve of stress concentration coefficient.

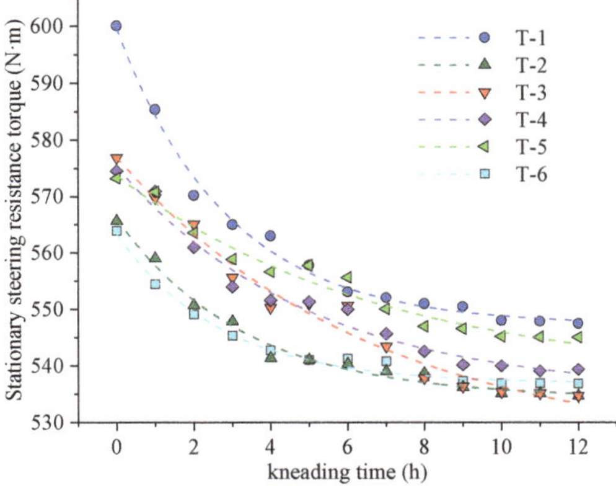

Figure 15. Fitting curve of stationary steering resistance torque.

Table 10. Attenuation parameter of stress concentration coefficient.

Parameter Combination	A	B	C	R^2
T-1	15.166	−0.238	45.184	0.983
T-2	13.428	−0.159	37.039	0.995
T-3	10.511	−0.270	45.835	0.965
T-4	11.421	−0.279	48.196	0.991
T-5	13.446	−0.240	46.467	0.995
T-6	16.420	−0.167	24.369	0.980

Table 11. Attenuation parameter of stationary steering resistance torque.

Parameter Combination	A	B	C	R^2
T-1	52.416	−0.347	547.305	0.992
T-2	31.743	−0.304	534.328	0.988
T-3	51.984	−0.152	525.004	0.977
T-4	40.137	−0.203	535.196	0.978
T-5	36.042	−0.146	537.679	0.970
T-6	26.407	−0.371	536.948	0.990

For different combinations of the texture parameters, the initial value, attenuation rate, and final value of the stress concentration factor are different. As shown in Figure 14 and Table 10, the indicator values of each sample decrease rapidly at the initial stage and then gradually tend to stabilize. Based on the value of the parameter B, the attenuation speed of T-1, T-3, T-4, and T-5 is higher than T-2 and T-6, which indicates that the greater the density of the curved groove, the more significant the stress concentration effect and the faster the abrasion of the pavement structure. The initial value order of the stress concentration coefficient is T-1 > T-5 > T-4 > T-3 > T-2 > T-6 and the values of T-1, T-5 and T-4 is similar. The initial values of curved groove interface are approximately in the range of 50–60%, while that of the rectangular grooved pavement is only 43%. The initial value of the evaluation indicator of the curved textured pavement is greater than that of the rectangular grooved pavement. Therefore, the meshing friction between the tire and the curved groove is more significant, which can prove that the texture roughness of the initial curved groove is better than that of the rectangular groove. The final values of T-1, T-3, T-4, and T-5 are significantly higher than those of T-2, indicating that the higher the groove density, the better the skid-resistance performance after abrasion. Meanwhile, the attenuation amplitude of T-3, T-4, and T-5 is less than that of T-1, which indicates that the spacing provided between the continuous curved grooves helps disperse the large stress, and the attenuation amplitude of the stress concentration coefficient can be effectively alleviated. Therefore, from the perspective of anti-skid performance, T-4 and T-5 are recommended, and the GGW is 50~70 mm.

As shown in Figure 15 and Table 11, the attenuation rate of the steering resistance torque in the early stage is higher than that in the later stage. The initial values of T-2, T-3, T-4, and T-5 are similar and greater than T-6 but significantly less than that of T-1. This proves that the groove density has a significant effect on the initial steering resistance moment. The greater the texture density, the greater the lateral force acting on the tire. Combined with the analysis of the friction mechanism between the textured pavement and the tire, the groove spacing has an interference effect at the tire–pavement contact interface, which reduces the effect of the lateral torque, yields a moderate initial index value, and reduces the driving shimmy on the curved grooved pavement. The order of final value is T-1 > T-5 > T-4 > T-3 > T-2, and T-6 is slightly higher than those of T-2, T-3, and T-4, which indicates that after abrasion, the appropriate texture parameters can help improve the driving stability of the textured pavement and maintain the steering resistance moment at a low level. Therefore, from the perspective of driving stability, T-2, T-3, and T-4 are recommended.

Considering the skid resistance performance and driving stability of the pavement, setting certain groove spacing can help effectively reduce the stationary steering resistance torque. In combination with the GGW, the skid resistance performance of the pavement can be improved. Therefore, a groove dimension scheme with a groove width of 8 mm, depth of 3 mm, spacing of 15 mm, and GGW of 50 mm is recommended.

7. Engineering Verification

The results of this study were applied in Wang Bei Ao Tunnel (3751 m left and 3713 m right) of Jiang Luo Highway (Jiangmen to Luoding Highway) in Guangdong Province, China. This project was completed in December 2016 and opened to traffic on 26 December 2016. In this project, P.O42.5 cement (initial setting time 235 min, final setting time 287 min, 7-day bending strength 4.95 MPa, 7-day compressive strength 38.1 MPa) was used for the tunnel concrete pavement, with a dosage of 378 kg. The granite aggregates (10–30 mm, 10–20 mm, and 5–10 mm aggregates with a weight ratio of 55%, 35%, and 10%, respectively) were used. The water–cement ratio was 0.37 and the external additive, CNF-3 water reducing agent (air-entraining, slow setting, and high efficiency water reduction agent) was used with a dosage of 7.6 kg. Considering the low design speed of the tunnel pavement, a higher level of stability and comfort was required for the project, so the texture parameter T-4 (W: 8 mm, D: 3 mm, GS: 15 mm, GGW: 50 mm) was chosen for the concrete pavement. The skid-resistance performance of highway cement pavements is mainly evaluated in terms of structure depth, the sideway-force coefficient, etc., of which the sideway-force coefficient is a more commonly used evaluation method. However, the test vehicle has strict requirements on the length of the road section and driving speed. As a post-evaluation, the sideway-force coefficient cannot provide guidance in the design stage. It is difficult obtain the sideway-force coefficients for some special road sections (such as cement ramps and toll plaza pavements). In addition, to guide the design of the texture parameters in the construction process, it is necessary to evaluate the skid resistance of the concrete pavement in the process of laboratory testing. In this study, the skid resistance performance of the optimized pavement (W: 8 mm, D: 3 mm, GS: 15 mm, GGW: 50 mm), built in 2017, was tracked, as shown in Figure 16, and the correlation between the stress concentration coefficient and the transverse force coefficient was studied.

Figure 16. Textured pavement.

A pressure film test system was used to carry out an annual inspection of highway pavements. A tire loading test was conducted at the same point every year to obtain the tire–pavement contact stress distribution. Figure 17 shows the results of the test conducted during the 2017–2020 period.

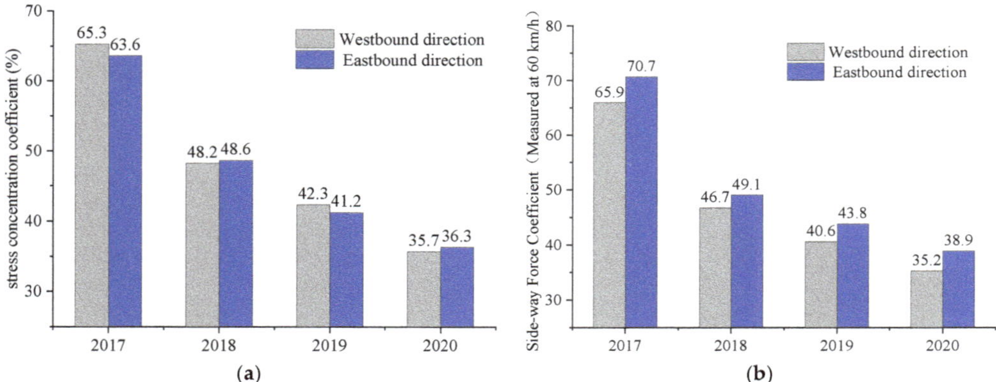

Figure 17. Tracking data: (**a**) stress concentration coefficient; (**b**) sideway-force coefficient (60 km/h).

Overall, the decrease percentages of the stress concentration coefficient of westbound and eastbound direction in the first year are 57.8% and 55.0%, respectively, whereas the corresponding values of the SFC are 62.1% and 68.0%. Under the effect of significant abrasion, the skid-resistance performance decreases rapidly in the initial stage, consistent with the laboratory test results.

Figure 18 shows the correlation analysis results of the stress concentration factor and transverse force coefficient. The R^2 of the test data is 0.871, indicating a good linear correlation between the tire contact stress concentration coefficient and sideway-force coefficient. Therefore, the stress concentration coefficient based on the friction mechanics can be used to characterize the skid-resistance of textured concrete pavements.

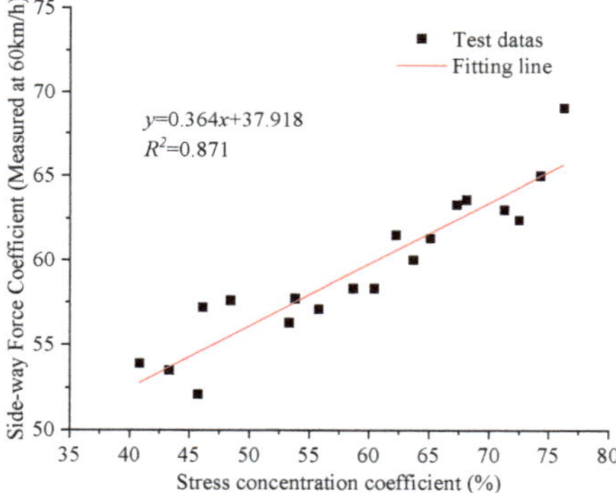

Figure 18. Relationship between stress concentration and SFC.

8. Conclusions

This study was conducted to optimize the texturing parameters of concrete pavement considering both the skid resistance and the driving stability. The work carried out by this study yields the following conclusions:

(1) The actual contact stress between the tire and the pavement can be characterized by the Weibull model. The discrete degree of the contact stress of different pavements can be ranked (from high to low) as follows: asphalt pavement (AC-16), curved grooved pavement, rectangular grooved pavement, and concrete pavements with no grooves.

(2) The compact texture structure of the textured pavement improved the friction resistance but reduced the driving stability. Theoretical and experimental analyses

showed that the stationary steering resistance torque based on the measured stress could effectively help evaluate the steering effect of textured pavements on tires.

(3) The most important factors influencing the texture depth, stress concentration coefficient, and steering resistance torque were the groove depth, groove width, and groove spacing. Through the analysis of variance, we found that each texture parameter had a significant influence on the stress concentration coefficient and stationary steering resistance torque.

(4) An asymptotic attenuation model successfully described the attenuation laws of the stress concentration coefficient and stationary steering resistance torque. Based on the results of orthogonal and abrasion resistance tests, we suggest a sample with optimal dimensions (8 mm in width, 3 mm in depth, 15 mm in spacing, and 50 mm in groove group width) for balancing skid-resistance performance and driving stability.

(5) The stress concentration coefficient and SFC (measured at 60 km/h) exhibited a good linear correlation, indicating that the stress concentration coefficient can effectively characterize the skid-resistance performance of textured concrete pavements.

Based on the requirements of the practical engineering project, this study is conducted to investigate the combination of texture parameters of cement pavement with longitudinal grooves for balancing skid-resistance performance and driving stability. The further study will be focused on the effects of groove direction, tire pattern form, tire structure type, tire load, and other factors on driving stability and braking performance. Moreover, the mix design of the wear resistance of concrete pavement is also an important factor that affects the durability of the skid resistance of concrete pavement and will also be systematically investigated in future study.

Author Contributions: Conceptualization, J.Y. and B.C.; methodology, J.Y.; investigation, B.Z. and P.L.; formal analysis, B.Z. and F.G.; writing—review and editing, B.C., P.L., and F.G. All authors have read and agreed to the published version of the manuscript.

Funding: The authors would like to acknowledge the financial support provided by the "National Natural Science Foundation of China" (51678251) and the "China Postdoctoral Science Foundation" (2020M672639).

Institutional Review Board Statement: Not applicable.

Informed Consent Statement: Informed consent was obtained from all subjects involved in the study.

Data Availability Statement: The data presented in this study are available on request from the corresponding author.

Conflicts of Interest: The authors declare no conflict of interest.

References

1. Rith, M.; Kim, Y.K.; Lee, S.W. Characterization of long-term skid resistance in exposed aggregate concrete pavement. *Constr. Build. Mater.* **2020**, *256*, 119423. [CrossRef]
2. Gao, L.; de Fortier Smit, A.; Prozzi, J.A.; Buddhavarapu, P.; Murphy, M.; Song, L. Milled pavement texturing to optimize skid improvements. *Constr. Build. Mater.* **2015**, *101*, 602–610. [CrossRef]
3. Wasilewska, M.; Gardziejczyk, W.; Gierasimiuk, P. Evaluation of skid resistance of exposed aggregate concrete pavement in the initial exploitation period. *Roads Bridges-Drog. I Mosty* **2017**, *16*, 295–308.
4. Chen, J.; Huang, X.; Zheng, B.; Zhao, R.; Liu, X.; Cao, Q.; Zhu, S. Real-time identification system of asphalt pavement texture based on the close-range photogrammetry. *Constr. Build. Mater.* **2019**, *226*, 910–919. [CrossRef]
5. Lin, C.; Tongjing, W. Effect of fine aggregate angularity on skid-resistance of asphalt pavement using accelerated pavement testing. *Constr. Build. Mater.* **2018**, *168*, 41–46. [CrossRef]
6. Yu, H.; Leng, Z.; Zhou, Z.; Shih, K.; Xiao, F.; Gao, Z. Optimization of preparation procedure of liquid warm mix additive modified asphalt rubber. *J. Clean. Prod.* **2017**, *141*, 336–345. [CrossRef]
7. Wang, K.; Lin, L.; Li, Q.J.; Nguyen, V.; Hayhoe, G.; Larkin, A. Runway groove identification and evaluation using 1 mm 3D image data. In Proceedings of the Airfield and Highway Pavement 2013: Sustainable and Efficient Pavements, Los Angeles, CA, USA, 9–12 June 2013; pp. 730–741.
8. de Oliveira, A.L.; Prudêncio, L.R., Jr. Evaluation of the superficial texture of concrete pavers using digital image processing. *J. Constr. Eng. Manag.* **2015**, *141*, 04015034. [CrossRef]

9. White, G. State of the art: Asphalt for airport pavement surfacing. *Int. J. Pavement Res. Technol.* **2018**, *11*, 77–98. [CrossRef]
10. Pranjić, I.; Deluka-Tibljaš, A.; Cuculić, M.; Šurdonja, S. Influence of pavement surface macrotexture on pavement skid resistance. *Transp. Res. Procedia* **2020**, *45*, 747–754. [CrossRef]
11. Fwa, T.; Ong, G.P. Transverse pavement grooving against hydroplaning. II: Design. *J. Transp. Eng.* **2006**, *132*, 449–457. [CrossRef]
12. Zhang, Z.; Luan, B.; Liu, X.; Zhang, M. Effects of surface texture on tire-pavement noise and skid resistance in long freeway tunnels: From field investigation to technical practice. *Appl. Acoust.* **2020**, *160*, 107120. [CrossRef]
13. Kogbara, R.B.; Masad, E.A.; Kassem, E.; Scarpas, A.; Anupam, K. A state-of-the-art review of parameters influencing measurement and modeling of skid resistance of asphalt pavements. *Constr. Build. Mater.* **2016**, *114*, 602–617. [CrossRef]
14. Plati, C. Sustainability factors in pavement materials, design, and preservation strategies: A literature review. *Constr. Build. Mater.* **2019**, *211*, 539–555. [CrossRef]
15. Chen, L.; Cong, L.; Dong, Y.; Yang, G.; Tang, B.; Wang, X.; Gong, H. Investigation of influential factors of tire/pavement noise: A multilevel Bayesian analysis of full-scale track testing data. *Constr. Build. Mater.* **2021**, *270*, 121484. [CrossRef]
16. Yu, H.; Zhu, Z.; Leng, Z.; Wu, C.; Zhang, Z.; Wang, D.; Oeser, M. Effect of mixing sequence on asphalt mixtures containing waste tire rubber and warm mix surfactants. *J. Clean. Prod.* **2020**, *246*, 119008. [CrossRef]
17. Zhang, S.; Wang, D.; Guo, F.; Deng, Y.; Feng, F.; Wu, Q.; Chen, Z.; Li, Y. Properties investigation of the SBS modified asphalt with a compound warm mix asphalt (WMA) fashion using the chemical additive and foaming procedure. *J. Clean. Prod.* **2021**, *319*, 128789. [CrossRef]
18. Guada, I.; Rezaei, A.; Harvey, J.; Spinner, D. *Evaluation of Grind and Groove (Next Generation Concrete Surface) Pilot Projects in California*; Working Paper Series; Institute of Transportation Studies: Berkeley, CA, USA, 2012.
19. Skarabis, J.; Stöckert, U. Noise emission of concrete pavement surfaces produced by diamond grinding. *J. Traffic Transp. Eng. Engl. Ed.* **2015**, *2*, 81–92. [CrossRef]
20. Ling, S.; Yu, F.; Sun, D.; Sun, G.; Xu, L. A comprehensive review of tire-pavement noise: Generation mechanism, measurement methods, and quiet asphalt pavement. *J. Clean. Prod.* **2021**, *287*, 125056. [CrossRef]
21. Beji, A.; Deboudt, K.; Khardi, S.; Muresan, B.; Lumière, L. Determinants of rear-of-wheel and tire-road wear particle emissions by light-duty vehicles using on-road and test track experiments. *Atmos. Pollut. Res.* **2021**, *12*, 278–291. [CrossRef]
22. Ran, S.; Besselink, I.; Nijmeijer, H. Energy Balance and Tyre Motions During Shimmy. In Proceedings of the 4th International Tyre Colloquium, Guilford, UK, 20–21 April 2015; University of Surrey: Guildford, UK, 2015; pp. 129–138.
23. Mi, T.; Stepan, G.; Takacs, D.; Chen, N.; Zhang, N. Model establishment and parameter analysis on shimmy of electric vehicle with independent suspensions. *Procedia IUTAM* **2017**, *22*, 259–266. [CrossRef]
24. Yu, M.; Xiao, B.; You, Z.; Wu, G.; Li, X.; Ding, Y. Dynamic friction coefficient between tire and compacted asphalt mixtures using tire-pavement dynamic friction analyzer. *Constr. Build. Mater.* **2020**, *258*, 119492. [CrossRef]
25. Zhuravlev, V.P.; Klimov, D.; Plotnikov, P. A new model of shimmy. *Mech. Solids* **2013**, *48*, 490–499. [CrossRef]
26. Araújo, J.P.C.; Palha, C.A.O.; Martins, F.F.; Silva, H.M.R.D.; Oliveira, J.R.M. Estimation of energy consumption on the tire-pavement interaction for asphalt mixtures with different surface properties using data mining techniques. *Transp. Res. Part D Transp. Environ.* **2019**, *67*, 421–432. [CrossRef]
27. Yu, M.; You, Z.; Wu, G.; Kong, L.; Liu, C.; Gao, J. Measurement and modeling of skid resistance of asphalt pavement: A review. *Constr. Build. Mater.* **2020**, *260*, 119878. [CrossRef]
28. Lee, Y.; Yurong, L.; Ying, L.; Fwa, T.; Choo, Y. Skid resistance prediction by computer simulation. In Proceedings of the Applications of Advanced Technologies in Transportation Engineering (2004), Beijing, China, 24–26 May 2004; pp. 465–469.
29. Bawono, A.A.; Lechner, B.; Yang, E.-H. Skid resistance and surface water drainage performance of engineered cementitious composites for pavement applications. *Cem. Concr. Compos.* **2019**, *104*, 103387. [CrossRef]
30. Yu, H.; Deng, G.; Zhang, Z.; Zhu, M.; Gong, M.; Oeser, M. Workability of rubberized asphalt from a perspective of particle effect. *Transp. Res. Part D: Transp. Environ.* **2021**, *91*, 102712. [CrossRef]
31. Yu, H.; Leng, Z.; Dong, Z.; Tan, Z.; Guo, F.; Yan, J. Workability and mechanical property characterization of asphalt rubber mixtures modified with various warm mix asphalt additives. *Constr. Build. Mater.* **2018**, *175*, 392–401. [CrossRef]
32. Xu, C.; Wang, D.; Zhang, S.; Guo, E.; Luo, H.; Zhang, Z.; Yu, H. Effect of Lignin Modifier on Engineering Performance of Bituminous Binder and Mixture. *Polymers* **2021**, *13*, 1083. [CrossRef]
33. Zhang, X.; Liu, T.; Liu, C.; Chen, Z. Research on skid resistance of asphalt pavement based on three-dimensional laser-scanning technology and pressure-sensitive film. *Constr. Build. Mater.* **2014**, *69*, 49–59. [CrossRef]
34. Araujo, V.M.C.; Bessa, I.S.; Castelo Branco, V.T.F. Measuring skid resistance of hot mix asphalt using the aggregate image measurement system (AIMS). *Constr. Build. Mater.* **2015**, *98*, 476–481. [CrossRef]
35. Liu, Y.; Tian, B.; Niu, K.-M. Research on skid resistance and noise reduction properties of cement concrete pavements with different surface textures. *J. Highw. Transp. Res. Dev. Engl. Ed.* **2013**, *7*, 22–27. [CrossRef]
36. Deng, Q.; Zhan, Y.; Liu, C.; Qiu, Y.; Zhang, A. Multiscale power spectrum analysis of 3D surface texture for prediction of asphalt pavement friction. *Constr. Build. Mater.* **2021**, *293*, 123506. [CrossRef]
37. Uz, V.E.; Gökalp, İ. Comparative laboratory evaluation of macro texture depth of surface coatings with standard volumetric test methods. *Constr. Build. Mater.* **2017**, *139*, 267–276. [CrossRef]
38. Chu, L.; Cui, X.; Zhang, K.; Fwa, T.F.; Han, S. Directional Skid Resistance Characteristics of Road Pavement: Implications for Friction Measurements by British Pendulum Tester and Dynamic Friction Tester. *Transp. Res. Rec.* **2019**, *2673*, 793–803. [CrossRef]

39. Guo, F.; Pei, J.; Zhang, J.; Li, R.; Zhou, B.; Chen, Z. Study on the skid resistance of asphalt pavement: A state-of-the-art review and future prospective. *Constr. Build. Mater.* **2021**, *303*, 124411. [CrossRef]
40. Liang, J.; Gu, X.; Deng, H.; Ni, F. Detecting device and technology of pavement texture depth based on high precision 3D laser scanning technology. *IOP Conf. Ser. Mater. Sci. Eng.* **2019**, *652*, 012063. [CrossRef]
41. Lee, M.-H.; Chou, C.-P.; Li, K.-H. Automatic measurement of runway grooving construction for pavement skid evaluation. *Autom. Constr.* **2009**, *18*, 856–863. [CrossRef]
42. Zheng, M.; Tian, Y.; Wang, X.; Peng, P. Research on Grooved Concrete Pavement Based on the Durability of Its Anti-Skid Performance. *Appl. Sci.* **2018**, *8*, 891. [CrossRef]
43. Dong, S.; Han, S.; Zhang, Q.; Han, X.; Zhang, Z.; Yao, T. Three-dimensional evaluation method for asphalt pavement texture characteristics. *Constr. Build. Mater.* **2021**, *287*, 122966. [CrossRef]
44. Persson, B. Theory of rubber friction and contact mechanics. *J. Chem. Phys.* **2001**, 115. [CrossRef]
45. Al-Assi, M.; Kassem, E. Evaluation of adhesion and hysteresis friction of rubber–pavement system. *Appl. Sci.* **2017**, *7*, 1029. [CrossRef]
46. Chen, B.; Zhang, X.; Yu, J.; Wang, Y. Impact of contact stress distribution on skid resistance of asphalt pavements. *Constr. Build. Mater.* **2017**, *133*, 330–339. [CrossRef]
47. Nicolosi, V.; D'Apuzzo, M.; Evangelisti, A. Cumulated frictional dissipated energy and pavement skid deterioration: Evaluation and correlation. *Constr. Build. Mater.* **2020**, *263*, 120020. [CrossRef]
48. Liang, K.; Tu, Q.-Z.; Shen, X.-M.; Fang, Z.-H.; Yang, X.; Zhang, Y.; Xiang, H.-Y. An improved LuGre model for calculating static steering torque of rubber tracked chassis. *Def. Technol.* **2021**. [CrossRef]
49. Yunchao, W.; Feng, P.; Pang, W.; Zhou, M. Pivot steering resistance torque based on tire torsion deformation. *J. Terramechanics* **2014**, *52*, 47–55. [CrossRef]
50. Hou, Y.; Zhang, H.; Wu, J.; Wang, L.; Xiong, H. Study on the microscopic friction between tire and asphalt pavement based on molecular dynamics simulation. *Int. J. Pavement Res. Technol.* **2018**, *11*, 205–212. [CrossRef]

Article

Study on Adhesion Property and Moisture Effect between SBS Modified Asphalt Binder and Aggregate Using Molecular Dynamics Simulation

Fucheng Guo [1], Jianzhong Pei [1,*], Jiupeng Zhang [1], Rui Li [1], Pengfei Liu [2] and Di Wang [3,4,*]

1 Key Laboratory for Special Area Highway Engineering of Ministry of Education, Chang'an University, Xi'an 710064, China
2 Institute of Highway Engineering, RWTH Aachen University, Mies-van-der-Rohe-Str. 1, 52074 Aachen, Germany
3 Department of Civil Engineering, Aalto University, Rakentajanaukio 4, 02150 Espoo, Finland
4 Hangzhou Telujie Transportation Technology Co., Ltd., Hangzhou 311121, China
* Correspondence: peijianzhong@126.com (J.P.); di.1.wang@aalto.fi (D.W.); Tel.: +86-29-82334681 (J.P.); +358-503422787 (D.W.)

Abstract: In this project, the adhesion property and moisture effect between styrene–butadiene–styrene (SBS) modified asphalt binder and aggregate were studied to reveal their interface adhesion mechanism. The influence of SBS contents on adhesion property and moisture effect between binder and aggregate phases were investigated using molecular dynamics simulation. Moreover, the double-layer adhesion models of asphalt binder–aggregate and triple-layer debonding models of asphalt binder–water–aggregate were constructed and equilibrated, and the adhesion property and the moisture effect were evaluated numerically. The results indicate that the built SBS-modified asphalt binder models show favorable reliability in representing the real one. The variation in the work of adhesion for SBS modified asphalt binder–quartz is not remarkable with the SBS content when its content is relatively low. However, the work of adhesion decreased significantly when the content was higher than 6 wt.%, which is consistent with the experimental results. The adhesion between SBS-modified asphalt binder and quartz is derived from Van der Waals energy. The modified asphalt binder with a high SBS modifier content (8 wt.% and 10 wt.%) shows much better moisture resistance (nearly 30% improved) than the unmodified asphalt binder from the work of debonding results. According to the Energy Ratio (*ER*) values, asphalt binders with high SBS content (8 wt.% and 10 wt.%) present a good moisture resistance performance. Therefore, the SBS content should be seriously selected by considering the dry and wet conditions that are used to balance the adhesion property and debonding properties. The content of 4 wt.% may be the optimal content under the dry adhesion and moisture resistance.

Keywords: SBS modified asphalt binder; aggregate; interface; adhesion property; moisture effect; molecular dynamics simulation

Citation: Guo, F.; Pei, J.; Zhang, J.; Li, R.; Liu, P.; Wang, D. Study on Adhesion Property and Moisture Effect between SBS Modified Asphalt Binder and Aggregate Using Molecular Dynamics Simulation. *Materials* **2022**, *15*, 6912. https://doi.org/10.3390/ma15196912

Academic Editors: Simon Hesp and Francesco Canestrari

Received: 5 August 2022
Accepted: 2 October 2022
Published: 5 October 2022

1. Introduction

Asphalt mixtures as complex multi-phase materials are mainly composed of binder, aggregate, and interface phases. The interface phase is one of the weakest phases, which is prone to damage and induces several types of pavement distress, including raveling, pothole, and fatigue cracking, under repeated traffic load and environmental effects [1–3]. The service life of the pavement is heavily shortened by these accumulated distresses caused by the weak interface phases. In order to gain better engineering performances, good adhesion strength between asphalt binder and aggregate is required to maintain the asphalt pavement structure as a stable system [4,5]. Asphalt, as the binder material, plays the dominant role in the adhesion process with aggregates [6]. Hence, most of the studies were conducted on the modified asphalt binder to improve its adhesion property and long-term road performance [7–9].

Styrene–butadiene–styrene (SBS), as a favorable modifier, has been verified and widely utilized in the construction of asphalt pavement to improve the performance properties of bituminous materials over a wide range of servicing temperatures [10]. Some studies validated that the addition of proper content SBS modifiers with asphalt binder could remarkably increase the viscosity of the asphalt binder and further improve the adhesion property [11]. However, most investigations focused on engineering performance using experimental methods, while the improvement mechanism is difficult to be revealed [12,13]. Therefore, it is vital to improve the adhesion property of asphalt binder purposefully when the mix design methods are gradually changed to the performance orientation.

The molecular dynamics (MD) simulation method, as an effective method, was introduced into the materials field to reveal the micro mechanism of the formation of macro behavior at the molecular level [13–15]. Different mechanical properties, interaction behaviors, diffusion behaviors, and effects of the aging process, together with the moisture effect, were modeled and evaluated properly [16–18]. The modification mechanism was favorably illustrated by using different parameters. Moreover, an important parameter, the work of adhesion, was proposed to characterize the adhesion behavior between asphalt binder and aggregate [17]. However, most MD-related investigations mainly focused on the effect of unmodified and aged binders on the adhesion between binder and aggregates; only limited studies considered the effect of the SBS modifier [19–21].

The objective of this study was to investigate the interface behavior between SBS-modified asphalt binder and aggregates, where the adhesion property and moisture effect were focused on revealing their interface adhesion mechanism. Firstly, the representative models of asphalt binder, styrene–butadiene–styrene (SBS) modifier, and aggregate were built based on molecular dynamics software Materials Studio 2021. Next, the double-layer model of SBS modified asphalt binder–aggregate with different SBS contents was constructed to evaluate the adhesion property. The geometry optimization and dynamics equilibration were conducted afterward, followed by the calculation of work of adhesion to better understand the influence of SBS content. Then, the triple-layer model of asphalt binder–water–aggregate was constructed, and the moisture effect was further analyzed by adding the 100 water molecules (about 10 wt.%) in the interlayer of asphalt binder–aggregate. Finally, the work of debonding and energy ratio was selected as the evaluation parameters to evaluate the moisture effect. The obtained result can be used for the selection of SBS modifier content and to understand the interface behavior.

2. Materials and Methods

2.1. Materials

In this study, the interface behavior between SBS-modified asphalt binder and aggregates was investigated. Therefore, the types of asphalt binder, SBS modifier, and aggregates should be selected. Asphalt binder as a mix consists of thousands of compounds [22]. Different models, such as the average molecular model and components constructed model, were used to represent the asphalt binder [23–26]. The average mode was proposed using Nuclear Magnetic Resonance (NMR) spectroscope, which can simulate asphalt binder accurately but fail to describe the fraction differences in asphalt binder. Multiple component models were firstly proposed by Zhang and Greenfield in 2007 to represent asphalt binders more accurately [27]. The components model was verified and widely used after it was developed [28,29]. In this study, the three-component constructed model was selected, as shown in Figure 1a–c, where 1,7-dimethylnapthalene and n-docosane (n-$C_{22}H_{46}$) were used to represent naphthene aromatic and saturate, respectively. The weight proportion of asphaltenes, naphthene aromatics, and saturates is approximately 20:20:60, respectively; the corresponding molecular number is 5, 27, and 41, respectively. The reliability of molecular structures and numbers has been verified by plenty of researchers [28,29].

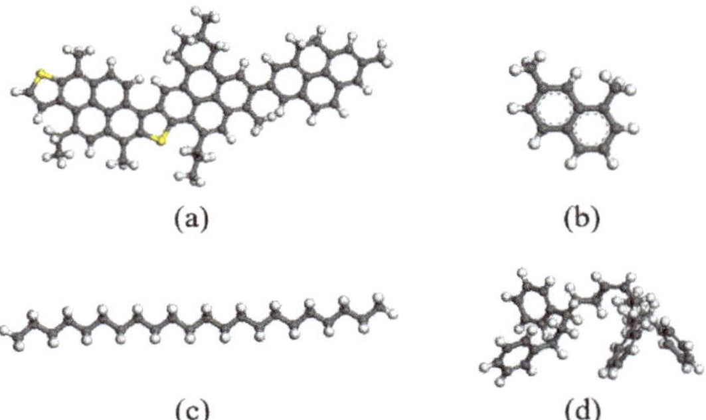

Figure 1. Molecular structures of SBS modified asphalt compositions for molecular simulations: (**a**) asphaltene; (**b**) naphthene aromatic; (**c**) saturate; (**d**) SBS.

After the constituent molecules of the asphalt binder were determined, the SBS molecular structure was then selected. In this study, a linear SBS structure with a molecular weight of 471 was selected (Figure 1d), where the weight proportion was nearly 2% by the weight of the asphalt binder.

Various types of aggregates, including basalt, granite, quartzite, limestone, etc., were used for asphalt pavement construction. Among them, quartz is the most common mineral with high proportions in mineral compositions [30]. Thus, quartz was selected in this study as the representative of aggregate.

2.2. Modeling of Asphalt Binder–Aggregate Interface

As for the adhesion simulation, the double-layer model of asphalt binder–aggregate needed to be built. The single molecular models of asphalt binder and aggregate had to be built separately. The asphalt binder model was built by assembling three components together with the corresponding molecular numbers. Similarly, the models of SBS modified asphalt binder with different SBS contents (0 wt.%, 2 wt.%, 4 wt.%, 6 wt.%, 8 wt.%, 10 wt.%) were built by assembling asphalt binder components with SBS structure of 1, 2, 3, 4, and 5, respectively.

As for the aggregate, the α-quartz was imported from the software database, and then the (1,1,0) plane was cleaved to obtain a quartz surface. Afterward, the two-dimensional (2D) structure of the quartz surface was created by increasing the unit cell to 4 and 6 for U and V directions, respectively. Then, a three-dimensional quartz crystal was obtained by adding a vacuum slab with a thickness of zero. Moreover, the hydroxylation of the surface silica was also considered.

After the confined SBS modified asphalt binder model and aggregate model were built, the asphalt binder–aggregate interface model could be constructed, as shown in Figure 2 (taking the interface model with 2 wt.% SBS modifiers as an example).

2.3. Modeling of Asphalt Binder–Water–Aggregate Interface

Water molecules of 100 were added to the interface between the SBS-modified asphalt binder layer and aggregate layer to characterize the moisture effect of adhesion, where the weight proportion was nearly 10 % of asphalt binder by weight. The built triple-layer interface model for asphalt binder–water–aggregate was shown in Figure 3 (taking the layer model with 2 wt.% SBS modified asphalt–aggregate–quartz as an example).

2.4. Simulation Details and Evaluation Index

All MD simulations were performed using Materials Studio 2021 with the COMPASS II force field to represent the atomistic interactions. COMPASS II force field is optimized ab initio force field and can be used to accurately simulate organic, inorganic, and polymer material [31]. The total potential energy (E_{total}) consists of valence and nonbonded interaction terms. The valence term (E_{val}) includes the interactions of bond stretching (E_b), angle bending (E_θ), internal torsion (E_φ), out-of-plane bending (E_χ), and the cross-coupling terms ($E_{bb'}$, $E_{b\theta}$, $E_{b\varphi}$, $E_{\theta\varphi'}$, $E_{\theta\theta'}$, and $E_{\theta\theta'\varphi}$). The non-bond interaction term ($E_{non\text{-}bond}$) quantifies the non-covalent contributions, including Coulomb electrostatic energy (E_{elec}) and the van der Waals energy (E_{LJ}) [32].

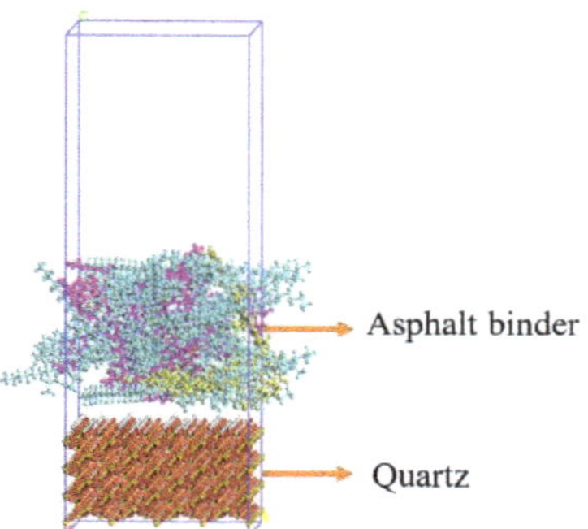

Figure 2. Double-layer interface model of SBS modified asphalt binder–aggregate.

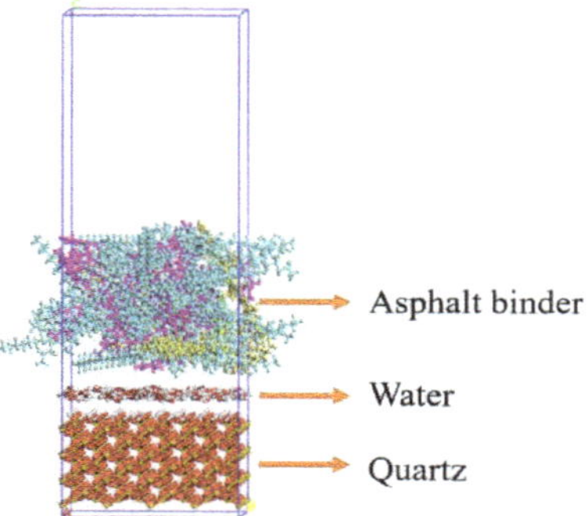

Figure 3. Triple-layer interface model of SBS modified asphalt binder–water–aggregate.

After all constituent molecules of asphalt binder and SBS structure were determined. The bulk models of SBS modified asphalt binder with different SBS contents were built. The bulk model is an amorphous cell with a three-dimensional periodic boundary. The geometry optimization with 10,000 iterations was conducted afterward to eliminate the highest energy overlap. The optimized configuration was further refined by performing Forcite dynamic calculations using Material Studio 2021 under the isothermal–isobaric (NPT) ensemble for 500 ps and 1 atm pressure and the canonical ensemble (NVT ensemble) for 500 ps to ensure that the system reached an equilibrium state. The temperature was set as 298 K, which is a common service temperature and has been widely used in this previous adhesion-related simulations [32,33]. The Ewald and the atom-based summation methods were used for electrostatic interactions and the Van der Waals interactions. The cutoff distance was set as 15.5 Å (this parameter was selected by considering the minimum image convention, the simulation cost, and the previous studies together).

After the bulk models were fully equilibrated, the density of SBS-modified asphalt binders was calculated. The experimental results were used to validate the built model's simulation reliability. The confined models were then further constructed to build the double-layer models. Different from the bulk model, the cell boundary for the confined model in the z-direction was confined by the hard repulsive wall, which stops any movements in the box beyond the wall. The confined single-layer models for the SBS modified asphalt binder were also optimized and equilibrated using the same procedures as the bulk one.

The double-layer models were then built to add the confined asphalt binder models to the top of the aggregate supercell. A vacuum of 40 Å was added above the top aggregate layer to avoid the interaction across the mirror image in the z-direction. The geometry optimization with 5000 iterations was conducted afterward for the built double-layer models, followed by the dynamic equilibration under the canonical ensemble (NVT ensemble) for 300 ps. It should be noted that the bottom aggregate layer was fixed during the optimization and dynamics simulation. The obtained stable configuration was selected for further adhesion strength calculation.

As previously stated, the work of adhesion, as a thermodynamic parameter, was selected to characterize the adhesion property between SBS-modified asphalt binder and aggregate [17]. This parameter is defined as the required energy to separate a unit area of an interface into two free surfaces in a vacuum. A better adhesion performance was presented with a larger work of adhesion. The work of adhesion was calculated by Equation (1) for the asphalt binder–aggregate layer model [34]. Specifically, the total potential energy of the asphalt binder–aggregate system under the equilibration state was calculated first. The potential energy for asphalt binder and aggregate could then be obtained by deleting the asphalt binder layer and aggregate layer from the system, respectively. Finally, the work of adhesion can be obtained using Equation (1).

$$W_{adhesion} = ((E_a + E_{agg}) - E_{total})/A, \tag{1}$$

where:

$W_{adhesion}$ is the work of adhesion between asphalt binder and aggregate;
E_{total} is the total potential energy of asphalt binder–aggregate interface model;
E_a is the potential energy of the asphalt binder;
E_{agg} is the potential energy of the aggregate;
A is the interface contact area between the asphalt binder and aggregate surface.

For the moisture effect, the optimization and dynamics simulation process for the asphalt binder–water–aggregate model was the same as the asphalt–aggregate model. After the stable configuration was obtained, the work of debonding for water to displace asphalt binder from its interface with aggregate was calculated according to Equation (2) [32]. Good moisture resistance was provided with the small work of debonding. Similar to the calculation procedures of the work of adhesion, the work of debonding was calculated after the triple-layer system reached the equilibration state. The interaction energy between two materials was calculated by deleting other materials.

$$W_{debonding} = ((\Delta E_{a\_w} + \Delta E_{agg\_w}) - \Delta E_{a\_agg})/A, \tag{2}$$

where:

$W_{debonding}$ is the work of debonding;
ΔE_{a_w} is the interaction energy between asphalt and water;
ΔE_{agg_w} is the interaction energy between aggregate and water;
ΔE_{a_agg} is the interaction energy between aggregate and asphalt.

Another parameter, energy ratio (*ER*), was also proposed to characterize the moisture resistance comprehensively [35]. *ER* is calculated by the ratio of the work of adhesion to the work of debonding, as shown in Equation (3). A higher *ER* value means better moisture resistance, which indicates better adhesion and low moisture damage susceptibility.

$$ER = W_{adhesion}/W_{debonding}, \tag{3}$$

3. Results and Discussion

3.1. Verification of the Built MD Models

The molecular dynamics (MD) simulation was conducted using the detailed procedures in Section 2.3. The energy and temperature were monitored during the dynamic equilibration process for the single-layer confined SBS modified asphalt binder models, double-layer interface models, and triple-layer interface models. Considering that the simulation was conducted under the NVT ensemble, the system should have constant total energy and temperature under the equilibration state. Thus, whether the system reaches the equilibration state can be judged by these two parameters. By taking SBS-modified asphalt binder–aggregate with 2 wt.% SBS content as the example, the variations in energy and temperature during the NVT dynamics equilibration are shown in Figures 4 and 5.

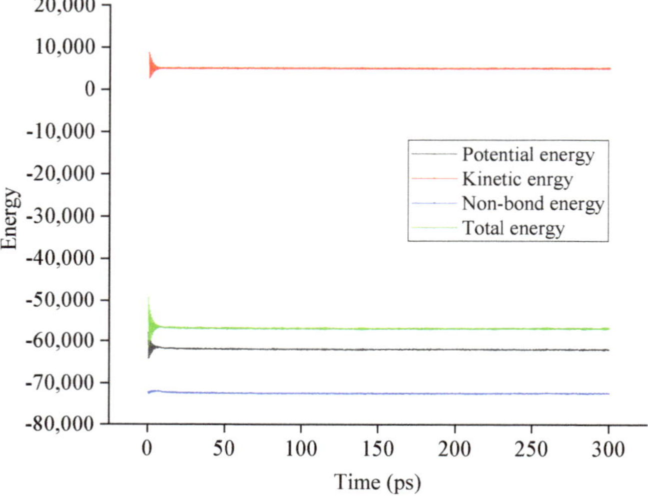

Figure 4. Variation in energy during the NVT dynamics equilibration.

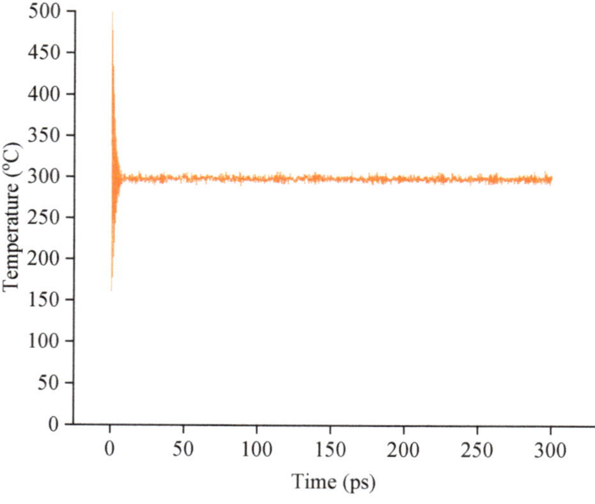

Figure 5. Variation in temperature during the NVT dynamics equilibration.

As shown in Figures 4 and 5, the energy and temperature fluctuated within an initial time of 10 ps while remaining at a relatively constant value with the increase in time. The constant value means that the built models are fully optimized and reach the equilibration state under the NVT ensemble. As for all single-layer confined SBS modified asphalt binder models, double-layer interface models, and triple-layer interface models, similar results are obtained, which verifies that the built models are under the equilibration state and can be used for further adhesion and debonding properties analysis.

After the models were fully optimized and equilibrated, the density of the SBS-modified asphalt binder under 298 K was calculated and compared with the previous studies to verify the reliability of the built model [36]. It should be noted that the SBS content of 0% means the unmodified asphalt binder, and its density was compared with the previous work, while the density for other SBS modified asphalt binder was estimated by the density of SBS modifier and asphalt binder using the hypothesis of volume additivity.

As shown in Figure 6, the density of SBS modified asphalt binder with different SBS contents from the simulation results shows good agreement with the previous study results with a maximum relative difference of 1%. Moreover, the addition of SBS into asphalt binder induces an increase in the density of asphalt binder from the simulation results, which is also consistent with the previous works [36]. This may be caused by the relatively larger density of SBS compared to the asphalt binder, with a density of 0.886. In addition, the molecular structure of SBS is more complex with the several benzene ring structures than the light components (naphthene aromatic and saturate) in asphalt binder with a large proportion. Therefore, the built SBS modified asphalt binder models reach the equilibration state and have the favorable reliability to represent the real one; thus, the models can be used for the further analysis of adhesion property and moisture effect.

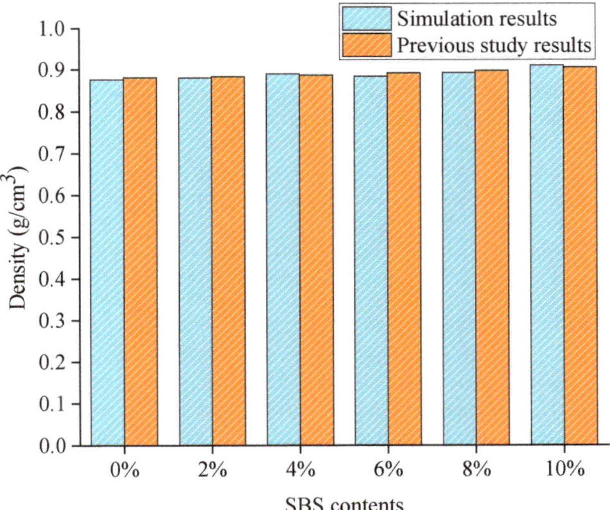

Figure 6. Density of SBS modified asphalt models from the simulation and previous study.

3.2. Variation in Work of Adhesion with the SBS Contents

For SBS modified asphalt binder with different SBS contents (0 wt.%, 2 wt.%, 4 wt.%, 6 wt.%, 8 wt.%, 10 wt.%), the work of adhesion between modified asphalt and aggregate can be calculated according to Equation (1). The adhesion results are shown in Figure 7.

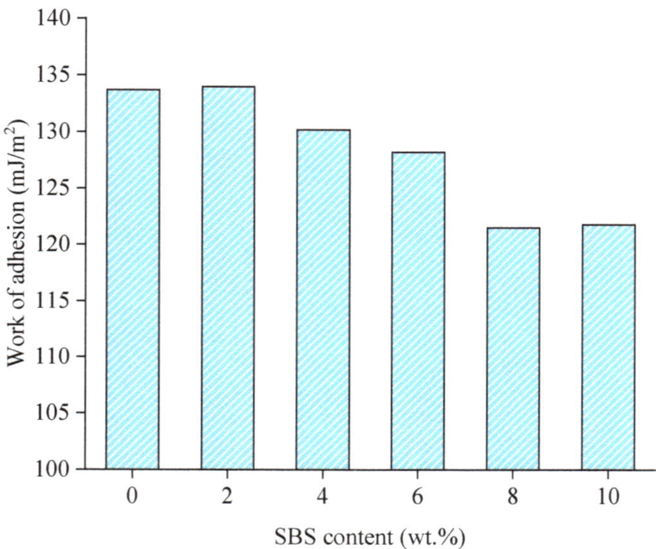

Figure 7. Variation in work of adhesion between SBS modified asphalt binder and quartz with SBS content.

As can be seen from Figure 7, the variation in the work of adhesion for SBS modified asphalt binder–quartz is not visible with the SBS content when its content is lower than 6 wt.%. However, the work of adhesion decreases significantly relatively when the content is high, which is consistent with the experimental results [36]. The former may attribute to the addition of SBS increased the benzene ring structure and cross-linked with asphalt binder to form a stable network structure, inducing the increase in adhesion within the small amount of SBS contents. The latter may be caused by poor compatibility when the content is relatively large (8% and 10 %), and the poor compatibility may result in a decrease in the adhesion property. Moreover, the work of adhesion between SBS modified asphalt binder and quartz is derived from the Van der Waals energy from the energy compositions, where the contribution of Van der Waals energy is minor and limited from the previous works [21]. Therefore, the variation in the work of adhesion is not remarkable.

3.3. Moisture Effect on the Interface of SBS Modified Asphalt Binder and Aggregate

The moisture effect was analyzed by adding water molecules into the interface of asphalt–aggregate. The same SBS contents were used to analyze the effect of water on SBS-modified asphalt binder–aggregate with different SBS contents. The work of debonding is calculated as shown in Figure 8. The *ER* values are calculated as well as shown in Figure 9.

As can be seen from Figure 5, the moisture effect for adhesion properties under different SBS contents is significant, especially for the asphalt binder–water–aggregate model with 6 wt.% SBS modifiers. The work of debonding for SBS modified asphalt binder–quartz increases firstly and decreases afterward with the SBS content except for the one with 4 wt.% SBS modifiers. This may be because the addition of SBS modifiers with the small percentage content into asphalt binder may increase the interaction energy of asphalt binder with water while decreasing the interaction energy of asphalt binder with the higher content. The larger interaction energy represents the strong interaction and the good affinity. Moreover, the modified asphalt binder with a high content (8 wt.% and 10 wt.%) of SBS modifiers shows much better moisture resistance than the unmodified asphalt binder, where the relative reduction ratio of work of working is about 30%. This means the addition of an SBS modifier has a positive effect on moisture resistance. Therefore, the asphalt binder with a high content of SBS modifiers may be more favorable from the debonding aspect.

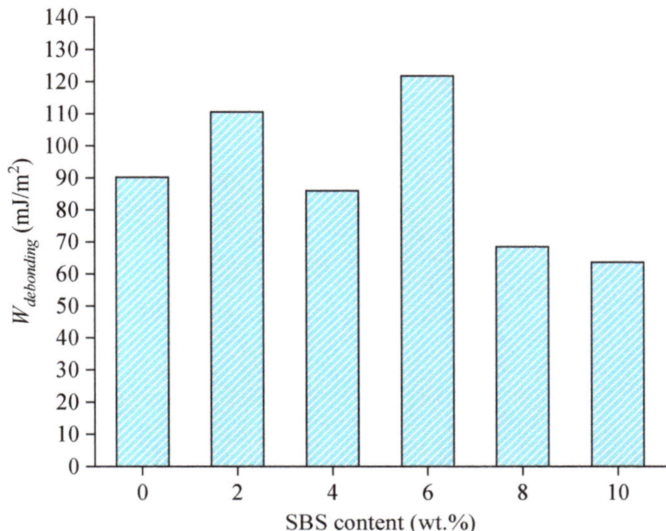

Figure 8. Variation in work of debonding with SBS content.

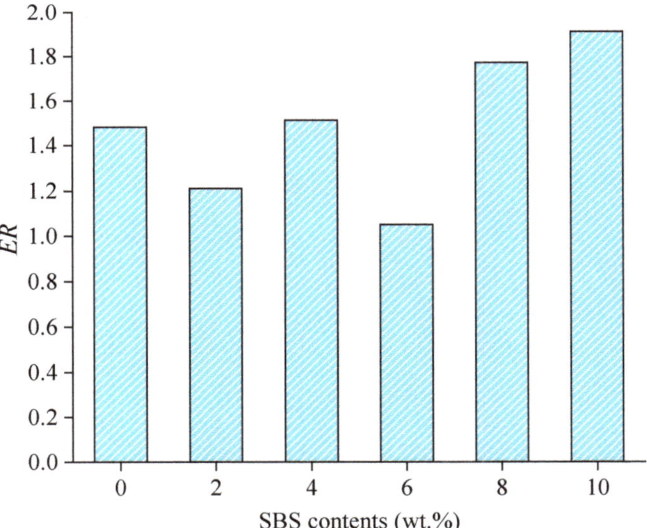

Figure 9. Variation in *ER* values with SBS content.

In order to comprehensively characterize the interface performance between SBS modified asphalt binder and aggregate, the *ER* values were applied. As shown in Figure 9, the *ER* values for all asphalt binder–aggregate models are larger than 1, which verifies that all modified asphalt binder has good moisture resistance. The asphalt binder with high content SBS presents a good performance in interface property for the higher *ER* values. This is mainly caused by the small value of the work of debonding. However, the large content of SBS modifiers means a high cost. The SBS content should be selected by considering the environment that is used to balance the adhesion property and debonding property.

4. Conclusions

In this study, the adhesion property and moisture effect between SBS-modified asphalt and aggregate were investigated using molecular dynamics simulation. The corresponding double-layer adhesion models and triple-layer debonding models were built. The variation in adhesion and debonding properties with SBS contents were investigated. The following conclusions were obtained:

(1) The built SBS modified asphalt binder models were verified to reach the equilibration state and have the favorable reliability to represent the real one from the dynamics simulation process parameters, e.g., energy and temperature, and density of SBS modified asphalt binder, thus, the models can be used for the further analysis of adhesion property and moisture effect;

(2) The variation in the work of adhesion for SBS modified asphalt binder–quartz is insignificant with the SBS content. The work of adhesion for SBS modified asphalt binder–quartz can be derived from the Van der Waals energy;

(3) The presence of water significantly reduces the interface adhesion. The modified asphalt binder with a high content (8 wt.% and 10 wt.%) of SBS modifiers shows much better moisture resistance than the unmodified asphalt binder from the results of work of debonding;

(4) The asphalt binder with high content SBS presents a good performance in moisture resistance according to *ER* values. Moreover, the SBS content should be selected by considering the environment that is used to balance the adhesion property and debonding property. The content of 4 wt.% may be the optimal content under the dry adhesion and moisture resistance.

This study used quartz to present the aggregate, and linear SBS modifiers represented SBS. Further studies should be conducted to investigate the impact of different types of aggregates and SBS modifiers on adhesion and debonding properties. Moreover, the experimental methods using AFM and surface free energy theory should be conducted to further reveal the modification mechanism and performance variations.

Author Contributions: Conceptualization, F.G., J.P. and D.W.; methodology, F.G. and J.P.; software, F.G.; validation, F.G. and J.Z.; formal analysis, F.G.; investigation, F.G.; resources, J.P. and D.W.; data curation, P.L. and D.W.; writing—original draft preparation, F.G.; writing—review and editing, J.P., P.L. and D.W.; visualization, R.L.; supervision, J.P. and D.W.; project administration, J.P.; funding acquisition, R.L. All authors have read and agreed to the published version of the manuscript.

Funding: This work was supported by the National Key R&D Program of China (grant number 2018YFE0103800); the National Natural Science Foundation of China (grant number 52178408); and the Fundamental Research Funds for the Central Universities, CHD (grant number 300102211702); and the German Research Foundation (No. OE 514/15-1(459436571)). The authors gratefully acknowledge their financial support. The first author would also like to acknowledge the China Scholarship Council (CSC) for supporting his visit to Aalto University, Finland (No. 202006560054).

Institutional Review Board Statement: Not applicable.

Informed Consent Statement: Not applicable.

Data Availability Statement: Not applicable.

Acknowledgments: The authors would like the technical support from Augusto Cannone Falchetto.

Conflicts of Interest: The authors declare no conflict of interest.

References

1. Cong, P.; Guo, X.; Ge, W. Effects of moisture on the bonding performance of asphalt-aggregate system. *Constr. Build. Mater.* **2021**, *295*, 123667. [CrossRef]
2. Guo, F.; Pei, J.; Zhang, J.; Xue, B.; Sun, G.; Li, R. Study on the adhesion property between asphalt binder and aggregate: A state-of-the-art review. *Constr. Build. Mater.* **2020**, *256*, 119474. [CrossRef]
3. Xing, C.; Jiang, W.; Li, M.; Wang, M.; Xiao, J.; Xu, Z. Application of atomic force microscopy in bitumen materials at the nanoscale: A review. *Constr. Build. Mater.* **2022**, *342*, 128059. [CrossRef]

4. Zhou, L.; Huang, W.; Zhang, Y.; Lv, Q.; Yan, C.; Jiao, Y. Evaluation of the adhesion and healing properties of modified asphalt binders. *Constr. Build. Mater.* **2020**, *251*, 119026. [CrossRef]
5. Júnior, J.L.L.; Babadopulos, L.F.; Soares, J.B. Moisture-induced damage resistance, stiffness and fatigue life of asphalt mixtures with different aggregate-binder adhesion properties. *Constr. Build. Mater.* **2019**, *216*, 166–175. [CrossRef]
6. Mohamed, A.S.; Xiao, F.; Hettiarachchi, C.; Abdel-Wahed, T. Bond strength in dry condition of reclaimed asphalt modified by crumb rubber modified binder. *J. Adhes.* **2022**, *99*, 1–30. [CrossRef]
7. Hu, M.; Sun, D.; Lu, T.; Ma, J.; Yu, F. Laboratory investigation of the adhesion and self-healing properties of high-viscosity modified asphalt binders. *Transp. Res. Rec.* **2020**, *2674*, 307–318. [CrossRef]
8. Office, J.E.; Chen, J.; Dan, H.; Ding, Y.; Gao, Y.; Guo, M.; Guo, S.; Han, B.; Hong, B.; Hou, Y.; et al. New innovations in pavement materials and engineering: A review on pavement engineering research 2021. *J. Traffic Transp. Eng.* **2021**, *8*, 815–999.
9. Mashaan, N.S.; Chegenizadeh, A.; Nikraz, H.; Rezagholilou, A. Investigating the engineering properties of asphalt binder modified with waste plastic polymer. *Ain Shams Eng. J.* **2021**, *12*, 1569–1574. [CrossRef]
10. Lin, P.; Huang, W.; Li, Y.; Tang, N.; Xiao, F. Investigation of influence factors on low temperature properties of SBS modified asphalt. *Constr. Build. Mater.* **2017**, *154*, 609–622. [CrossRef]
11. Tian, Z.; Zhang, Z.; Zhang, K.; Huang, S.; Luo, Y. Preparation and properties of high viscosity and elasticity asphalt by styrene–butadiene–styrene/polyurethane prepolymer composite modification. *J. Appl. Polym. Sci.* **2020**, *137*, 49123. [CrossRef]
12. Wang, T.; Wei, X.; Zhang, D.; Shi, H.; Cheng, Z. Evaluation for low temperature performance of SBS modified asphalt by dynamic shear rheometer method. *Buildings* **2021**, *11*, 408. [CrossRef]
13. Chen, M.; Geng, J.; Xia, C.; He, L.; Liu, Z. A review of phase structure of SBS modified asphalt: Affecting factors, analytical methods, phase models and improvements. *Constr. Build. Mater.* **2021**, *294*, 123610. [CrossRef]
14. Sun, D.; Lin, T.; Zhu, X.; Tian, Y.; Liu, F. Indices for self-healing performance assessments based on molecular dynamics simulation of asphalt binders. *Comput. Mater. Sci.* **2016**, *114*, 86–93. [CrossRef]
15. Yao, H.; Liu, J.; Xu, M.; Ji, J.; Dai, Q.; You, Z. Discussion on molecular dynamics (MD) simulations of the asphalt materials. *Adv. Colloid Interface Sci.* **2022**, *299*, 102565. [CrossRef]
16. Guo, F.; Zhang, J.; Chen, Z.; Zhang, M.; Pei, J.; Li, R. Investigation of friction behavior between tire and pavement by molecular dynamics simulations. *Constr. Build. Mater.* **2021**, *300*, 124037. [CrossRef]
17. Chen, Z.; Pei, J.; Li, R.; Xiao, F. Performance characteristics of asphalt materials based on molecular dynamics simulation–A review. *Constr. Build. Mater.* **2018**, *189*, 695–710. [CrossRef]
18. Xu, G.; Wang, H. Study of cohesion and adhesion properties of asphalt concrete with molecular dynamics simulation. *Comput. Mater. Sci.* **2016**, *112*, 161–169. [CrossRef]
19. Luo, L.; Chu, L.; Fwa, T.F. Molecular dynamics analysis of moisture effect on asphalt-aggregate adhesion considering anisotropic mineral surfaces. *Appl. Surf. Sci.* **2020**, *527*, 146830. [CrossRef]
20. Xu, G.; Wang, H. Molecular dynamics study of interfacial mechanical behavior between asphalt binder and mineral aggregate. *Constr. Build. Mater.* **2016**, *121*, 246–254. [CrossRef]
21. Gao, Y.; Zhang, Y.; Yang, Y.; Zhang, J.; Gu, F. Molecular dynamics investigation of interfacial adhesion between oxidised bitumen and mineral surfaces. *Appl. Surf. Sci.* **2019**, *479*, 449–462. [CrossRef]
22. Ding, Y.; Tang, B.; Zhang, Y.; Wei, J.; Cao, X. Molecular dynamics simulation to investigate the influence of SBS on molecular agglomeration behavior of asphalt. *J. Mater. Civ. Eng.* **2015**, *27*, C4014004. [CrossRef]
23. Hossain, K.; Hossain, Z. A synthesis of computational and experimental approaches of evaluating chemical, physical, and mechanistic properties of asphalt binders. *Adv. Civ. Eng.* **2019**, *2019*. [CrossRef]
24. Guo, F.; Zhang, J.; Pei, J.; Zhou, B.; Hu, Z. Study on the mechanical properties of rubber asphalt by molecular dynamics simulation. *J. Mol. Model.* **2019**, *25*, 365. [CrossRef]
25. Zhan, Y.; Wu, H.; Song, W.; Zhu, L. Molecular dynamics study of the diffusion between virgin and aged asphalt binder. *Coatings* **2022**, *12*, 403. [CrossRef]
26. Ren, S.; Liu, X.; Lin, P.; Gao, Y.; Erkens, S. Molecular dynamics simulation on bulk bitumen systems and its potential connections to macroscale performance: Review and discussion. *Fuel* **2022**, *328*, 125382. [CrossRef]
27. Zhang, L.; Greenfield, M.L. Analyzing properties of model asphalts using molecular simulation. *Energy Fuels* **2007**, *21*, 1712–1716. [CrossRef]
28. Guo, F.; Zhang, J.; Pei, J.; Zhou, B.; Cannone Falchetto, A.; Hu, Z. Investigating the interaction behavior between asphalt binder and rubber in rubber asphalt by molecular dynamics simulation. *Constr. Build. Mater.* **2020**, *252*, 118956. [CrossRef]
29. Hu, D.; Gu, X.; Cui, B.; Pei, J.; Zhang, Q. Modeling the oxidative aging kinetics and pathways of asphalt: A ReaxFF molecular dynamics study. *Energy Fuels* **2020**, *34*, 3601–3613. [CrossRef]
30. Xu, S.; Xiao, F.; Amirkhanian, S.; Singh, D. Moisture characteristics of mixtures with warm mix asphalt technologies–A review. *Constr. Build. Mater.* **2017**, *142*, 148–161. [CrossRef]
31. Sun, H.; Jin, Z.; Yang, C.; Akkermans, R.L.; Robertson, S.H.; Spenley, N.A.; Miller, S.; Todd, S.M. COMPASS II: Extended coverage for polymer and drug-like molecule databases. *J. Mol. Model.* **2016**, *22*, 47. [CrossRef] [PubMed]
32. Gao, Y.; Zhang, Y.; Gu, F.; Xu, T.; Wang, H. Impact of minerals and water on bitumen-mineral adhesion and debonding behaviours using molecular dynamics simulations. *Constr. Build. Mater.* **2018**, *171*, 214–222. [CrossRef]

33. Dong, Z.; Liu, Z.; Wang, P.; Gong, X. Nanostructure characterization of asphalt-aggregate interface through molecular dynamics simulation and atomic force microscopy. *Fuel* **2017**, *189*, 155–163. [CrossRef]
34. Long, Z.; You, L.; Tang, X.; Ma, W.; Ding, Y.; Xu, F. Analysis of interfacial adhesion properties of nano-silica modified asphalt mixtures using molecular dynamics simulation. *Constr. Build. Mater.* **2020**, *255*, 119354. [CrossRef]
35. Cui, B.; Wang, H. Molecular interaction of Asphalt-Aggregate interface modified by silane coupling agents at dry and wet conditions. *Appl. Surf. Sci.* **2022**, *572*, 151365. [CrossRef]
36. Hu, D.; Pei, J.; Li, R.; Zhang, J.; Jia, Y.; Fan, Z. Using thermodynamic parameters to study self-healing and interface properties of crumb rubber modified asphalt based on molecular dynamics simulation. *Front. Struct. Civ. Eng.* **2020**, *14*, 109–122. [CrossRef]

 *materials*

MDPI

Article

Research on Performance of SBS-PPA and SBR-PPA Compound Modified Asphalts

Jianguo Wei [1], Song Shi [1,2], Yuming Zhou [1,3,*], Zhiyuan Chen [1], Fan Yu [1], Zhuyi Peng [1] and Xurui Duan [1]

1 School of Traffic and Transportation Engineering, Changsha University of Science and Technology, Changsha 410004, China; jianguowei9969@126.com (J.W.); songs20222022@163.com (S.S.); chenzycsust@163.com (Z.C.); 15673102532@163.com (F.Y.); p19375175428@163.com (Z.P.); duanxuruile@163.com (X.D.)

2 Henan Railway Construction & Investment Group Co., Ltd., Zhengzhou 450018, China

3 National Engineering Laboratory of Highway Maintenance Technology, Changsha University of Science and Technology, Changsha 410004, China

* Correspondence: zym_2015@csust.edu.cn; Tel.: +86-188-7499-8425

Abstract: Although several studies indicated that the addition of Styrene-Butadiene-Styrene (SBS) and Styrene-Butadiene Rubber (SBR) bring a lot of benefits on properties of asphalt binders, high production costs and poor storage stability confine the manufacture of better modified asphalt. To reduce the production costs, polyphosphoric acid (PPA) was applied to prepare better compound modified asphalt binders. In this research, five PPA (0.5%, 0.75%, 1.0%, 1.25% and 1.5%) and two SBR/SBS (4% and 6%) concentrations were selected. Dynamic shear rheometer (DSR) and Bending Beam Rheometer (BBR) tests were performed to evaluate the rheological properties of the compound modified asphalt. Rolling Thin Film Oven (RTFO) test was performed to evaluate the aging properties of the compound modified asphalts. The results indicate that SBS/SBR modified asphalts with the addition of PPA show better high-temperature properties significantly, the ability of asphalt to resist rutting is improved, and the elastic recovery is increased. However, the low-temperature properties of the compound modified asphalts are degraded by increasing the creep stiffness (S) and decreasing the creep rate (m). At the same time, RTFO tests results show that PPA was less prone to oxidation to improve the anti-aging ability of modified asphalts. Overall, the combination of 4% SBS and 0.75–1.0% PPA, the combination of 4% SBR and 0.5–0.75% PPA is recommended based on a comprehensive analysis of the performance of compound modified asphalt, respectively, which can be equivalent to 6% SBS/SBR modified asphalt with high-temperature properties, low-temperature properties, temperature sensitivity and aging properties.

Keywords: road engineering; polyphosphoric acid; dynamic shear rheological test; temperature sensing performance; aging performance

Citation: Wei, J.; Shi, S.; Zhou, Y.; Chen, Z.; Yu, F.; Peng, Z.; Duan, X. Research on Performance of SBS-PPA and SBR-PPA Compound Modified Asphalts. *Materials* **2022**, *15*, 2112. https://doi.org/10.3390/ma15062112

Academic Editors: Markus Oeser, Michael Wistuba, Pengfei Liu and Di Wang

Received: 4 January 2022
Accepted: 9 March 2022
Published: 13 March 2022

1. Introduction

Bituminous pavement has become the main highway pavement structure due to its characteristics of good smoothness, low noise, safe driving and ease of mechanized construction. With growing traffic volumes and various natural environments, a variety of pavement stresses such as ruts, cracks and water damage have emerged [1–3]. Many researchers came to the conclusion that it is needed modified asphalt binders to resist against distresses and improve the performance and longevity of bituminous pavement [4–7].

In recent years, the most commonly used asphalt modifiers are polymers such as styrene-butadiene-styrene (SBS), styrene-butadiene rubber (SBR), crumb rubber (CR) and polyethylene (PE). Among polymers, SBS polymer modification could be the most accepted and wide application all over the world. Many studies showed that the addition of SBS to the asphalt increases the softening point, decreases the penetration slightly, lessens the thermal susceptibility, increases the viscosity and decreases the Frass breaking point [8].

However, the SBS polymer is not chemically stable and vulnerable to aging due to a number of unsaturated double carbon bonds inside SBS molecular [9]. Meanwhile, SBR is a kind of synthetic rubber that can improve the elastic recovery and low-temperature ductility of asphalt binders. Furthermore, it could apply as a modifier along with other modifiers to produce compound modified binders [10]. According to Zhang and Yu, their results indicated that the addition of SBR could improve the thermal cracking resistance of the Polyphosphoric acid (PPA) modified binders [11]. However, the limited effect of SBR on asphalt high temperature performance makes it difficult to meet the demand for applying in hot areas. In addition, the great discrepancy in structure and solubility parameter between SBR and asphalt leads to potential instability for blends [12].

Polyphosphoric acid (PPA) is a polymer of orthophosphoric acid (H_3PO_4) which can be used alone and also with a combination of other additives in the modification of neat binders [4,13]. Past experience has shown PPA increases the high temperature stiffness of an asphalt binders with only minor effects on the intermediate and low temperature properties [14]. One of the first patents involved adding PPA to the asphalt binders to increase viscosity without increasing the penetration for asphalt modification was in 1973 [13]. Nunez et al. Their indicated that the addition of PPA into asphalt could increase its fatigue and rutting resistance [15]. Ho et al. studied the temperature sensitivity of PPA modified asphalt with different contents. The results showed that PPA has little effect on the low-temperature properties of asphalt [16]. Zhou Yan et al. evaluated the influence of PPA on the low-temperature properties of asphalt through 5 °C ductility test and beam bending creep test, and these results demonstrated that the modification effect of PPA on the low-temperature properties of asphalt was affected by the chemical composition of asphalt [17]. Ramasamy, B.S. et al. compared the rheological properties and aging properties of polymer modified asphalt and neat asphalt, their results showed that PPA can significantly improve the stiffness modulus and aging resistance of asphalt [18]. Based on the low price of PPA, the utilization of PPA to make SBR-modified binders of lower amount of SBR percentage. Peng Liang et al. comprehensively analyzed the influence of PPA on the rheological properties of SBR modified asphalt at high and low temperatures. The outcomes revealed that the gelation of PPA can improve the bonding between asphalts and enhance their elastic properties [12]. According to John D'Angelo, the multistress creep and recovery test were conducted to evaluate the interaction of PPA and polymer modification. The testing demonstrated that there is an interaction between PPA and SBS polymers which improved the high-temperature stiffness of the SBS-PPA compound modified asphalt [13]. Darrell Fee et al. studied the rheological properties of PPA compound SBS modified asphalt at low and high temperatures. The results showed that the performance was not obvious at low temperatures [19]. In a word, researchers have reached a relatively unified understanding of the high-temperature properties of PPA modified asphalt.

Considering the characteristics of PPA, SBS and SBR modified asphalt binders, the primary objective of this study is to investigate partially replacing SBS/SBR with PPA as asphalt modifiers, which can not only achieve lower cost, but also show better performances such as rutting resistance, high-temperature properties and anti-aging. To achieve the objective, the specific work is to investigate the effects of different PPA content on high-temperature properties, low-temperature properties, temperature sensitivity and aging properties of SBS/SBR modified asphalt binders using Dynamic Shear Rheometer (DSR), Bending Beam Rheometer (BBR) and Rolling Thin Film Oven (RTFO) testing.

2. Materials and Methods

2.1. Materials

2.1.1. Asphalt

In order to analyze the effect of PPA on the properties of different neat asphalt, two types of neat asphalt widely used in China were selected in this study: DH-70# produced by SINOPEC (Beijing, China) and LH-90# produced by Liaohe Petrochemical Asphalt

Corporation (Panjin, China) (the number represents the penetration grade). The asphalt technical indexes were tested according to the relevant test methods in the "*Standard Test Methods of Bitumen and Bituminous Mixtures for Highway Engineering*" (JTG E20-2011).The results are listed in Table 1.

Table 1. Technical performance of asphalt.

	Technical Indexes	Units	Results	
			DH-70#	LH-90#
Unaged	Softening point(R&B)	°C	53.1	49.2
	Ductility (50 mm/min, 10 °C)	cm	39.7	>100
	Penetration (25 °C, 5 s, 100 g)	0.1 mm	64.2	84.6
	Solubility	%	99.9	99.7
	Density(15 °C)	g/cm^3	1.033	1.042
	Flash point	°C	279	316
RTFO aged	Mass variation	%	0.47	−0.16
	Retained penetration ratio (25 °C)	%	70.2	73.1
	Retained ductility (10 °C)	cm	14.1	8.7

2.1.2. PPA

The 110% PPA (PPA-110) was produced by Nanjing Chemical Reagent Corporation (Nanjing, China) and was selected for the preparation of modified asphalt.

2.1.3. Polymer Modifier

Two types of polymer modifiers were selected: SBS1401(YH-792, linear) produced by SINOPEC (Beijing, China) and SBR produced by Beijing Yudahang Industry and Trade Corporation (Beijng, China). The basic technical indexes are listed in Table 2.

Table 2. Basic technical indexes of SBS and SBR.

SBS1401			SBR		
Technical Index	Units	Measured Value	Technical Index	Units	Measured Value
S/B ratio	-	4/6	Granularity	Mesh No.	10~80
Oil filling rate	%	0	Bound styrene content	%	10~50
Volatile matter	%	≤0.7	Mooney viscosity	ML	45~65
Ash content	%	≤0.2	300% Constant tensile stress	MPa	15
300% Constant tensile stress	Mpa	≥3.5	Tensile strength	Mpa	≥20
Tensile strength	Mpa	24			
Elongation at break	%	730			
Shore hardness	A	85			
Melt flow rate	g/10 min	0.1~5.0			

2.1.4. Schemes of Modified Asphalt with PPA and SBS/SBR Compound Ratio

The SBS/SBR content (by weight of neat asphalt) of SBS/SBR modified asphalt is usually 3–6%, and the PPA content of polymer modified asphalt is generally 0.5–1.5% [20,21]. Different contents of SBS/SBR and PPA were used to prepare different modified asphalt binders. Based on the recommendations of the past studies, the polymer content was set as 4%, and the PPA content was 0.5%, 0.75%, 1.0%, 1.25% and 1.5% respectively. Seven test schemes are shown in Table 3.

Table 3. Compound modification schemes of PPA.

No	Schemes
S_6/R_6	6% SBS/SBR Modification
S_4/R_4	4% SBS/SBR Modification
$S_4/R_4 + P_{0.5}$	4% SBS/SBR + 0.5% PPA Modification
$S_4/R_4 + P_{0.75}$	4% SBS/SBR + 0.75% PPA Modification
$S_4/R_4 + P_{1.0}$	4% SBS/SBR + 1.0% PPA Modification
$S_4/R_4 + P_{1.25}$	4% SBS/SBR + 1.25% PPA Modification
$S_4/R_4 + P_{1.5}$	4% SBS/SBR + 1.5% PPA Modification

2.2. Specimen Preparation

The following process is adopted for the preparation of PPA and SBS/SBR compound modified asphalt:

(1) The neat asphalt was heated to 150 °C until it became fluid before mixing;
(2) The weighed SBS/SBR particles were added into the asphalt and conducted high-speed shear at the speed of 4500 rpm for 30 min;
(3) Then, the weighed PPA was added into the prepared modified asphalt and conducted high-speed shear at the speed of 4500 rpm for 30 min;
(4) The sheared PPA and SBS/SBR compound modified asphalt were put into an oven at 180 °C for swelling and developing for 1 h; therefore, PPA and SBS/SBR compound modified asphalt was prepared, then the prepared PPA and SBS/SBR compound modified asphalt was poured into corresponding containers for storage.

2.3. Experimental Program

2.3.1. DSR Test

The Physica MCR 301 DSR was used to investigate the high-temperature properties of the compound modified asphalt binders. The test modes include temperature sweep test and frequency sweep test to obtain the phase angle (δ) reflecting the viscoelasticity of asphalt and the complex shear modulus (G*) reflecting the high temperature stability of asphalt [22]. The loading frequency of temperature sweep test was 1.59 Hz [23]. Considering that in the frequency sweep test, the driving speeds of 8~16 and 80~100 km/h correspond to the frequencies of 0.15 and 1.5 Hz, respectively, in order to cover the driving speed range of vehicles specified on roads, the frequency are set at 0.01~16.0 Hz [24]. The test temperature of frequency sweep test is 60 °C. The viscosity temperature index (VTS) method and complex modulus index (GTS) method can be used to evaluate asphalt temperature sensitivity based on DSR test. In this study, G*, δ and other rheological parameters of each sample was tested to evaluate the temperature sensing performance of modified asphalt binders.

2.3.2. BBR Test

The TE-BBR was used to evaluate the low-temperature performance of asphalt binders. Small beam specimens were prepared and tested to obtain the creep rate m-value and creep stiffness (S) under constant load at different temperatures. The m-value and S were used as the evaluation indexes of asphalt low-temperature performance. The asphalt binder with low S value and high m-value showed better low-temperature properties [25].

2.3.3. RTFO Test

The Rolling thin film oven (RTFO) test was used to simulate the short-term aging of asphalt. The heating temperature was kept at 163 ± 0.5 °C, and the penetration, softening point and asphalt quality of PPA modified asphalt with different contents are tested before and after aging.

3. Results and Discussion

3.1. High-Temperature Properties

3.1.1. Temperature Sweep Test

SHRP proposes to adopt rutting factor G*/sin δ to characterize the ability of asphalt to resist high temperature rutting deformation, the larger G*/sin δ, the stronger the rutting resistance of asphalt. G* is the complex shear modulus, which is the ratio of the maximum shear stress to the maximum shear strain of the material, and represents the measurement parameter of the total deformation resistance of the material under repeated shear load; δ is the phase angle, which represents the viscoelastic characteristics of the material, most materials are between ideal elasticity and ideal viscosity. The smaller δ, the stronger the elasticity of the material.

This can be seen in Figures 1–4.

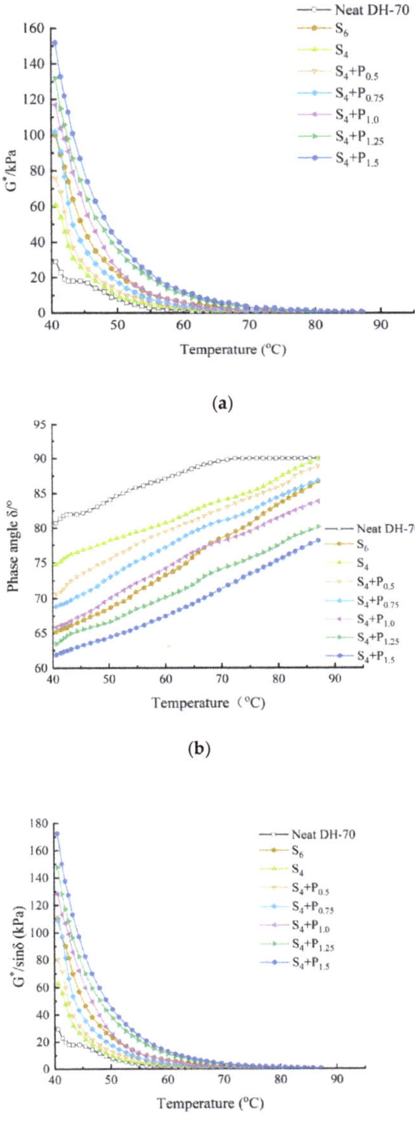

Figure 1. High-temperature properties of compound ratio of PPA–SBS and DH–70# neat asphalt: (**a**) Complex modulus-temperature results; (**b**) Phase angle-temperature results; (**c**) G*/sinδ-tenperature results.

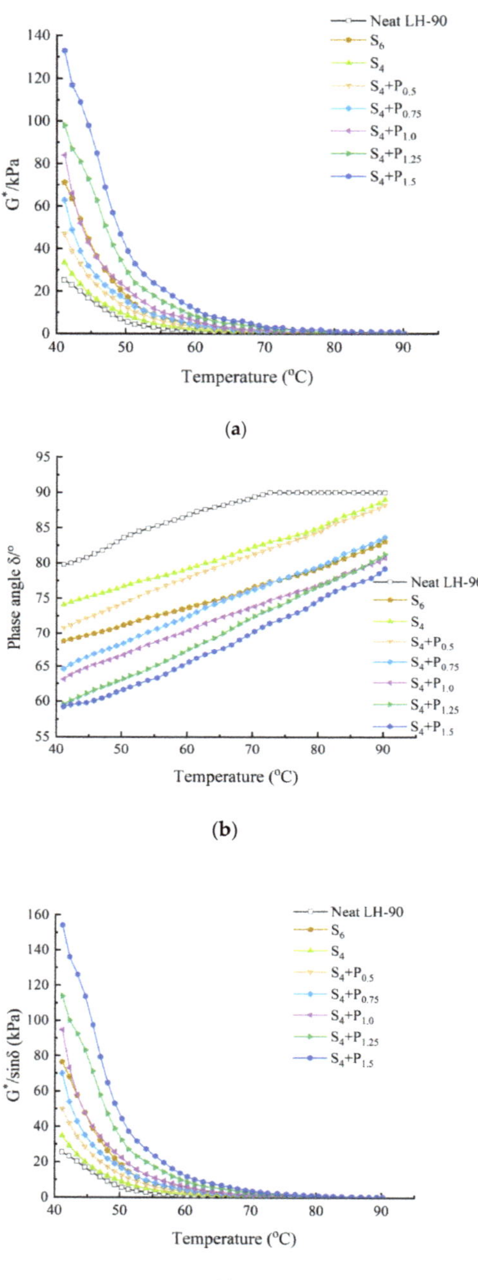

Figure 2. High-temperature properties of compound ratio of PPA–SBS and LH–90# neat asphalt: (**a**) Complex modulus-temperature results; (**b**) Phase angle-temperature results; (**c**) G*/sinδ-tenperature results.

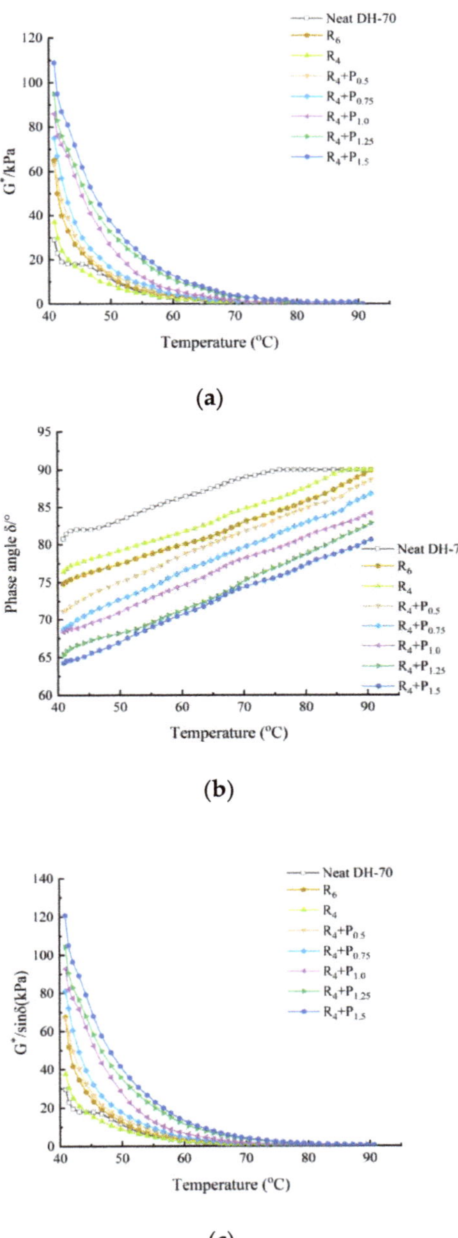

Figure 3. High-temperature properties of compound ratio of PPA–SBR and DH–70# neat asphalt: (**a**) Complex modulus-temperature results; (**b**) Phase angle-temperature results; (**c**) G*/sinδ-tenperature results.

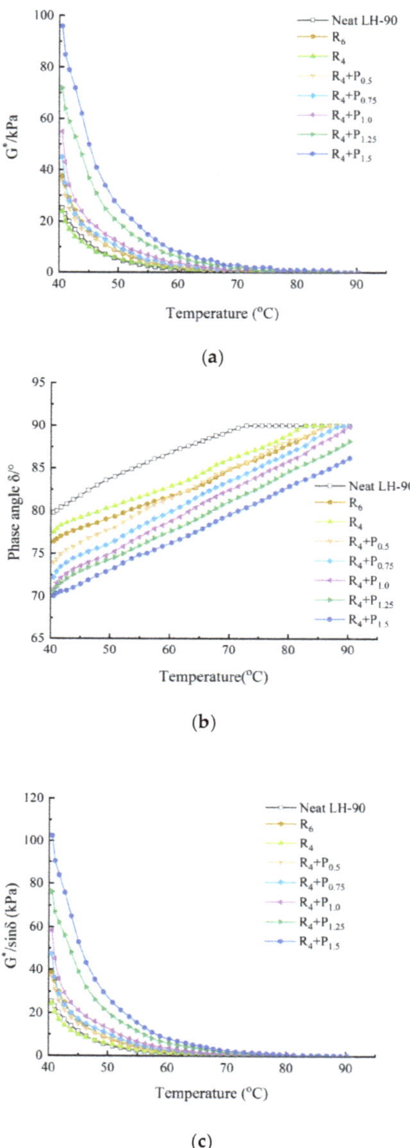

Figure 4. High-temperature properties of compound ratio of PPA–SBR and LH–90# neat asphalt: (**a**) Complex modulus-temperature results; (**b**) Phase angle-temperature results; (**c**) G*/sinδ-tenperature results.

Under the same neat asphalt type and temperature, adding PPA modifier can significantly improve the G* of modified asphalt, reduce the phase angle δ, it shows that PPA can reduce the viscosity characteristics at the same temperature, improve the elastic characteristics of modified asphalt and enhance the shear deformation resistance of asphalt at high temperature. The larger the content of PPA, the more obvious the modification effect is.

The rutting factor G*/sinδ decreases with the increase of temperature.S_6 shows higher G*/sinδ value than S_4 and R_6 shows higher G*/sinδ value than R_4. At the same temperature, by comparing the G*/sinδ values of SBS-PPA and SBR-PPA compound modified asphalt binders, it can be found that the G*/sinδ increased with the increase of PPA concentration, indicating that PPA can improve the rutting resistance ability of the SBS/SBR modified asphalt binders

The same high-temperature properties can be achieved through replacing 2% SBS with 0.75–1.0% PPA or replacing 2% SBR with 0.5% PPA, which also cost less.

3.1.2. Frequency Sweep Test

Asphalt will show different viscoelastic characteristics and mechanical states when the temperature and load frequency change. Asphalt with more stiff and elastic characteristics under high temperature conditions can have better rutting resistance.

The frequency sweep test was performed to identify the mechanical and viscoelastic behavior of asphalt binder. Figures 5–8 illustrate the complex modulus and phase angle of compound modified binders with different SBS/SBR and PPA concentrations. The G* studied by Superpave can be decomposed into two components: storage modulus G′ and loss modulus G″. The G′ represents the elastic part of asphalt binder, and the G″ represents the viscous part of asphalt binder [26]. It reflects the energy loss part of asphalt during deformation. At high temperature, the greater the G′, the stronger the high temperature stability of the binder is. At low temperature, the greater the G″, the stronger the crack resistance of the binder is [27].

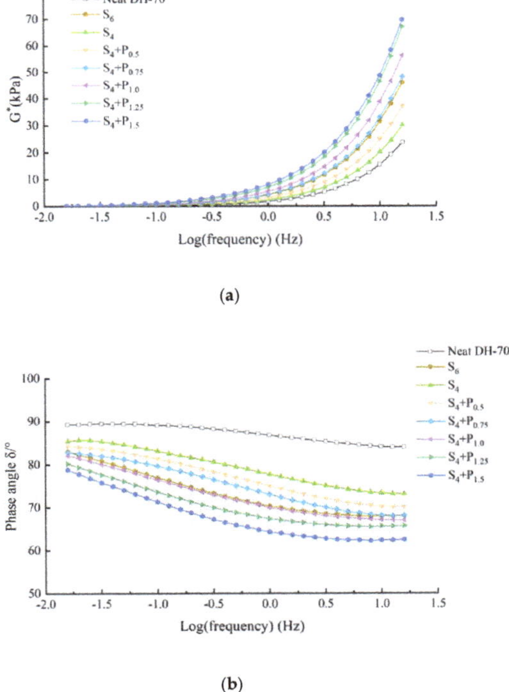

Figure 5. Complex modulus and phase angle of compound PPA–SBS modified binders of DH–70# asphalt: (**a**) Complex modulus-Log(frequency) results; (**b**) Phase angle-Log(frequency) results.

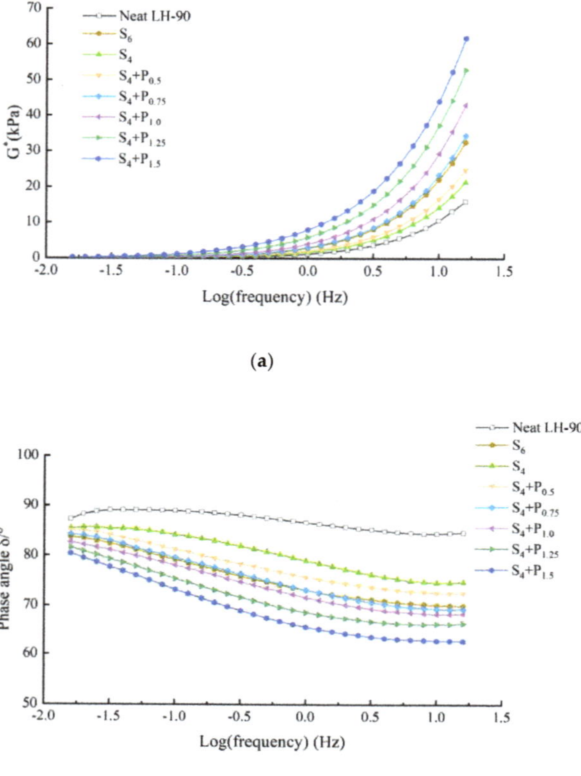

Figure 6. Complex modulus and phase angle master of compound PPA–SBS modified binders of LH–90# asphalt: (**a**) Complex modulus-Log(frequency) results; (**b**) Phase angle-Log(frequency) results.

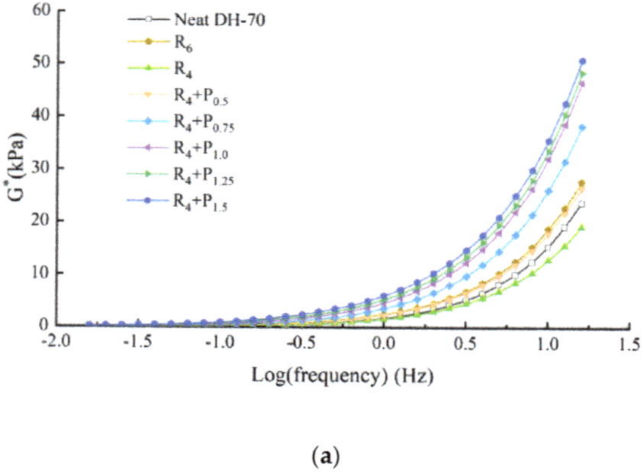

(**a**)

Figure 7. *Cont.*

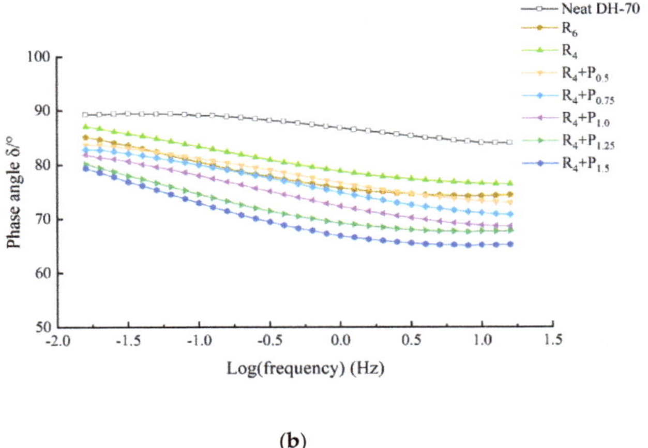

(**b**)

Figure 7. Complex modulus and phase angle master of compound PPA–SBR modified binders of DH-70# asphalt: (**a**) Complex modulus-Log(frequency) results; (**b**) Phase angle-Log(frequency) results.

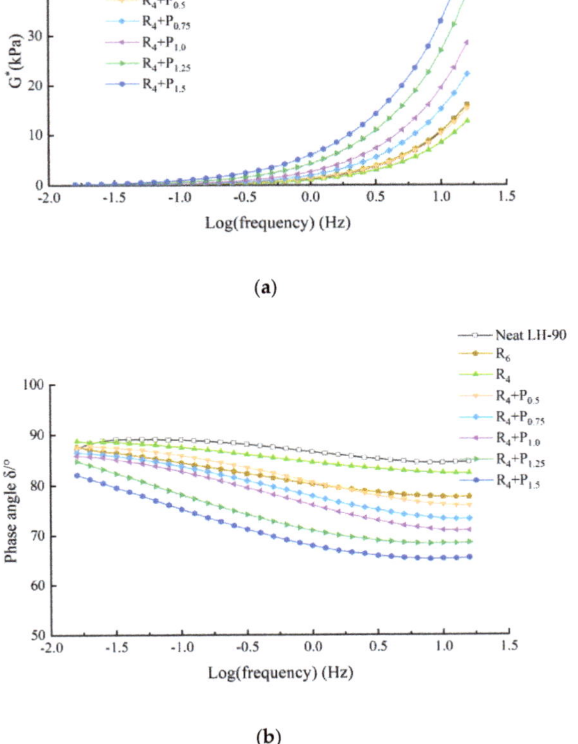

(**b**)

Figure 8. Complex modulus and phase angle master of compound PPA–SBR modified binders of LH-90# asphalt: (**a**) Complex modulus-Log(frequency) results; (**b**) Phase angle-Log(frequency) results.

This can be seen in Figures 5–8.

Complex modulus values increasing and phase angle values decreasing can be observed for all the tested asphalt binders. The curves of the two binders show increasing complex modulus and decreasing phase angle for higher content of PPA addition, which indicates transition of asphalt binder from a viscoelastic to elastic state.

At a specific frequency, by comparing the complex modulus and phase angle of SBS-PPA or SBR-PPA compound modified asphalt binders with that of 6% SBS/SBR single modified asphalt, the compound modified asphalt schemes with similar performance to these single modified asphalts can be obtained, which are listed in Table 4.

Table 4. Schemes of compound modified asphalt with similar performance to S6/R6.

Asphalt Type	Polymer Modifier Type	Complex Modulus G*/kPa	Phase Angleδ/°	Recommended Scheme
DH-70#	SBS	$S_4 + P_{0.75}$	$S_4 + P_{1.0}$	$S_4 + P_{0.75}$
	SBR	$R_4 + P_{0.5}$	$R_4 + P_{0.5\sim0.75}$	$R_4 + P_{0.5}$
LH-90#	SBS	$S_4 + P_{0.75}$	$S_4 + P_{0.75}$	$S_4 + P_{0.75}$
	SBR	$R_4 + P_{0.5}$	$R_4 + P_{0.5\sim0.75}$	$R_4 + P_{0.5}$

Compared with the base asphalt, the addition of modifier effectively improves the elastic characteristics and the total deformation resistance of asphalt under the same load frequency. Considering the price of PPA, it is more recommended to replace 2% SBS/SBR with 0.75/0.5% PPA for DH-70# asphalt, and replace 2% SBS/SBR with 0.5% PPA for LH-90# asphalt.

3.2. Low-Temperature Properties

BBR tests were conducted to evaluate the low-temperature cracking resistance of the studied modified asphalt binders. The tests were carried out at different temperatures (−12 °C, −18 °C and −24 °C) and the creep stiffness(S) and m-value were evaluated. Creep stiffness(S) represents the resistance load performance of asphalt. The creep rate m-value is the absolute value of the slope of the logarithm of the stiffness curve versus the logarithm of time and reflects the stress relaxation ability of asphalt at low temperature. The test results are shown in Figures 9–16.

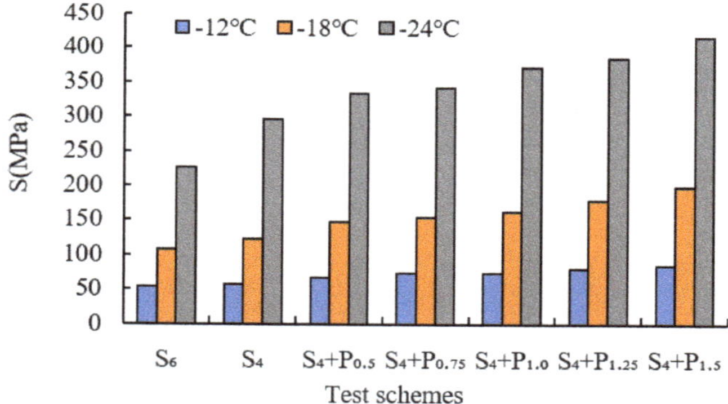

Figure 9. Creep stiffness of PPA–SBS and DH–70# asphalt.

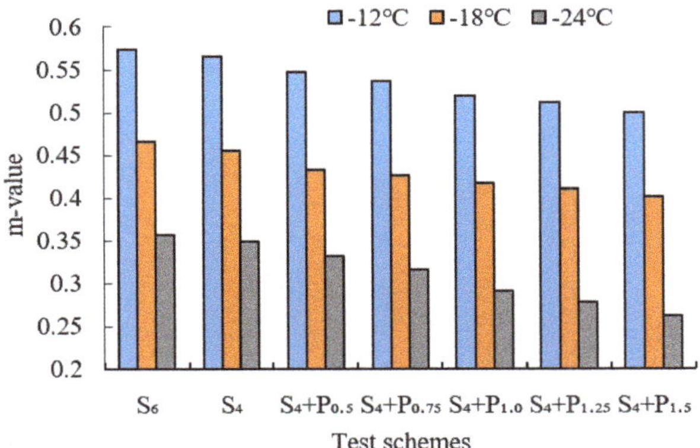

Figure 10. m-value of PPA–SBS and DH–70# asphalt.

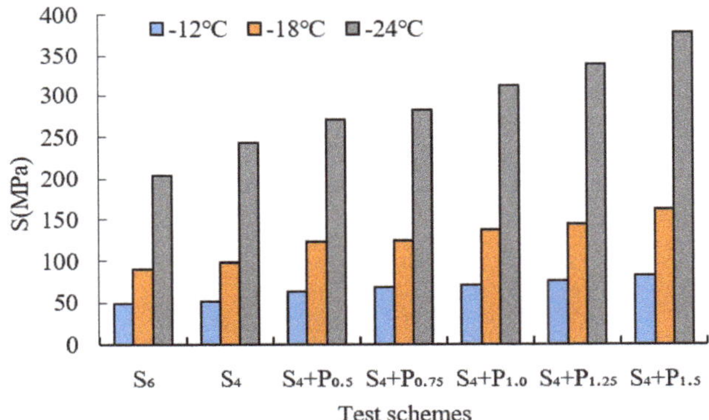

Figure 11. Creep stiffness of PPA–SBS and LH–90# asphalt.

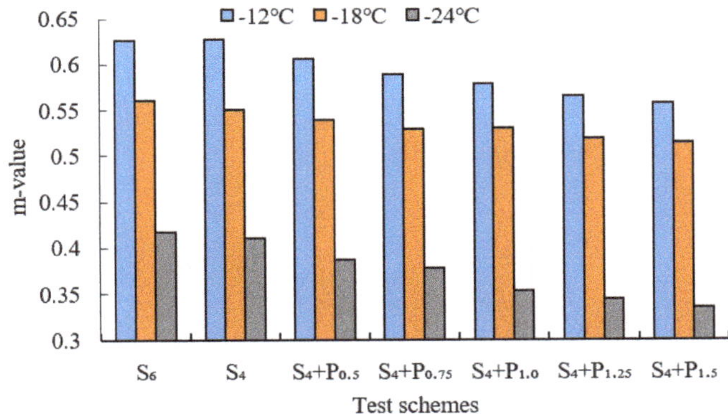

Figure 12. m-value of PPA–SBS and LH–90# asphalt.

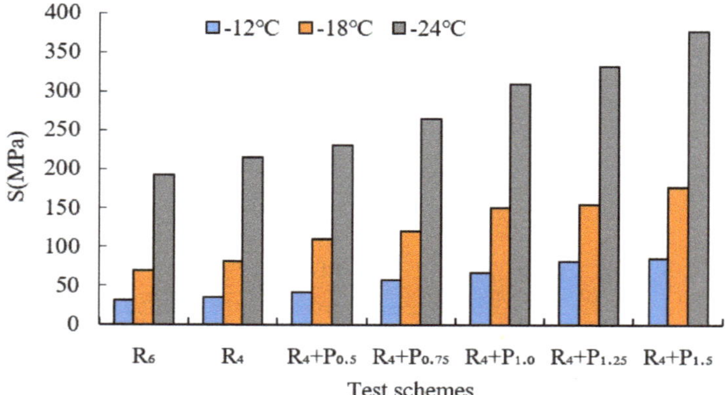

Figure 13. Creep stiffness of PPA–SBR and DH–70# asphalt.

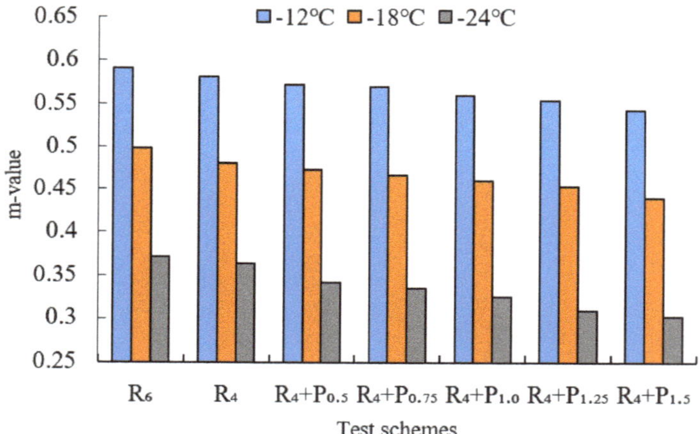

Figure 14. m-value of PPA–SBR and DH–70# asphalt.

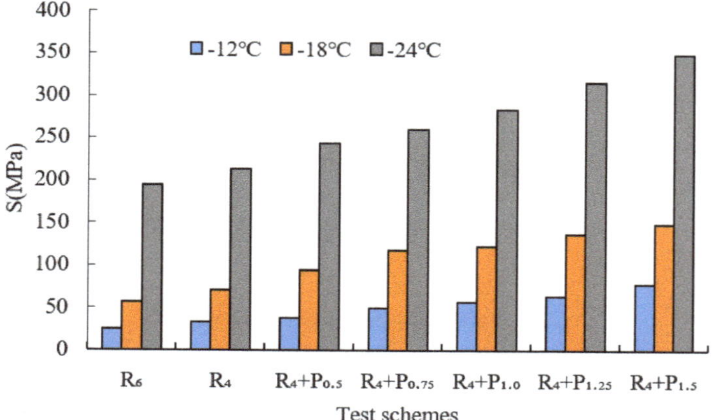

Figure 15. Creep stiffness of PPA–SBR and LH–90# asphalt.

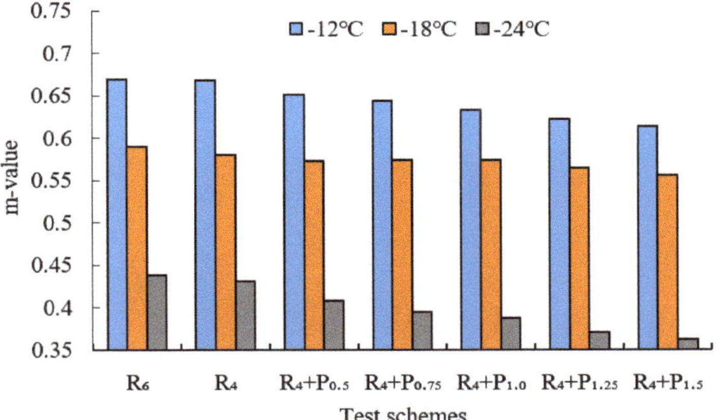

Figure 16. m-value of PPA–SBR and LH–90# asphalt.

For grading of the low-temperature performance of asphalt binders, the Superpave limits that the specified maximum creep stiffness value at 60 s is 300 MPa, i.e., S < 300 MPa, and the creep rate to a minimum of 0.30, i.e., m value > 0.3, at 60 s. As shown in Figures 9–16, The S of 6% SBS is lower (or similar) than 4% SBS, the m-value of 6% SBS is higher (or similar) than 4% SBS, and SBR has the same results. However, with the addition of PPA, the S gradually increases and the m-value gradually decreases. The results indicate that PPA has a negative impact on the low-temperature properties and stress relaxation ability of SBS/SBR modified asphalt, which significantly reduces the low-temperature ductility of SBS/SBR modified asphalt.

For SBR-PPA compound modified asphalt binders of LH-90# asphalt at −24 °C, the m-value is higher than 0.3, which indicates that the low-temperature performance grade decreased one level. The S of some asphalts are higher than 300 MPa at −24 °C with the addition of PPA. In order to guarantee the low-temperature properties of asphalt, PPA content should be no more than 1.0%, and service temperature cannot less than −18 °C.

3.3. *Temperature Sensitivity*

3.3.1. Complex Modulus Index (GTS) Method

SHRP proposes to use complex modulus index (GTS) to test the temperature sensitivity of asphalt under medium temperature to high temperature conditions [28]. GTS is calculated by fitting the double logarithm of complex modulus G* with the logarithm of test temperature T. Equation (1) presents the definition of GTS.

$$\lg[\lg(G^*)] = GTS \times \lg(T) + C \quad (1)$$

where G* is the complex shear modulus, T is the test temperature, C is the constant term in regression fitting, GTS is the complex modulus index.

Considering that the pavement surface temperature is usually up to 60 °C or higher in summer, temperature sweep tests were conducted at 58 °C, 64 °C, 70 °C and 76 °C, and the GTS results are shown in Figures 17 and 18.

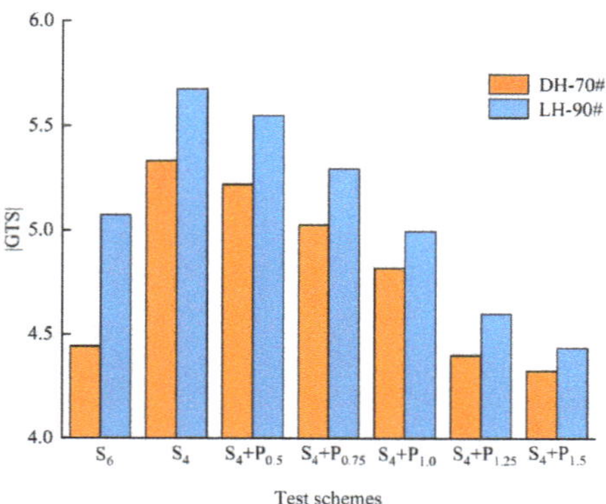

Figure 17. Effect of different PPA–SBS compound ratio on temperature sensitivity of asphalt.

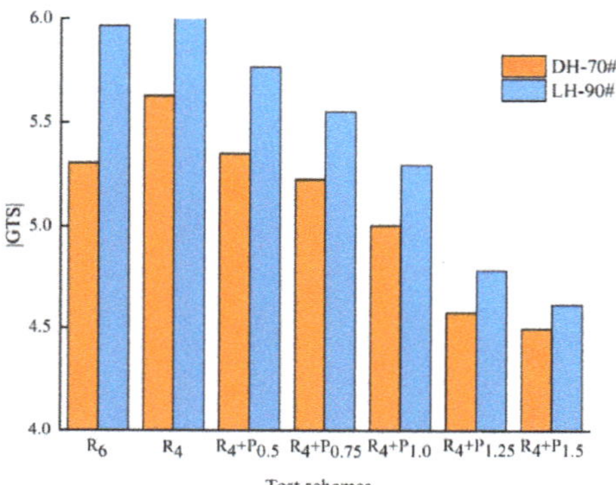

Figure 18. Effect of different PPA–SBR compound ratio on temperature sensitivity of asphalt.

As shown in Figures 17 and 18, the |GTS| of S_6 is lower than S_4, and SBR has the same results. With the addition of PPA, the |GTS| of modified asphalt gradually decreases. The results indicate that the temperature sensitivity of compound modified asphalt is improved. For DH-70 # asphalt, the | GTS | values of S_4 + $P_{1.25}$ shows similar to S_6 and R_4 + $P_{0.5}$ shows similar to R_6. For LH-90# asphalt, the |GTS| values of S_4 + $P_{1.0}$ shows similar to S_6 and R_4 +$P_{0.5}$ shows similar to R_6.

In order to decrease costs and reduce the temperature sensitivity of asphalt binder, replacing 2% SBS/SBR with 1.25%/0.5% PPA for DH-70# asphalt, and replacing 2% SBS/SBR with 1.0%/0.5% PPA for LH-90# asphalt is more recommended, respectively.

3.3.2. Viscosity Temperature Index (VTS) Method

The results of many domestic scholars show that [29], it is more appropriate to use Kelvin temperature scale to calculate viscosity temperature index VTS to evaluate the temperature sensitivity of modified asphalt. Low temperature sensitivity of asphalt binder indicates better asphalt pavement performance. The viscosity of asphalt could be obtained by Equation (2), it refers to the ability of asphalt material to resist shear deformation under the action of external force. and VTS could be obtained by Equation (3). Curve fitting is performed to show the correlation between viscosity and temperature and, from which, the slopes (VTS) and coefficient of determinations (R^2) are obtained.

$$\eta = \frac{G^*}{\omega}\left(\frac{1}{\sin\delta}\right)^{4.8628} \tag{2}$$

$$\mathrm{VTS} = \frac{\lg(\lg\eta_1) - \lg(\lg\eta_2)}{\lg T_1 - \lg T_2} \tag{3}$$

where η is the viscosity, ω is the loading frequency, ω = 10 Hz, η1 and η2 is the viscosity corresponding to the adjacent temperature, T is Kelvin temperature, T = t + 273.13, t is Celsius temperature, which is 58 °C, 64 °C, 70 °C and 76 °C.

According to Equation (2), logarithmic regression is carried out for different compound asphalt at different temperatures and corresponding viscosity. The slope of the regression equation is the viscosity temperature change rate. The greater its absolute value, the worse the temperature sensitivity of asphalt. The regression curve and regression results are shown in Figures 19–21 and Table 5.

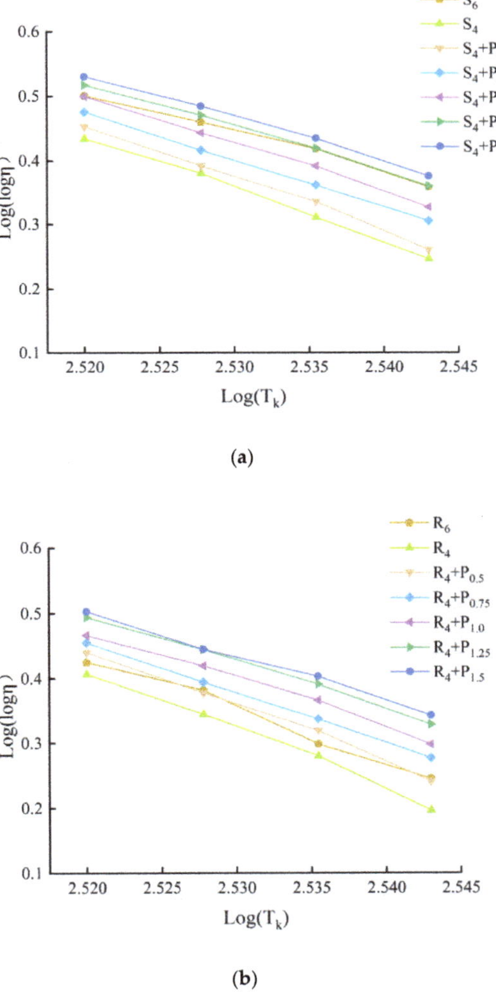

Figure 19. Double logarithm regression viscosity temperature curve of DH–70# asphalt: (**a**) PPA–SBS compound ratio; (**b**) PPA–SBR compound ratio.

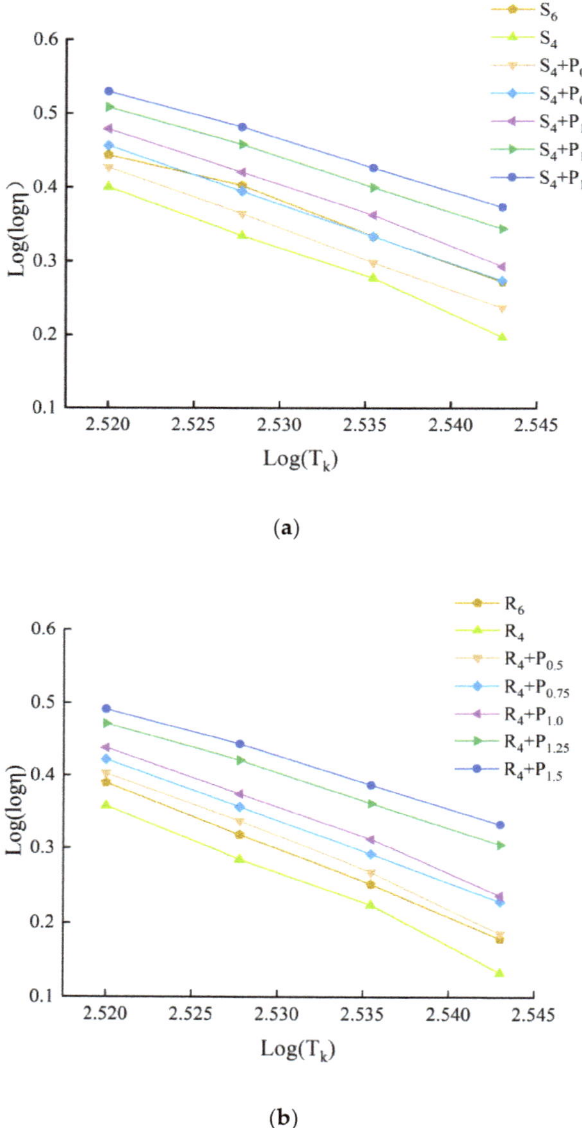

Figure 20. Double logarithm regression viscosity temperature curve of LH–90# asphalt: (**a**) PPA–SBS compound ratio; (**b**) PPA–SBR compound ratio.

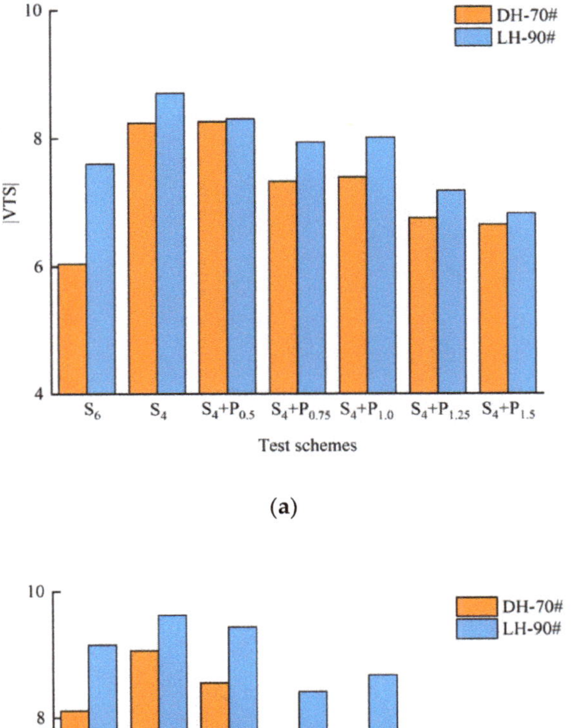

(**a**)

DH-70#
LH-90#
|VTS|
10
8
6
4
R_6
R_4
R_4+$P_{0.5}$
R_4+$P_{0.75}$
R_4+$P_{1.0}$
R_4+$P_{1.25}$
R_4+$P_{1.5}$
Test schemes

(**b**)

Figure 21. SBS/SBR-PPA variation curve of |VTS|: (**a**) |VTS| of PPA–SBS with DH–70# and LH–90#; (**b**) |VTS| of PPA–SBR with DH–70# and LH–90#.

Table 5. Double logarithm regression results of different asphalt mixtures.

Asphalt Type	No	Regression Equation	Correlation Coefficient R^2	ǀVTSǀ
DH-70#	S_6	y = −6.0467x + 15.743	$R^2 = 0.9895$	6.0467
	S_4	y = −8.2404x + 21.204	$R^2 = 0.9966$	8.2404
	$S_4 + P_{0.5}$	y = −8.2611x + 21.274	$R^2 = 0.9953$	8.2611
	$S_4 + P_{0.75}$	y = −7.3347x + 18.959	$R^2 = 0.9999$	7.3347
	$S_4 + P_{1.0}$	y = −7.4027x + 19.156	$R^2 = 0.9971$	7.4027
	$S_4 + P_{1.25}$	y = −6.7658x + 17.571	$R^2 = 0.9964$	6.7658
	$S_4 + P_{1.5}$	y = −6.6601x + 17.317	$R^2 = 0.9958$	6.6601
	R_6	y = −8.1127x + 20.876	$R^2 = 0.9853$	8.1127
	R_4	y = −9.0659x + 23.258	$R^2 = 0.9931$	9.0659
	$R_4 + P_{0.5}$	y = −8.5545x + 22.001	$R^2 = 0.995$	8.5545
	$R_4 + P_{0.75}$	y = −7.7112x + 19.888	$R^2 = 0.9998$	7.7112
	$R_4 + P_{1.0}$	y = −7.2983x + 18.864	$R^2 = 0.9911$	7.2983
	$R_4 + P_{1.25}$	y = −7.1512x + 18.519	$R^2 = 0.996$	7.1512
	$R_4 + P_{1.5}$	y = −6.8034x + 17.647	$R^2 = 0.9953$	6.8034
LH-90#	S_6	y = −7.6056x + 19.617	$R^2 = 0.9894$	7.6056
	S_4	y = −8.7073x + 22.345	$R^2 = 0.9947$	8.7073
	$S_4 + P_{0.5}$	y = −8.3059x + 21.359	$R^2 = 0.9998$	8.3059
	$S_4 + P_{0.75}$	y = −7.9426x + 20.472	$R^2 = 1$	7.9426
	$S_4 + P_{1.0}$	y = −8.017x + 20.684	$R^2 = 0.9974$	8.017
	$S_4 + P_{1.25}$	y = −7.1883x + 18.625	$R^2 = 0.9988$	7.1883
	$S_4 + P_{1.5}$	y = −6.8357x + 17.758	$R^2 = 0.9988$	6.8357
	R_6	y = −9.1534x + 23.457	$R^2 = 0.9995$	9.1534
	R_4	y = −9.6161x + 24.594	$R^2 = 0.9926$	9.6161
	$R_4 + P_{0.5}$	y = −9.4301x + 24.171	$R^2 = 0.9962$	9.4301
	$R_4 + P_{0.75}$	y = −8.4108x + 21.618	$R^2 = 1$	8.4108
	$R_4 + P_{1.0}$	y = −8.6712x + 22.292	$R^2 = 0.9969$	8.6712
	$R_4 + P_{1.25}$	y = −7.2864x + 18.836	$R^2 = 0.9986$	7.2864
	$R_4 + P_{1.5}$	y = −6.9272x + 17.95	$R^2 = 0.9987$	6.9272

Table 5 illustrates the slopes ǀVTSǀ and the coefficient of determinations (R^2), respectively. The R^2 value is greater than 0.99, which indicates that lg(lg(viscosity)) has a good linear correlation with lg(T). The larger the magnitude of the VTS value is found to be, the more susceptible the binder is to changes in viscosity with temperature [30]. It can be seen that the slope value of modified asphalt binders integrally decreases with the more PPA concentration, indicating that PPA contributes to the reduction in temperature sensitivity.

To obtain similar sensitive properties, replacing 2% SBS/SBR with 1.5%/0.5% PPA for DH-70# asphalt, and replacing 2% SBS/SBR with 0.75%/0.5% PPA for LH-90# asphalt is more recommended, respectively.

3.4. *Aging Properties*

In this study, RTFO test was used to simulate the short-term aging effect and test the performance changes of PPA compound modified asphalt before and after aging. The heating temperature was 163 ± 0.5 °C. The Retained penetration ratio, Softening point increment and Mass loss rate results are shown in Tables 6 and 7.

Table 6. Performance changes of PPA–SBS modified asphalt before and after aging.

Asphalt Type	No	Retained Penetration Ratio (%)	Softening Point Increment (°C)	Mass Loss Rate (%)
DH-70#	S_6	77.32	4.5	0.379
	S_4	73.17	5.8	0.442
	$S_4 + P_{0.5}$	75.44	4.7	0.412
	$S_4 + P_{0.75}$	76.81	4.5	0.391
	$S_4 + P_{1.0}$	78.24	4.2	0.357
	$S_4 + P_{1.25}$	79.25	3.8	0.321
	$S_4 + P_{1.5}$	80.64	3.1	0.271
LH-90#	S_6	71.25	5.1	0.418
	S_4	66.15	6.9	0.533
	$S_4 + P_{0.5}$	69.03	5.4	0.435
	$S_4 + P_{0.75}$	71.39	5.2	0.401
	$S_4 + P_{1.0}$	73.51	4.8	0.379
	$S_4 + P_{1.25}$	75.28	4.0	0.314
	$S_4 + P_{1.5}$	76.78	3.3	0.258

Table 7. Performance changes of PPA–SBR modified asphalt before and after aging.

Asphalt Type	No	Retained Penetration Ratio (%)	Softening Point Increment (°C)	Mass Loss Rate (%)
DH-70#	R_6	68.68	8.8	0.385
	R_4	64.94	10.5	0.467
	$R_4 + P_{0.5}$	67.24	9.5	0.426
	$R_4 + P_{0.75}$	68.93	8.6	0.374
	$R_4 + P_{1.0}$	71.11	6.1	0.318
	$R_4 + P_{1.25}$	74.62	4.4	0.243
	$R_4 + P_{1.5}$	75.65	3.7	0.221
LH-90#	R_6	61.26	10.4	0.422
	R_4	57.19	12.0	0.486
	$R_4 + P_{0.5}$	60.24	10.2	0.452
	$R_4 + P_{0.75}$	61.87	8.9	0.371
	$R_4 + P_{1.0}$	65.02	6.6	0.309
	$R_4 + P_{1.25}$	68.75	4.3	0.227
	$R_4 + P_{1.5}$	71.96	3.3	0.172

As shown in Tables 6 and 7, the compound modified binders with S_6 shows higher retained penetration ratio (RP) value than those with S_4, and the binder with higher SBS concentration shows lower value softening point increment (ΔS) and mass loss rate (ΔM) than that with lower concentration. Furthermore, SBR has the same results.PPA shows a significant effect on RP, ΔS and ΔM. With RTFO aging, RP value increases with the increase of PPA, but ΔS and ΔM decrease. This indicates that PPA is less prone to oxidation to improve the anti-aging properties of SBS/SBR modified asphalt.

In addition, for DH-70 # asphalt, the aging indices of $S_4 + P_{1.0}$ and S_6 show similar values and the aging indices of $R_4 + P_{0.75}$ and R_6 show similar values. For LH-90 # asphalt, the aging indices of $S_4 + P_{0.75}$ and S_6 show similar values and the aging indices of $R_4 + P_{0.75}$ and R_6 show similar values.

Based on the above statements, to obtain the similar anti-aging properties, replacing 2% SBS/SBR with 1.0%/0.75% PPA for DH-70# asphalt, and replacing 2% SBS/SBR with 0.75% PPA for LH-90# asphalt is more recommended, respectively.

4. Conclusions

To investigate the feasibility of using PPA to partially replace SBS/SBR as asphalt modifier, through a series of tests, including DSR, BBR and RTFO tests, the performance of compound asphalt is improved compared with that of 4% SBS/SBR single modified asphalt. The results show that appropriate PPA content can improve the elastic performance of SBS/SBR modified asphalt, reduce its viscosity, and improve the high-temperature deformation resistance, therefore, the rutting resistance of compound asphalt is improved. After adding PPA, |GTS| decreases significantly and |VTS| decreases gradually, indicating that PPA reduces the temperature sensitivity of modified asphalt.

In addition, PPA can inhibit the thermal oxidative aging of polymer modified asphalt, thereby enhancing the anti-aging properties, and the greater the amount of PPA, the more significant the modification effect is.

Overall, the high temperature, temperature sensitivity and anti-aging properties of polymer modified asphalt were improved after adding PPA, and because of its low price, the polymer modified asphalt can reduce some polymer content by adding PPA. PPA may have an adverse effect on the stiffness and stress relaxation of asphalt binder, but it can meet the specification requirements in the area where the average temperature in winter is not lower than −18 °C. Considering those performance and economic benefits, the recommended scheme of PPA-SBS compound modified asphalt is 4% SBS + 0.75–1.0% PPA, and the recommended scheme of PPA-SBR compound modified asphalt is 4% SBR + 0.5–0.75% PPA, which is equivalent to the performance of 6% SBS/SBR single modified asphalt. Following research work should include investigating the combinations of PPA with other modifiers and neat asphalt, as well as the verification of field correlation, so as to prepare asphalt with better performance and wider application range.

Author Contributions: Conceptualization, J.W.; methodology, J.W. and Y.Z.; validation, S.S.; investigation, S.S. and Z.C.; data curation, Z.C.; writing—original draft preparation, Y.Z. and X.D.; writing—review and editing, F.Y. and Z.P. All authors have read and agreed to the published version of the manuscript.

Funding: The research was supported by National Natural Science Foundation of China (grant numbers, 52108396); and Scientific Research Project of Hunan Education Department (grant numbers, 19B032) and the Natural Science Foundation of Hunan Province (grant numbers, 2020JJ4615), and postgraduate Scientific Research Innovation Project of CSUST (SJCX202104).

Institutional Review Board Statement: Not applicable.

Informed Consent Statement: Not applicable.

Data Availability Statement: Data is contained within the article.

Conflicts of Interest: The authors declare no conflict of interest.

References

1. Pei, X.; Fan, W. Effects of Amorphous Poly Alpha Olefin (APAO) and Polyphosphoric Acid (PPA) on the Rheological Properties, Compatibility and Stability of Asphalt Binder. *Materials* **2021**, *14*, 2458. [CrossRef] [PubMed]
2. Li, B.; Li, X.; Kundwa, M.J.; Li, Z.; Wei, D. Evaluation of the adhesion characteristics of material composition for polyphosphoric acid and SBS modified bitumen based on surface free energy theory. *Constr. Build. Mater.* **2021**, *266*, 121022. [CrossRef]
3. Behnood, A.; Olek, J. Rheological properties of asphalt binders modified with styrene-butadiene-styrene (SBS), ground tire rubber (GTR), or polyphosphoric acid (PPA). *Constr. Build. Mater.* **2017**, *151*, 464–478. [CrossRef]
4. Ameli, A.; Khabbaz, E.H.; Babagoli, R.; Norouzi, N.; Valipourian, K. Evaluation of the effect of carbon nano tube on water damage resistance of Stone matrix asphalt mixtures containing polyphosphoric acid and styrene butadiene rubber. *Constr. Build. Mater.* **2020**, *261*, 119946. [CrossRef]
5. Babagoli, R.; Ameli, A.; Shahriari, H. Laboratory evaluation of rutting performance of cold recycling asphalt mixtures containing SBS modified asphalt emulsion. *Pet. Sci. Technol.* **2016**, *34*, 309–313. [CrossRef]

6. Imaninasab, R. Rutting resistance and resilient modulus evaluation of polymer-modified SMA mixtures. *Pet. Sci. Technol.* **2016**, *34*, 1483–1489. [CrossRef]
7. Li, Y.; Hao, P.; Zhao, C.; Ling, J.; Wu, T.; Li, D.; Liu, J.; Sun, B. Anti-rutting performance evaluation of modified asphalt binders: A review. *J. Traffic Transp. Eng. Engl. Ed.* **2021**, *8*, 339–355. [CrossRef]
8. Porto, M.; Caputo, P.; Loise, V.; Eskandarsefat, S.; Teltayev, B.; Oliviero Rossi, C. Bitumen and Bitumen Modification: A Review on Latest Advances. *Appl. Sci.* **2019**, *9*, 742. [CrossRef]
9. Xu, J.; Pei, J.; Cai, J.; Liu, T.; Wen, Y. Performance improvement and aging property of oil/SBS modified asphalt. *Constr. Build. Mater.* **2021**, *300*, 23735. [CrossRef]
10. Zhang, Q.; Fan, W.; Wang, T.; Nan, G. Studies on the temperature performance of SBR modified asphalt emulsion. In Proceedings of the 2011 International Conference on Electric Technology and Civil Engineering (ICETCE), Lushan, China, 22–24 April 2011; IEEE: Manhattan, NY, USA, 2011.
11. Zhang, F.; Yu, J. The research for high-performance SBR compound modified asphalt. *Constr. Build. Mater.* **2010**, *24*, 410–418. [CrossRef]
12. Liang, P.; Liang, M.; Fan, W.; Zhang, Y.; Qian, C.; Ren, S. Improving thermo-rheological behavior and compatibility of SBR modified asphalt by addition of polyphosphoric acid (PPA). *Constr. Build. Mater.* **2017**, *139*, 183–192. [CrossRef]
13. D'Angelo, J.A. Workshop Summary. In *Polyphosphoric Acid Modification of Asphalt Binders*; Transportation Research Board: Washington, DC, USA, 2009.
14. Baumgardner, G.L.; Masson, J.F.; Hardee, J.R.; Menapace, A.M.; Williams, A.G. Polyphosphoric Acid Modified Asphalt: Proposed Mechanisms. *Road Mater. Pavement Des.* **2005**, *74*, 283–305.
15. Nuñez, J.Y.; Domingos, M.D.; Faxina, A.L. Susceptibility of low-density polyethylene and polyphosphoric acid-modified asphalt binders to rutting and fatigue cracking. *Constr. Build. Mater.* **2014**, *73*, 509–514. [CrossRef]
16. Man Sze Ho, S.; Zanzotto, L.; MacLeod, D. Impact of Different Types of Modification on Low-Temperature Tensile Strength and Tcritical of Asphalt Binders. *Transp. Res. Rec.* **2002**, *1810*, 1–8. [CrossRef]
17. Baldino, N.; Gabriele, D.; Rossi, C.O.; Seta, L.; Lupi, F.R.; Caputo, P. Low temperature rheology of polyphosphoric acid (PPA) added bitumen. *Constr. Build. Mater.* **2012**, *36*, 592–596. [CrossRef]
18. Naresh Baboo, R. *Effect of Polyphosphoric Acid on Aging Characteristics of PG 64-22 Asphalt Binder*; University of North Texas: Denton, TX, USA, 2010.
19. Fee, D.; Maldonado, R.; Reinke, G.; Romagosa, H. Polyphosphoric Acid Modification of Asphalt. *Transp. Res. Rec. J. Transp. Res. Board* **2010**, *2179*, 49–57. [CrossRef]
20. Fini, E.H.; Al-Qadi, I.L.; You, Z.; Zada, B.; Mills-Beale, J. Partial replacement of asphalt binder with bio-binder: Characterisation and modification. *Int. J. Pavement Eng.* **2012**, *13*, 515–522. [CrossRef]
21. Sengoz, B.; Topal, A.; Isikyakar, G. Morphology and image analysis of polymer modified bitumens. *Constr. Build. Mater.* **2009**, *23*, 1986–1992. [CrossRef]
22. Rossi, C.O.; Spadafora, A.; Teltayev, B.; Izmailova, G.; Amerbayev, Y.; Bortolotti, V. Polymer modified bitumen: Rheological properties and structural characterization. *Colloids Surf. A Physicochem. Eng. Asp.* **2015**, *480*, 390–397. [CrossRef]
23. *JTG E20-2011*; Standard Test Methods of Bitumen and Bituminous Mixtures for Highway Engineering. Ministry of Transport of the People's Republic of China: Beijing, China, 2011.
24. Bahia, H.U.; Hanson, D.I.; Zeng, M.; Zhai, H.; Khatri, M.A.; Anderson, R.M. Characterization of modified asphalt binders in Superpave mix design. In *Characterization of Modified Asphalt Binders in Superpave Mix Design*; Transportation Research Board: Washington, DC, USA, 2001.
25. Wang, T.; Xiao, F.; Amirkhanian, S.; Huang, W.; Zheng, M. A review on low temperature performances of rubberized asphalt materials. *Constr. Build. Mater.* **2017**, *145*, 483–505. [CrossRef]
26. Strategic Highway Research Program (SHRP). *Superior Performing Asphalt Pavements (Superpave): The Product of the SHRP Asphalt Research Program (SHRP-A-410)*; Strategic Highway Research Program: Washington, DC, USA, 1994.
27. YongNing, W. Effect of admixtures on properties of SBS modified asphalt mixed with polyphosphate. *J. China Foreign Highw.* **2018**, 264–268. [CrossRef]
28. Local Standards of Hebei Province. *Technical Standard for Tire Crumb Asphalt Rubber and Asphalt Rubber Hot Mixture (DB13/T 1013-2009)*; Local Standards of Hebei Province: HuBei, China, 2009.
29. Chen, H.-X.; Li, N.-L.; Zhang, Z.-Q. Temperature Susceptibility Analysis of Asphalt Binders. *J. Chang. Univ. Nat. Sci. Ed.* **2009**, *1*, 505–509.
30. Rasmussen, R.O.; Lytton, R.L.; Chang, G.K. Method to Predict Temperature Susceptibility of an Asphalt Binder. *J. Mater. Civ. Eng.* **2002**, *14*, 246–252. [CrossRef]

materials

MDPI

Article

Performance Evaluation and Structure Optimization of Low-Emission Mixed Epoxy Asphalt Pavement

Yulou Fan , You Wu, Huimin Chen, Shinan Liu, Wei Huang, Houzhi Wang and Jun Yang *

School of Transportation, Southeast University, Nanjing 211189, China
* Correspondence: yangjun@seu.edu.cn

Abstract: Epoxy asphalt concrete (EAC) has excellent properties such as high strength, outstanding thermal stability, and great fatigue resistance, and is considered to be a long-life pavement material. Meanwhile, the low initial viscosity of the epoxy components provides the possibility to reduce the mixing temperature of SBS-modified asphalt. The purpose of this study is to verify the feasibility of low-emission mixing of SBS-modified epoxy asphalt and to compare the mechanical responses in several typical structures with EAC, in order to perform structure optimization for practical applications of EAC. In this paper, the Brookfield rotational viscosity test was conducted to investigate the feasibility of mixing SBS-modified epoxy asphalt at a reduced temperature. Subsequently, the dynamic modulus tests were carried out on EAC to obtain the Prony series in order to provide viscoelastic parameters for the finite element model. Six feasible pavement structures with EAC were proposed, and a finite element method (FEM) model was developed to analyze and compare the mechanical responses with the conventional pavement structure. Additionally, the design life was predicted and compared to comprehensively evaluate the performance of EAC structures. Finally, life cycle assessment (LCA) on carbon emissions was developed to explore the emission reduction effect of the epoxy asphalt pavement. The results indicate that the addition of epoxy components could reduce the mixing temperature of SBS-modified asphalt by 30 °C. The proper use of EAC can significantly improve the mechanical condition of the pavement and improve its performance and service life. It is recommended to choose S5 (with EAC applied in the middle-lower layer) as the optimal pavement structure, whose allowable load repetitions to limit fatigue cracking were more than 1.7 times that of conventional pavements and it has favorable rutting resistance as well. The LCA results show that in a 25-year life cycle, the carbon emissions of epoxy asphalt pavements could be reduced by 29.8% in comparison to conventional pavements.

Keywords: epoxy asphalt concrete (EAC); low-emission mixed asphalt; pavement structure; finite element method (FEM); mechanical response; life cycle assessment (LCA)

Citation: Fan, Y.; Wu, Y.; Chen, H.; Liu, S.; Huang, W.; Wang, H.; Yang, J. Performance Evaluation and Structure Optimization of Low-Emission Mixed Epoxy Asphalt Pavement. *Materials* **2022**, *15*, 6472. https://doi.org/10.3390/ma15186472

Academic Editors: Di Wang, Pengfei Liu, Michael Wistuba and Markus Oeser

Received: 17 August 2022
Accepted: 15 September 2022
Published: 18 September 2022

1. Introduction

With the increasing traffic volume, heavier traffic loads, and higher requirements for pavement, the conventional ordinary asphalt pavement has gradually failed to meet the demands of roads. Numerous pavement distresses, such as rutting, fatigue failure, and thermal cracking, frequently occur before the expected service life; thus, there is a demand for new durable pavement materials. Epoxy asphalt (EA), known as a thermosetting polymer-modified asphalt, consists of thermosetting epoxy resin, curing agent, asphalt binder, and other additives [1]. After mixing and curing, the epoxy resin and the curing agent undergo a cross-linking reaction, forming irreversible cross-linked networks, which greatly restrict the mobility of asphalt molecules, giving the epoxy asphalt thermosetting characteristics [2–4]. The cured epoxy asphalt concrete has changed the thermoplastic nature of conventional asphalt and has outstanding physical and mechanical properties such as high strength and stiffness, excellent thermal stability, good corrosion resistance, and superior fatigue resistance [5,6]. Consequently, EAC has been widely used as paving materials for airfields and orthotropic steel bridge decks around the world [7–9]. In China, epoxy asphalt was used for the first time on the Nanjing Second Yangtze River Bridge in 2000, where a 50 mm thick dense-graded EAC was paved on the deck [10,11]. In the over 20 years of service so far, the EAC on the Nanjing Second Yangtze River Bridge has shown an amazingly outstanding performance, which has led to the widespread application of EAC on orthotropic steel bridge decks built in China, including the Nansha Bridge (main span 1688 m), the Sutong Yangtze River Bridge (main span 1088 m), and the Xihoumen Bridge (main span 1650 m) [12,13].

Meanwhile, due to the outstanding properties and satisfactory performance in bridge deck paving, researchers have started to experiment with EAC in roadway pavement applications, in which EAC is considered to possess the potential to be a long-life pavement material [14–16]. In most performance indicators, epoxy asphalt outperforms ordinary asphalt by at least two times and even more than 10 times in some cases. Lu and Bors [13] reviewed the potential application of epoxy asphalt in roadway engineering, and pointed out that EAC has good fatigue resistance, high-temperature stability, and strong resistance to aging and chemical attack. Xie et al. [17] summarized the factors affecting the mechanical properties, bonding properties, and thermal stability of epoxy asphalt and analyzed the mechanism of phase separation and its effect on the mechanical properties of epoxy asphalt. Chen et al. [16] investigated the physical, chemical, and mechanical properties of EAC and found that it exhibited remarkably higher compressive strength, rutting resistance, durability, and water resistance; however, its cracking resistance at low temperatures was slightly reduced. In other studies, rheological characteristics were studied as an important indicator affecting deformation and construction, and it was found that epoxy asphalt has better rheological properties than conventional modified asphalt, with favorable fluidity and low viscosity before curing, which can effectively fill the voids in the mixture and make it more uniform [7,14,18]. Additionally, the fatigue properties of EAC reflect the durability of the road and is directly related to the service life of asphalt pavements [19]. Due to the high stiffness and bonding properties of epoxy asphalt, EAC has superior fatigue resistance when compared to conventional asphalt concrete, especially at high epoxy resin dosages [20,21]. Li et al. [22] explored the fatigue performance of porous epoxy asphalt mixtures and found that epoxy asphalt binders can significantly improve the resistance to fatigue damage of porous asphalt mixtures but it is relatively sensitive to strain levels. Furthermore, the addition of glass or mineral fibers can further strengthen the fatigue resistance of EAC [23].

Additionally, because the initial viscosity of epoxy components is relatively low compared to asphalt, the viscosity of the epoxy asphalt will gradually increase from a lower value during the curing reaction [14]. This property provides the epoxy asphalt with better workability, especially for SBS-modified asphalt, which allows EAC to be mixed at a lower temperature, thus reducing emissions in the production process. Moreover, due to its excellent performance, EAC generally has a much longer life than the conventional asphalt mixtures and can meet the increasing requirements of low-emission, "green", and long-life pavements [13]. Furthermore, the most effective measure for emission reduction is to improve the service life of the pavement. From this perspective, epoxy asphalt pavement may also have a more favorable emission reduction effect. In summary, the feasibility of epoxy asphalt for roadway pavement has been demonstrated by numerous researchers based on extensive research and test results.

However, besides the superb properties of the epoxy asphalt itself, pavement structure is also crucial for the application and field performance of new pavement materials [24]. Recently, the finite element method (FEM) has become a popular numerical simulation method for calculating the mechanical response of numerous pavement structures, which can adapt to complex geometries and boundary conditions and is applicable to solve a variety of linear and nonlinear problems [25,26]. Guo et al. [27] explored the dynamic response of five typical pavement structures based on ABAQUS and found that the stresses and strains in the full-depth pavements were less than those in semi-rigid base pavements, but the rutting distress was the opposite. Yu-Shan and Shakiba [28] explored the performance of asphalt pavement under different asphalt concrete and base thicknesses, base and subgrade permeability, and various saturations by using a three-dimensional FEM model and found that the pavement structure has a significant effect on the peeling phenomenon of the asphalt layer. In other studies, researchers have investigated various pavement performance qualities under different structures using FEM, such as mechanical responses, rutting, top-down cracking, and reflective cracking [29–32]. In summary, the structure combination will greatly influence the mechanical responses and the field performance of the pavement. Thus, the pavement performance of EAC should be evaluated in order to determine the optimal structure before it is widely used in the field. Nevertheless, there are research gaps for the performance evaluation and structure optimization of EAC, which limits its application in engineering practice.

The objective of this study is to verify the low-emission mixed characteristics of epoxy asphalt through viscosity tests, compare the performance of EAC in several typical structures, and select the optimal pavement structure considering the mechanical responses and the design life, in order to provide a reference for the practical application of EAC. Meanwhile, the emission reduction effect of the epoxy asphalt pavement is explored using life cycle assessment (LCA).

2. Materials and Methods

2.1. Experimental Design

2.1.1. Materials

Styrene–butadiene–styrene (SBS)-modified epoxy asphalt was used in this study. The matrix asphalt was a 60/80 asphalt with the performance grade of PG 64–22, and the SBS was SBS (I-D) produced in China, with a dosage of 5.0%. The epoxy resin and curing agent were produced in China, and the ratio of epoxy resin to curing agent is 1:1. The basic properties and technical indicators of asphalt binder, epoxy resin, and curing agent were tested according to the Chinese specification "Standard Test Methods of Bitumen and Bituminous Mixtures for Highway Engineering" (JTG E20-2011) [33] and shown in Table 1. The ratio of the mass of the epoxy resin components (including the epoxy resin and curing agent) to that of the asphalt was chosen as 3:7, that is, the dosage of epoxy resin components was 30% of the entire epoxy asphalt binder by weight. The gradation of EAC-13, i.e., epoxy asphalt concrete with a nominal maximum aggregate size (NMAS) of 13 mm, is the same as the typical gradation of AC-13 (asphalt concrete with an NMAS of 13 mm), as shown in Table 2. According to the "Technical Specifications for Construction of Highway Asphalt Pavements" (JTG F40-2004) [34], the optimal binder–aggregate ratio for EAC-13 was determined to be 6.3% based on the Marshall method. Two EAC-13 specimens with a height of 170 mm and a diameter of 150 mm were prepared on a shear gyratory compactor, which were cored and cut for dynamic modulus testing with a height of 150 mm and a diameter of 100 mm.

Table 1. Properties of asphalt binder, epoxy resin, and curing agent [33].

Materials	Indicators	Values	Materials	Indicators	Values
Asphalt binder	Penetration (0.1 mm)	51.5	Epoxy resin	Density (g/cm^3)	1.13
	Softening point (°C)	81.7		Viscosity (Pa•s, 25 °C)	4.1
	Ductility (cm, 5 °C)	29.2	Curing agent	Density (g/cm^3)	0.92
	Kinematic Viscosity (Pa•s, 135 °C)	2		Viscosity (Pa•s, 25 °C)	0.2

Table 2. Gradation of EAC-13.

Sieve size (mm)	16	13.2	9.5	4.75	2.36	1.18	0.6	0.3	0.15	0.075
Passing Percentage (%)	100	95	76.5	53	37	26.5	19	13.5	10	6

2.1.2. Viscosity Tests

The viscosity–temperature relationship of the asphalt binder is the key to control the workability of the asphalt mixture. The initial viscosity of the epoxy asphalt is reduced after mixing due to the lower viscosity of the epoxy components compared to the asphalt binder. However, it will gradually increase as the curing reaction proceeds. Therefore, the viscosity of epoxy asphalt as a function of time and temperature needs to be investigated to determine its rheological properties and mixing temperatures.

The viscosity of SBS-modified epoxy asphalt was tested at different temperatures (120, 130, 140, and 150 °C) using a Brookfield rotary viscometer (Model DV2T, Brookfield Engineering Inc., Middleboro, MA, USA). The test time was 120 min at a speed of 100 rpm, which was changed to 50 rpm for 120 °C and 130 °C after 80 min in order to keep the rotational torque within the normal range.

2.1.3. Dynamic Modulus Tests

As a typical viscoelastic material, the mechanical properties of asphalt mixtures exhibit dependency on time and temperature. In order to evaluate the viscoelastic properties of EAC, dynamic modulus tests were carried out by the universal testing machine (UTM-25, IPC Global, CONTROLS, Milan, Italy) using the semi-sine wave loading form, according to the "Standard Test Methods of Bitumen and Bituminous Mixtures for Highway Engineering" (JTG E20-2011) [33]. The dynamic moduli were calculated under 6 test frequencies (0.1, 0.5, 1, 5, 10, and 25 Hz) and 5 test temperatures (0, 10, 25, 40, and 55 °C). The dynamic modulus master curve at reference temperature of 25 °C was constructed using the Sigmoid model (Figure 1), and the time–temperature shift factor α_T was calculated using the Williams–Landel–Ferry (WLF) equation.

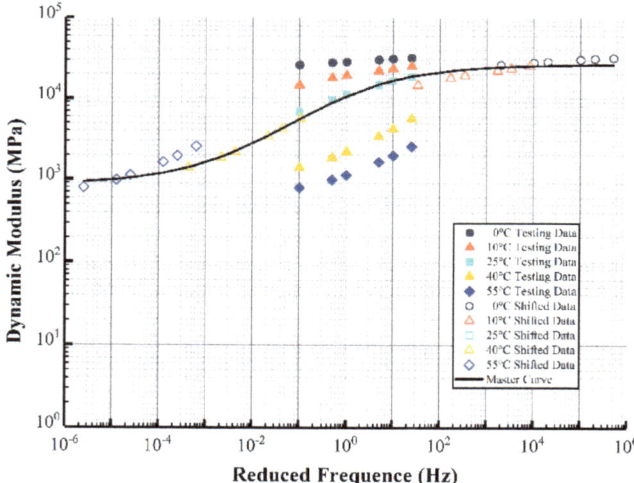

Figure 1. The dynamic modulus master curve of EAC-13.

The viscoelastic properties of asphalt mixtures in ABAQUS can be characterized by the Prony series. The fitted Prony series parameters of EAC at 25 °C with 11 relaxation components were used for modeling, as shown in Table 3. Due to the strong bonding of epoxy asphalt, gradation has a relatively reduced effect on the properties of EAC, so the same Prony series was applied for EAC-13 and EAC-16. The Prony series of the other asphalt mixtures used in the simulation were determined according to reference [31].

Table 3. Prony series parameters of EAC and other asphalt mixtures.

EAC-13/16		SMA-13 *		SUP-20 **		SUP-25 **	
τ_i	G_i or K_i	τ_i	G_i or K_i	τ_i	G_i or K_i	τ_i	G_i or K_i
0.00001	0.02538	0.00001	0.17312	0.00001	0.16223	0.00001	0.16223
0.0001	0.01741	0.0001	0.21719	0.0001	0.20595	0.0001	0.20595
0.001	0.05832	0.001	0.20891	0.001	0.20636	0.001	0.20636
0.01	0.11526	0.01	0.14639	0.01	0.15432	0.01	0.15432
0.1	0.19804	0.1	0.07766	0.1	0.08746	0.1	0.08746
1	0.25276	1	0.03453	1	0.04104	1	0.04104
10	0.17329	10	0.01363	10	0.01682	10	0.01682
100	0.07532	100	0.00527	100	0.00671	100	0.00671
1000	0.03208	1000	0.00179	1000	0.00234	1000	0.00234
10,000	0.01114	10,000	0.00078	10,000	0.00092	10,000	0.00092
100,000	0.00575						

* SMA-13 = stone mastic asphalt with NMAS of 13 mm. ** SUP-20/25 = Superpave with NMAS of 20/25 mm.

2.2. *Finite Element Modeling of Different Structures*

2.2.1. Pavement Structures

Based on the previous experience, 6 feasible pavement structures with EAC (S1–S6) were proposed in this paper, and 1 conventional structure was selected as the control group (S0). The pavement structure is composed of an asphalt surface layer with a total thickness of 26 cm, a cement-treated base (CTB, 38 cm), a lime-fly ash stabilized soil sub-base (LFS, 20 cm), and soil (SG, 8 m). The pavement width was selected as 42 m, based on a two-way eight-lane highway with a designated speed of 120 km/h (each lane is 3.75 m in width with a 4.5 m central reservation and 3.75 m shoulders on both sides). All pavement structures are summarized in Table 4.

Table 4. Summary of 7 pavement structures.

Structure Number	Pavement Structure
S0 (control group)	4 cm SMA-13 + 6 cm SUP-20 + 6 cm SUP-20 + 10 cm SUP-25
S1	4 cm SMA-13 + ***6 cm EAC-16*** + 6 cm SUP-20 + 9 cm SUP-25 + 1 cm AR-SAMI *
S2	4 cm SMA-13 + ***6 cm EAC-16*** + 6 cm SUP-20 + 10 cm SUP-25
S3	4 cm SMA-13 + ***6 cm EAC-16*** + ***6 cm EAC-16*** + 9 cm SUP-25 + 1 cm AR-SAMI
S4	***4 cm EAC-13*** + ***6 cm EAC-16*** + 6 cm SUP-20 + 10 cm SUP-25
S5	4 cm SMA-13 + 6 cm SUP-20 + ***6 cm EAC-16*** + 9 cm SUP-25 + 1 cm AR-SAMI
S6	***4 cm EAC-13*** + 6 cm SUP-20 + 6 cm SUP-20 + 10 cm SUP-25

* AR-SAMI = asphalt rubber stress-absorbing membrane interlayer, applied between SUP-25 and CTB.

2.2.2. Simulation Method

The two-dimensional ABAQUS models of 7 pavement structures with a size of 42 m × 8.84 m were established, and the material parameters are shown in Table 5. Additionally, the viscoelastic properties were taken into account for the asphalt layers according to the Prony series fitted in Section 2.1.2. The standard axle load BZZ-100 specified in Chinese "Specifications for Design of Highway Asphalt Pavement" (JTG D50-2017) [35] was used to convert the loading area. The double-circle vertical uniform load was set at the top of the pavement, with the equivalent circle radius obtained by Equation (1), and the vertical compressive stress was 0.7 MPa. The distance between the two equivalent circles is 0.1065 m; thus, the width of the loading area, d, is 0.213 m, with an interval of 0.1065 m. The traffic loads were applied at the center of the 8 lanes.

$$d = \sqrt{\frac{4P}{\pi p}} = \sqrt{\frac{4 \times \frac{100}{4}}{\pi \times 700000}} = 0.213 \text{ (m)} \tag{1}$$

where P = axle load on each wheel and p = tire internal pressure, i.e., the contact pressure.

Table 5. Material parameters of ABAQUS model.

Materials	Thickness (cm)	Modulus (MPa)	Poisson's Ratio	Density (kg/m^3)
EAC-13	4/6	2500	0.25	2400
EAC-16	4/6	2300	0.25	2400
SMA-13	4	1500	0.25	2400
SUP-20	6	1400	0.25	2400
SUP-25	9/10	1150	0.25	2400
AR-SAMI	1	200	0.3	1150
CTB	38	1500	0.25	2300
LFS	20	700	0.3	1800
SG	800	80	0.4	1800

In this study, the pavement structure was simplified to a two-dimensional plane strain problem, so the standard load of 0.7 MPa should be converted into a linear load of 117,371 Pa based on the equivalent static principle. To simulate the dynamic response of the pavement, a half-sine dynamic cyclic load with a period of 0.2 s was applied, according to Equation (2):

$$y = 0.11737 sin 10\pi t \text{ (MPa)} \quad (2)$$

The boundary conditions are: completely fixed at the bottom edge of the soil and no lateral displacement on both sides of the whole structure. The mesh of the loading area was refined, and the element type is CPE4R, i.e., 4-node continuum bilinear plane strain with reduced integration (Figure 2). The mechanical responses of 7 structures were captured, including the tensile strain at the bottom of the asphalt layer, the tensile stress at the bottom of the base, the vertical compressive strain on the top of the subgrade, and the shear strain of the asphalt layer.

Figure 2. Pavement modeling in ABAQUS: (**a**) loads and boundary conditions; (**b**) meshes.

2.3. Calculation of Design Life

In order to compare the design life of different pavement structures, load repetitions were calculated using the functions applicable to the response of various types of structures proposed by JTG D50-2017 [35]. The fatigue life of the asphalt layer N_f was obtained based on the tensile strain at the bottom of the asphalt layer according to Equation (3):

$$N_f = 6.32 \times 10^{15.96-0.29\beta} k_a k_b k_T^{-1} \left(\frac{1}{\varepsilon_a}\right)^{3.97} \left(\frac{1}{E_a}\right)^{1.58} (VFA)^{2.72} \quad (3)$$

where N_f is the fatigue cracking life of asphalt layer (times), β is the target reliability index (1.65 for freeway), k_a is the coefficient for seasonal permafrost areas, and k_b is the coefficient of the fatigue loading mode, calculated based on Equation (4):

$$k_b = \left[\frac{1+0.03{E_a}^{0.43}(VFA)^{-0.85}e^{0.024h_a-5.41}}{1+e^{0.024h_a-5.41}}\right]^{3.33} \tag{4}$$

where E_a is the dynamic modulus of asphalt mixture at 20 °C (MPa), VFA is the voids filled with asphalt (%), h_a is the thickness of the asphalt layer, k_T is the temperature adjustment coefficient, and ε_a is the tensile strain at the bottom of asphalt layer (10^{-6}).

Additionally, in this study, to limit the effects of rutting and subgrade deformation, the load repetitions N_d were calculated based on the allowable vertical compressive strain on the top of the subgrade according to Equation (5):

$$[\varepsilon_z] = 1.25 \times 10^{4-0.1\beta}(k_T N_d)^{-0.21} \tag{5}$$

where $[\varepsilon_z]$ is the allowable vertical compressive strain on the top of the subgrade (10^{-6}). Based on the principle that the calculated strain should be less than the allowable strain ($\varepsilon_z < [\varepsilon_z]$), N_d can be determined.

In this paper, the material parameters E_a and VFA were determined by laboratory tests (Section 2.1.3), while the mechanical parameters ε_a and ε_z were obtained from simulation results in Section 3.2. The load repetitions N_f and N_d calculated above were used to compare the service life of epoxy asphalt pavement with that of conventional asphalt pavement to verify its long-life characteristics.

2.4. Life Cycle Assessment (LCA)

To further investigate the carbon reduction effect of epoxy asphalt pavement, a life cycle assessment (LCA) on the carbon emissions of epoxy asphalt pavement and conventional pavement was conducted based on the calculated fatigue life. The service life of conventional asphalt pavement is usually 15 years, while that of epoxy asphalt pavement is 1.7 times longer, according to the results in Section 3.3, up to 25 years or more, so a calculation period of 25 years was used for LCA.

The LCA model on carbon emissions of asphalt pavements can be divided into two main parts, the construction period and the operation period. The construction period includes material production, mixing, transportation, paving, and rolling, while the operation period consists of milling, maintenance and repair, and traffic delays. The basic energy consumption and emissions were referenced from the Chinese Life Cycle Database (CLCD) [36], where the calorific value of the energy consumption process referred to "China Energy Statistical Yearbook 2020" [37], and the carbon emission factors referred to "IPCC Guidelines for National Greenhouse Gas Inventories" [38]. The carbon emissions of material production were available through the references, and the carbon emission factors of equipment used in other processes were obtained by converting from the total power.

In this paper, it is assumed that in the 25-year life cycle for conventional asphalt pavement, a total of 3 major repairs are needed with a milling thickness of 10 cm, as well as 4 minor repairs by adding an ultra-thin overlay with a thickness of 1 cm. For epoxy asphalt pavement, because of its excellent performance and longer fatigue life, only 1 major repair and 2 minor repairs are required. Taking a 1000 m*0.1 m*30 m pavement section as an example, the carbon emissions in a life cycle of epoxy asphalt pavements and conventional asphalt pavements in the construction period and operation period were calculated and compared.

3. Results and Discussions

3.1. Viscosity Tests

The viscosity–time curves of SBS-modified epoxy asphalt (with a 30% dosage of epoxy components) at different temperatures are shown in Figure 3. It illustrates that the viscosity of epoxy asphalt has a typical dependence on time and temperature. As time increases and the curing reaction develops, the epoxy asphalt gradually acquires thermosetting properties and therefore, the viscosity increases. The curves of viscosity and curing time are consistent with the trends observed in the studies of Cong [39] and Luo [40], where they pointed that the viscosity of hot-mixed epoxy asphalt showed a logarithmic relationship with time. Meanwhile, the initial viscosity of the epoxy asphalt binder reduces with an increase in temperature, and it is 615, 450, 400, and 265 mPa•s at 120, 130, 140, and 150 °C, respectively. This characteristic indicates the importance of temperature in the mixing and construction process.

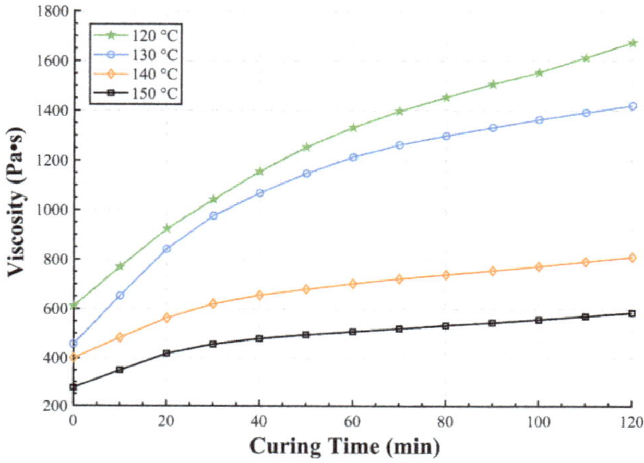

Figure 3. Viscosity test results of SBS-modified epoxy asphalt.

Additionally, the low initial viscosity of the epoxy component opens up the possibility of lower mixing and compaction temperatures. According to Zhang et al. [41], the suitable viscosities for mixing and compaction of SBS-modified asphalt are (0.32 ± 0.03) Pa•s and (0.45 ± 0.05) Pa•s, respectively. To achieve this viscosity, it is usually necessary to mix the SBS-modified asphalt at around 180 °C, which might result in numerous emissions and the deterioration of the asphalt binder. Nevertheless, from the viscosity test results in Figure 3, the addition of epoxy components provides the SBS-modified epoxy asphalt with a lower viscosity, and the requirements described above can be achieved at 150 °C (0–40 min), thus allowing the mixing process to proceed at a lower temperature. It can also be noted that a mixing temperature below 150 °C might be undesirable, because the viscosity has exceeded 1.0 Pa•s after 30 min of curing at 120 °C or 130 °C, which will cause difficulties in construction and compaction. Additives such as warm mix agents are needed if there is a demand for much lower mixing temperatures.

In summary, it is pleasing to find that the addition of the epoxy components allows the SBS-modified asphalt to be mixed at a lower temperature. The mixing temperature of SBS-modified epoxy asphalt is reduced by 30 °C (from 180 °C to 150 °C), thus achieving a low-emission mixed-asphalt mixture, which has a better temperature reduction effect and meets the low-carbon and emission reduction requirements.

3.2. Mechanical Responses of Different Structures

3.2.1. Tensile Strain at the Bottom of Asphalt Layer

The tensile strain at the bottom of the asphalt layer is closely correlated with the fatigue life of the pavement. According to JTG D50-2017 [35], the mechanical response of the asphalt pavement structure can be calculated in the elastic layer system (Figure 4a), so the point A in the inner lane was selected as the calculation point based on the maximum principle (Figure 4b). In this model, point A corresponds to Node 2650, Element 5170 for structures S0, S2, S4, and S6. For structures S1, S3, and S5, point A is Node 2028 and Element 3674. The tensile strains at the bottom of the asphalt layers of seven pavement structures under dynamic loading were extracted and initially filtered, as shown in Figure 5. It demonstrates that under dynamic loading, the mechanical response of the asphalt pavement also varies dynamically; however, it exhibits a hysteresis to some extent. According to Equation (2), the traffic load achieved its maximum value at 0.05 s, but the tensile strain reached the maximum value at approximately 0.06 s. This is because the asphalt mixture was given viscoelastic properties in the simulation, so the external loads would be transferred gradually in the depth direction of the pavement, reflecting the time-delayed effect, which is consistent with the actual situation.

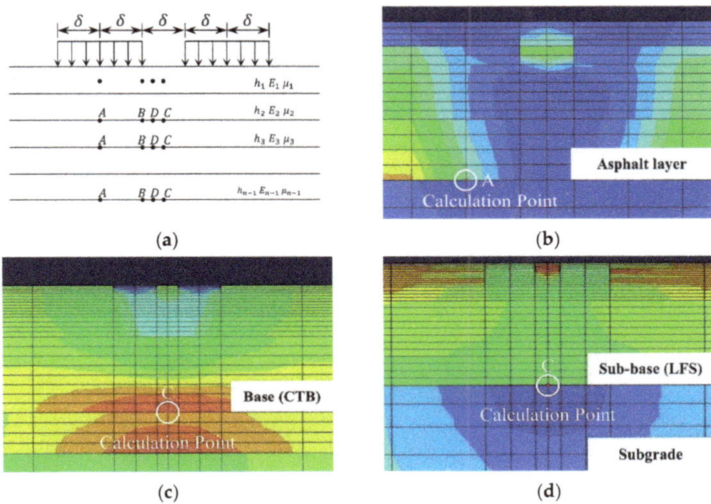

Figure 4. Calculation basis: (**a**) elastic layer system; (**b**) calculation point of tensile strain; (**c**) calculation point of tensile stress; (**d**) calculation point of vertical compressive strain.

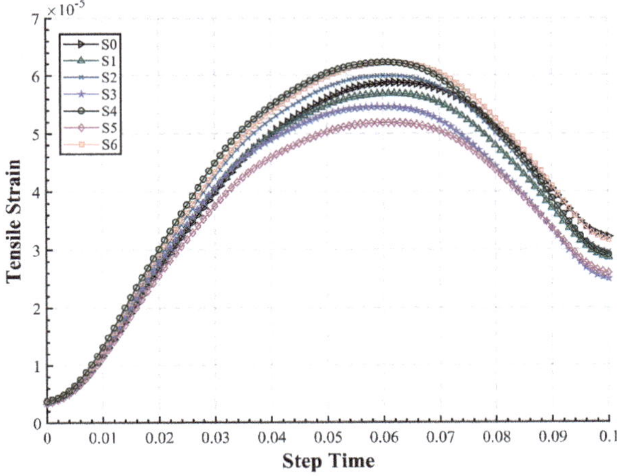

Figure 5. Time–history curve of tensile strain at the bottom of asphalt layer.

It can be discovered from Figure 5 that structure S5 has the lowest maximum tensile strain at the bottom of asphalt layer of 5.21×10^{-5}, while S4 and S6 are similar and highest, exceeding 6.25×10^{-5}. The maximum tensile strain and its occurrence time of the seven structures are listed in Table 6. Three conditions (S1, S3, and S5) have significantly lower tensile strains than the conventional pavement S0 without EAC, decreasing by 3.13%, 6.96%, and 11.81%, respectively, showing a better mechanical response, which indicates that structures with EAC might present a longer fatigue life. The structural mechanics study by Chen et al. also obtained fatigue life by controlling the maximum tensile strain at the bottom of the asphalt layer [42]. The values of epoxy asphalt pavements in this paper are much smaller than the simulation results of conventional asphalt pavements of Jiang et al. (around 9×10^{-5} to 12×10^{-5}) [43]. Therefore, the placement of EAC can prevent the structure from generating excessive flexural tensile strains, thus delaying the fatigue-induced structural failure of the pavement. Furthermore, the maximum value appears earliest in S0, and the other structures with EAC exhibit a longer time delay, which might suggest that the new structures with EAC have better viscoelasticity and ductility. Meanwhile, the results also indicate that setting a 1 cm AR-SAMI as a stress-absorbing layer can significantly reduce the maximum tensile strain at the bottom of the asphalt layer, which in turn can optimize the stress state and improve the fatigue life of the pavement to a certain extent.

Table 6. Summary of maximum indicators and their occurrence time.

Structure	S0	S1	S2	S3	S4	S5	S6
Max tensile strain at the bottom of asphalt layer	5.90×10^{-5}	5.72×10^{-5}	6.03×10^{-5}	5.49×10^{-5}	6.28×10^{-5}	5.21×10^{-5}	6.26×10^{-5}
Occurrence time	0.060	0.062	0.063	0.062	0.063	0.062	0.064
Max tensile stress at the bottom of base (Pa)	71190	70330	74412	71828	75352	70095	73367
Occurrence time	0.052	0.051	0.051	0.051	0.05	0.051	0.051
Max vertical compressive strain on the top of subgrade	1.84×10^{-4}	1.78×10^{-4}	1.79×10^{-4}	1.75×10^{-4}	1.75×10^{-4}	1.80×10^{-4}	1.78×10^{-4}
Occurrence time	0.055	0.053	0.055	0.053	0.056	0.053	0.055

3.2.2. Tensile Stress at the Bottom of Base

As the main bearing layer of the semi-rigid base pavement, the mechanical characteristics of the base are critical to the service life of asphalt pavements. According to the principle of maximum value, the center point in the outermost lane has the maximum tensile stress, so the point C at the bottom of CTB in the outermost lane was chosen as the calculation point of tensile stress (Figure 4c), which was Node 74, Element 1448 for S0, S2, S4, and S6 and Node 423, Element 5371 for S1, S3, and S5. The tensile stress values at the bottom of the base of seven structures were extracted and filtered, as shown in Figure 6. Since the traffic load is a half-sine load, the tensile stress at the bottom of the base also demonstrates a sinusoidal trend with time. It is interesting that the time-delayed effect of tensile stress is much less compared to the tensile strain, which is in line with the viscoelastic characteristics of asphalt pavements, i.e., the deformation will exhibit a significant time delay under continuous loading rather than stress. The maximum tensile stress and its occurrence time in the seven structures are also listed in Table 6.

The maximum tensile stress at the bottom of CTB ranges from 59.7% to 64.2% of the traffic load, which indicates that there is an overall attenuation in the downward transfer of the surface loads (which might first increase before decreasing in the asphalt layer), consistent with accepted perceptions. The analysis in Figure 6 illustrates that S5 still has the minimum tensile stress at the bottom of base with a maximum value of 70,094 Pa, which is similar to S1, and both of them reduce by 1.54% and 1.21% compared to the conventional structure S0. However, the structure S4 shows the largest tensile strain of about 75,352 Pa.

This might be due to the fact that the high modulus of EAC applied to the upper layer causes a change in the stress distribution of the pavement, which makes the stress more dispersed in the lateral direction and influenced by the adjacent lanes, thus increasing the tensile stress of the base.

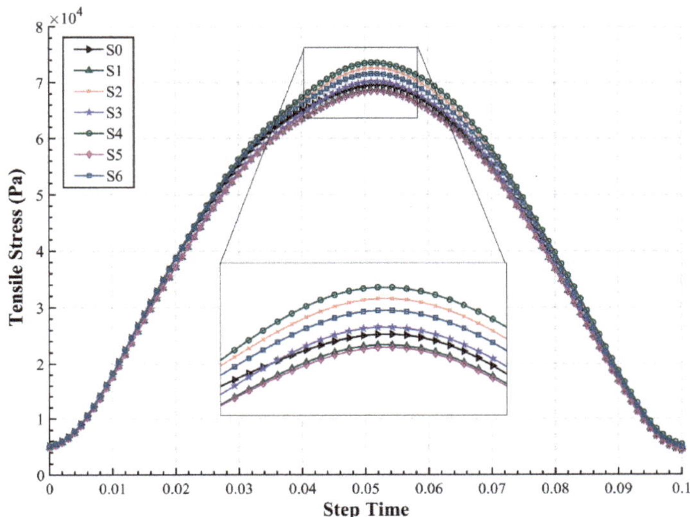

Figure 6. Time–history curve of tensile stress at the bottom of the base (CTB).

3.2.3. Maximum Strain and Stress along Depth of the Pavement

In addition to analysis of the maximum tensile strain at the bottom of the asphalt layer and tensile stress at the bottom of the base as design control indicators, the strain and stress characteristics of the whole asphalt pavement at loading peak are also of concern. As for strain characteristics, still using point A as the calculation point, the strain of the asphalt layer along depth was evaluated and classified with and without SAMI (Figure 7). It can be seen that the compressive strain in the upper 4 cm of S4 and S6 with EAC applied to the upper layer is significantly reduced and there is a more uniform strain distribution when compared to other structures. As the EAC layer transformed into an SUP layer, S4 and S6 underwent a sudden change in strain at 10 cm and 4 cm, respectively, from compressive strain to tensile strain. In the upper-middle layer (4–10 cm in depth), the same phenomenon occurs, where S0 and S6 without the application of EAC generate large tensile strains, while S2 and S4 still present a low compressive strain. Especially for S2, the original tensile strain in the upper layer increased sharply, while the strain in the upper-middle layer suddenly changed from tensile to compressive after the application of EAC. From this point of view, although S4 has a greater tensile strain at the bottom of the asphalt layer than the conventional structure, it might be argued that the EAC application solution for S4 is not unreasonable, because S4 is at a significantly lower strain level throughout the asphalt layer, which might be a considerable contribution to improving the overall fatigue life. Figure 7b shows a similar trend in that the tensile strain in the layer with EAC applied decreases remarkably, allowing for more uniform forces and deformations, which might lead to an alleviation of fatigue failure in the asphalt layer in some cases.

Figure 8 shows the stress distribution in the pavement structure, calculated using point C. It can be seen that different structures have little effect on the stress in the subgrade below the sub-base, and the impact is mainly distributed in the upper part of the pavement, especially in the asphalt layer. The trend of stress along depth is almost the same for all structures due to the fact that the changes in materials will significantly affect the stress–strain characteristics near its location (the asphalt layer), while having little impact on the overall structure at a distance. Meanwhile, the difference in the distribution of stress and strain in the asphalt layer might be caused by the variation of the calculation point locations. Compared to S0, the structures S2, S4, and S6 with EAC have less compressive stress in the asphalt layer, which is beneficial for the pavement. However, they have greater tensile stress in the lower part of the base, especially for S4, so when applying EAC, the issue of the fatigue life of the base should be considered and enhanced, and several methods such as strengthening the base materials and laying a bedding layer may be feasible.

Based on the previous calculations, more attention should be paid to structures with SAMI layers, as they may serve as a more reasonable solution. Figure 8b demonstrates that S5 not only has the lowest tensile strain at the bottom of the asphalt layer and tensile stress at the bottom of the base, but also has a lower value of tensile stress in the whole tensile region. In addition, its compressive stress is not excessively large, so S5 has a more balanced response and may possess the potential to be an engineering-suitable structure.

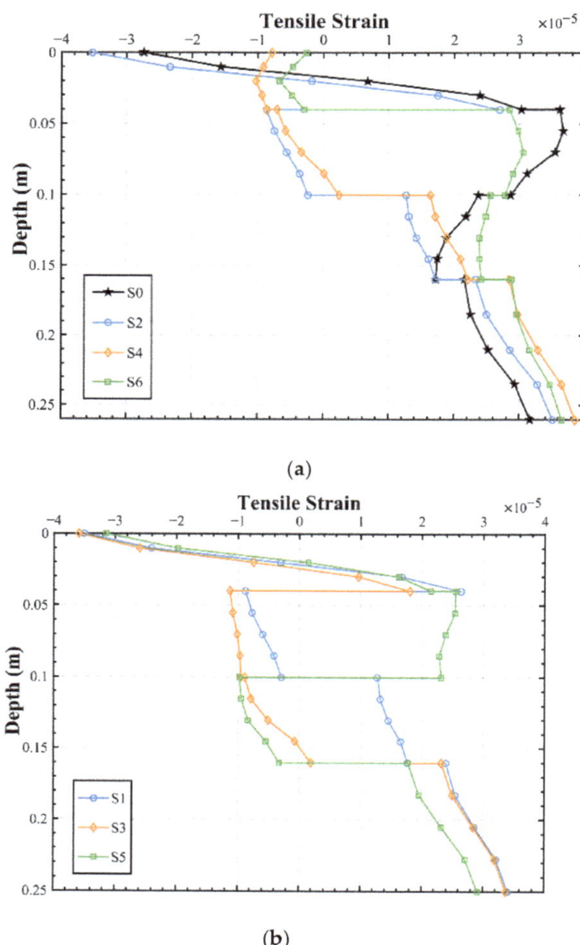

Figure 7. Pavement strain along depth: (**a**) S0, S2, S4, and S6; (**b**) S1, S3, and S5.

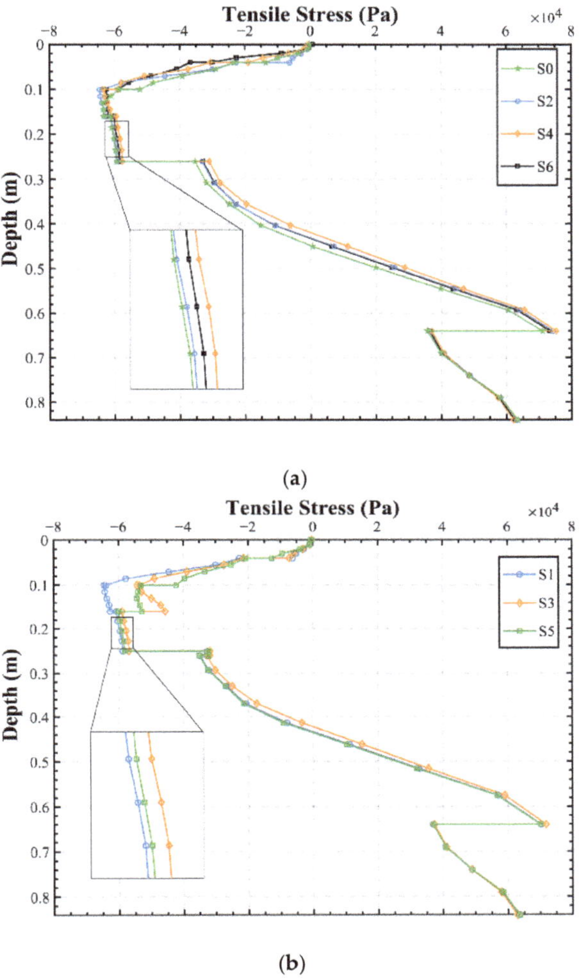

Figure 8. Pavement stress along depth: (**a**) S0, S2, S4, and S6; (**b**) S1, S3, and S5.

3.2.4. Vertical Compressive Strain on the Top of Subgrade

The vertical compressive strain on the top of the subgrade plays an important role in the deformation of the soil and the generation of rutting. In this study, the point C on the top of the subgrade in the outermost lane was chosen as the calculation point for the vertical compressive strain (Figure 4d). For S0, S2, S4, and S6, it was Node 59 and Element 1027 and for S1, S3, and S5, it was Node 411 and Element 5111. The time–history curves and maximum values of vertical compressive strain on the top of the subgrade for different pavement structures are plotted and listed in Figure 9 and Table 6, respectively. The time–history curve is still identical to that of the loading pattern and is distributed as a sinusoidal curve. However, due to viscoelasticity, the vertical compressive strain also has a certain hysteresis effect, reaching its peak around 0.055 s, consistent with previous conclusions. Figure 9 illustrates that the vertical compressive strain presents three levels, with the conventional pavement S0 having the largest compressive strain, and the other two levels decrease by around 2.17% and 4.89%, respectively. S3 and S4 have the lowest value located in the first level, while S1, S2, and S5 are similar and lower than S0, lying in the middle level.

It is interesting that all structures with EAC applied exhibited significantly smaller vertical compressive strains on the top of subgrade. This indicates that the application of EAC could substantially improve the rutting resistance of the pavement, which is in agreement with the findings of previous laboratory tests [44,45]. According to the laboratory results of wheel tracking tests, the dynamic stability of epoxy asphalt mixtures could be as high as 30,000 times/mm or more, presenting an extremely strong rutting resistance. This can also be verified by the three value levels in the figure. Compared with S1, S2, and S5, the use of double layers of EAC in S3 and S4 further reduces the vertical compressive strain on the top of subgrade, which further delays the generation of rutting on the pavement. This suggests that the proper use of EAC in pavement structures might be a way to reduce rutting distress, especially in high-temperature areas.

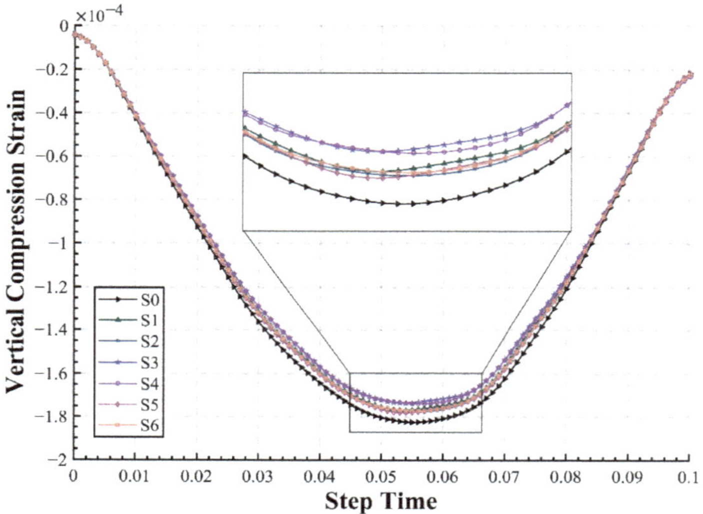

Figure 9. Time–history curve of vertical compressive strain on the top of subgrade.

3.2.5. Shear Stress in the Asphalt Layer

The shear stress in the asphalt layer has a non-negligible effect on the service life of asphalt pavements, although it is not currently used as a design indicator in China. At the same time, the shear stress in the upper and middle asphalt layers is also closely related to the appearance of rutting distresses, so the effect of shear stress in the asphalt layer should be investigated. The shear stress within the upper and upper-middle layers (0–10 cm) were mainly considered, and for the seven structures, the locations where the maximum values occur were selected as the calculation points.

The time–history curves, maximum values and their locations of the shear stress were calculated, as shown in Figure 10 and Table 7. It can be seen that the shear stress varies greatly among different structures. S6 has a shear stress 1.6 times greater than the shear stress of S0, and S1, S2, and S4 have the same value of about 1.35 times the shear stress of S0. The increase in shear stress in S3 and S5, which have stress-absorbing layers with EAC in the lower-middle layers, is relatively less significant. The shear stress values of all structures with EAC applied are larger than S0, which may be due to the fact that the high stiffness of EAC changes the stress distribution, making the maximum value occur mostly in the layers where EAC is applied; this can also be verified from the vertical depths in Table 7. Therefore, even though EAC has the advantages of high strength and fatigue life, its shear resistance deserves heightened attention.

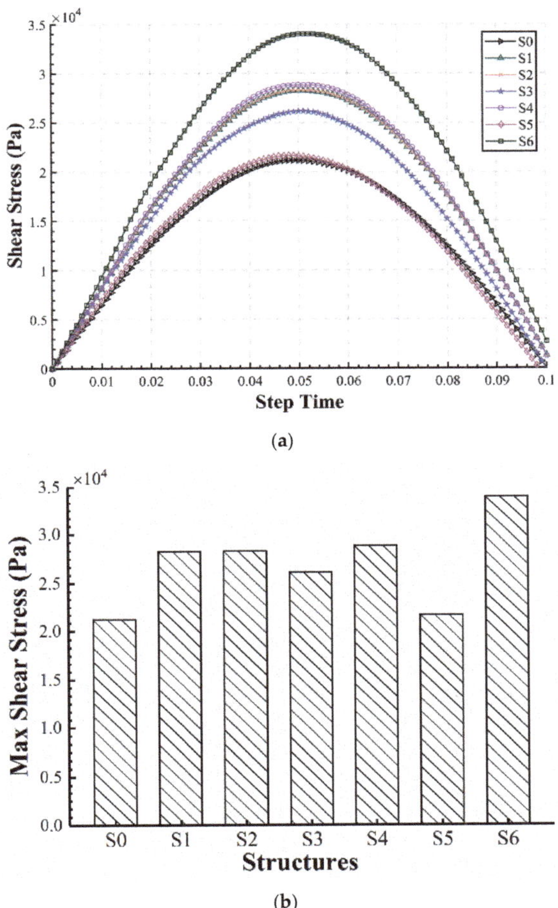

Figure 10. Shear stress in the asphalt layer: (**a**) Time–history curve; (**b**) maximum shear stress.

Table 7. Maximum values and locations of shear stress in the asphalt layer.

Structure	S0	S1	S2	S3	S4	S5	S6
Maximum shear stress (Pa)	21,190.9	28,266.6	28,303.3	26,153.7	28,823.6	21,655.2	34,016.1
Element number	1918	5738	1891	5741	1715	5761	1714
Vertical depth (cm)	4	5.5	5.5	7	4	8.5	3
Horizontal points	Lane Center (Point C)	Center of the loading area (Point A)					

Additionally, the large value in S6 might be attributed to the application of EAC in the upper layer that is directly interacting with traffic loads, generating high local stress in the case of discontinuous pavement stiffness. A continuously applied EAC or an additional transition layer can be used to minimize this phenomenon. Among all the structures, the upside is that S5 has similar shear stress to S0, in spite of a slight increase. In S5, the EAC was applied to the lower layers near the base, so it has less of an effect on the stiffness continuity and the shear stress in the upper layers. From this perspective, it is more reasonable for EAC to be applied in the lower layer.

3.3. Prediction and Comparison of Design Life

Some researchers have used the empirical formulas from the America Asphalt Institute (AI) in previous studies [43], but only took into account two parameters, strain and modulus, leading to inaccuracy and large variability. In this paper, formulas in the Chinese specification JTG D50-2017 were used to calculate the design life, which incorporated more comprehensive parameters to achieve greater precision in the calculation. The allowable number of load repetitions to limit fatigue cracking and rutting of the seven structures is shown in Figure 11. With regard to the allowable number N_f to limit fatigue cracking, S0 has a similar value to S2, while S4 and S6 are relatively lower. This indicates that EAC applied to the upper layer might affect the fatigue of the underlying asphalt mixtures, so the life of the overall structure needs to be taken into account, even though EAC has higher resistance. The structures with the SAMI layer (S1, S3, and S5) have significantly higher allowable load repetitions, especially S5, which allows for more than 1.7 times the fatigue number of conventional pavement S0, indicating that the epoxy asphalt pavement might have the potential to be a long-life pavement.

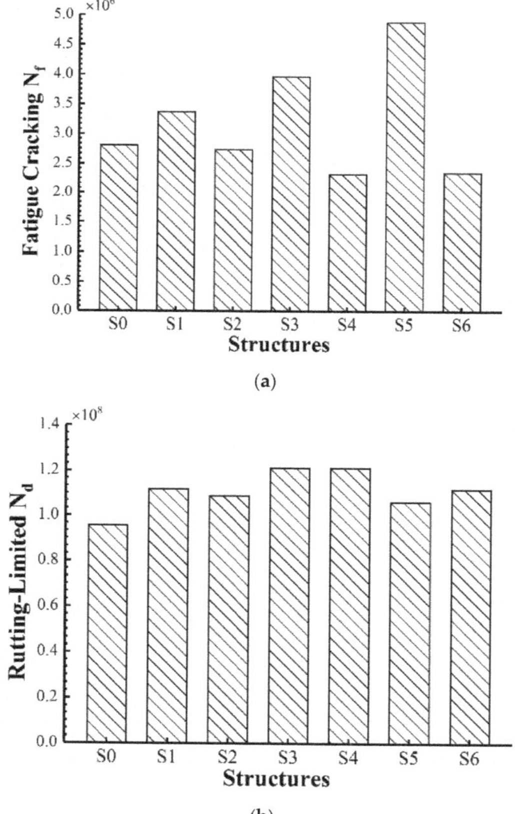

Figure 11. The allowable load repetitions: (**a**) N_f to limit fatigue cracking; (**b**) N_d to limit rutting.

Meanwhile, the aging of the asphalt binder is also related to the expected service life of the asphalt mixture. In fact, the aging properties of the epoxy asphalt have been studied by several researchers, and it has been proven that the addition of epoxy components can provide epoxy asphalt with much better aging resistance. Apostolidis et al. [46] tracked the formation of carbonyl compounds by Fourier transform infrared (FTIR) spectroscopy and found that its reaction rate decreased when the epoxy components were added to the asphalt, which indicates that oxidative aging in epoxy asphalt occurs relatively slowly. Si et al. [47] proposed that aging changed the structure of the epoxy asphalt and improved its stress tolerance, increasing the tensile strength to 5.16 MPa. Additionally, after long-term aging, the fatigue life of EAC-13 is still three to four times that of AC-13, which is far superior to that of conventional asphalt concrete [22]. In terms of UV aging, the addition of the epoxy components gives the epoxy asphalt strong anti-aging properties, with a 12% increase in the residual penetration ratio [48]. From this point of view, the aging resistance of epoxy asphalt might be an advantage, or at least not much of a concern. As for the changes of material properties during use, a feasible approach in practical cases is to install sensors in the pavement and adjust the calculation of service life according to the real-time stress–strain data.

As for the allowable load repetitions N_d to limit rutting, it can be seen that all structures with EAC (S0–S6) can withstand a higher number of loads than conventional pavement S0, indicating that they have much better rutting resistance, which is consistent with the results of laboratory wheel tracking tests on epoxy asphalt mixtures. Meanwhile, the load repetitions of rutting limits greatly exceed the fatigue cracking limits; therefore, the high-temperature performance of EAC might require less concern. However, despite the high resistance to fatigue cracking, S5 has a relatively weak rutting resistance, which could be further improved, though it still exceeds that of conventional pavement and the specification requirements.

3.4. *Life Cycle Assessment (LCA) of Epoxy Asphalt Pavements*

The LCA results on carbon emissions of epoxy asphalt pavements and conventional asphalt pavements are shown in Figure 12. Due to the high production emissions of epoxy resin, the carbon emissions in the construction period of epoxy asphalt pavements are higher than those of conventional pavements but could already be significantly reduced by the lower mixing temperatures studied in this paper. More importantly, due to the longer service life and better performance, the epoxy asphalt pavements exhibit much lower carbon emissions from maintenance and repair during the operation period, about 48% of that of conventional pavements. As mentioned above, the extension of the service life of the pavement is a remarkably effective way to reduce emissions. It is critical to reduce pavement distresses and maintain good service condition, thereby significantly decreasing the carbon emissions generated by repairs. Thus, over the whole life cycle, epoxy asphalt pavements have an excellent emission reducing effect, reducing carbon emissions by 29.8% compared to conventional pavements. From the LCA results, the low-emission mixed EAC has strong feasibility as a long-life, "green", and low-carbon pavement material.

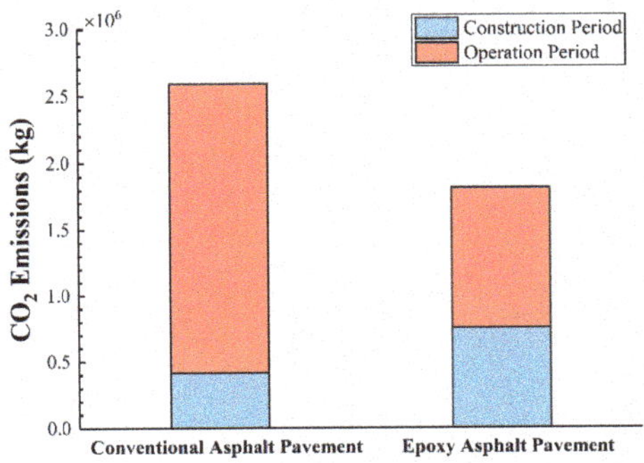

Figure 12. LCA of carbon emissions for epoxy asphalt pavement and conventional asphalt pavement.

4. Conclusions

In this study, the feasibility of low-emission mixed epoxy asphalt mixtures was verified, and six pavement structures with EAC were proposed for performance evaluation and compared to conventional pavements. Based on the viscoelastic parameters obtained from the dynamic modulus tests, an FEM model were developed to analyze the mechanical responses and the design life in order to conduct structure optimization. Finally, life cycle assessment on the carbon emissions of the recommended epoxy asphalt pavement structures was conducted. The following conclusions can be drawn:

1. Through the viscosity tests, the SBS-modified epoxy asphalt has a suitable mixing and compaction viscosity of 150 °C, so the mixing temperature of SBS-modified asphalt can be reduced by 30 °C after adding epoxy components. Thus, the SBS-modified epoxy asphalt can be considered to be a low-emission mixed material.
2. S5 with EAC applied in the middle-lower layers has the lowest tensile strain at the bottom of the asphalt layer and lowest tensile stress at the bottom of the base, decreasing by 11.81% and 1.54% compared to S0, respectively; however, S4 and S6 with EAC applied have an increasing trend in the upper layer.
3. The application of AR-SAMI as a stress-absorbing layer can significantly reduce the maximum tensile strain and stress in the asphalt layer, which might improve the fatigue life of the pavement. Additionally, the layer with EAC applied showed a remarkable decrease in tensile strain, resulting in more uniform stresses and deformations in the depth direction.
4. S3 and S4 reduced the vertical compressive strain on the top of the subgrade by 4.89%, indicating that the reasonable use of EAC could be a way reduce rutting distresses. The application of EAC increases the shear stress in the asphalt layer, especially when applied to the upper layer, while it has little increase in S5 compared to the conventional structure.
5. The allowable number of load repetitions to limit fatigue cracking of S5 is 1.7 times higher than that of the conventional structure, suggesting that it might have the potential to act as a long-life pavement. All structures with EAC have improved the allowable number to limit rutting, indicating that EAC has excellent rutting resistance.
6. The LCA results proved that in the 25 year life cycle, the carbon emissions of epoxy asphalt pavements could be reduced by 29.8% compared with the conventional pavements, showing a superior emission reduction effect.

In conclusion, the use of EAC, which allows for low-emission mixing, could enhance the performance and service life of the pavement, and has a favorable carbon emission reduction effect. Considering the performance and cost, structure S5 with EAC in the middle-lower layer is recommended. Future research can focus on simulations of fatigue cracking, top-down cracking, and rutting in different structures with EAC, while test sections in the field are also needed to obtain field stress–strain data to calibrate the accuracy of the simulation.

Author Contributions: Conceptualization, Y.F., H.W., W.H. and J.Y.; methodology, Y.F., H.C. and Y.W.; software, Y.F.; validation, Y.W. and S.L.; formal analysis, H.C.; investigation, Y.F.; resources, W.H. and J.Y.; data curation, H.W.; writing—original draft preparation, Y.F.; writing—review and editing, J.Y.; visualization, Y.F. and Y.W.; supervision, W.H.; project administration, J.Y.; funding acquisition, J.Y. and W.H. All authors have read and agreed to the published version of the manuscript.

Funding: This research was funded by the National Natural Science Foundation of China, grant number 51778140, 52078130; the Technology Research and Development Program of China State Railway Group Co., Ltd., grant number P2019G030; and the Major Technology Demonstration of Epoxy Asphalt in Pavement as a Green Highly-Efficient Carbon Mitigation Technology, grant number BE2022615.

Institutional Review Board Statement: Not applicable.

Informed Consent Statement: Not applicable.

Data Availability Statement: The data presented in this study are available on request from the corresponding author.

Conflicts of Interest: The authors declare no conflict of interest.

References

1. Jing, R.; Apostolidis, P.; Liu, X.; Naus, R.; Erkens, S.; Skarpas, T. Effect of recycling agents on rheological properties of epoxy bitumen. *Road Mater. Pavement Des.* **2021**, 1–15. [CrossRef]
2. Zhang, F.; Zhang, L.; Muhammad, Y.; Cai, Z.; Guo, X.; Guo, Y.; Huang, K. Study on preparation and properties of new thermosetting epoxy asphalt. *Constr. Build. Mater.* **2021**, *311*, 125307. [CrossRef]
3. Wang, J.; Yu, X.; Ding, G.; Si, J.; Ruan, W.; Zou, X. Influence of asphalt solvents on the rheological and mechanical properties of cold-mixed epoxy asphalt. *Constr. Build. Mater.* **2021**, *310*, 125245. [CrossRef]
4. Apostolidis, P.; Liu, X.; Erkens, S.; Scarpas, A. Evaluation of epoxy modification in bitumen. *Constr. Build. Mater.* **2019**, *208*, 361–368. [CrossRef]
5. Luo, S.; Qian, Z.; Harvey, J. Research on Dynamic Modulus for Epoxy Asphalt Mixtures and Its Master Curve. *China J. Highw. Transp.* **2010**, *23*, 16–20.
6. Luo, S.; Qian, Z.; Harvey, J. Experiment on Fatigue Damage Characteristics of Epoxy Asphalt Mixture. *China J. Highw. Transp.* **2013**, *26*, 20–25.
7. Nie, W.; Wang, D.; Sun, Y.; Xu, W.; Xiao, X. Integrated Design of Structure and Material of Epoxy Asphalt Mixture Used in Steel Bridge Deck Pavement. *Buildings* **2021**, *12*, 10009. [CrossRef]
8. Bocci, E.; Graziani, A.; Canestrari, F. Mechanical 3D characterization of epoxy asphalt concrete for pavement layers of orthotropic steel decks. *Constr. Build. Mater.* **2015**, *79*, 145–152. [CrossRef]
9. Liu, Y.; Qian, Z.; Wang, Y.; Xue, Y. Development and Laboratory Evaluation of a Cold Mix High-Early-Strength Epoxy Asphalt Concrete for Steel Bridge Deck Pavements. *Materials* **2021**, *14*, 164555. [CrossRef]
10. Huang, W. Epoxy asphalt concrete paving on the deck of long-span steel bridges. *Chin. Sci. Bull.* **2003**, *48*, 2391–2394. [CrossRef]
11. Huang, W.; Pei, M.; Liu, X.; Wei, Y. Design and construction of super-long span bridges in China: Review and future perspectives. *Front. Struct. Civ. Eng.* **2020**, *14*, 803–838. [CrossRef]
12. Huang, W. Integrated Design Procedure for Epoxy Asphalt Concrete–Based Wearing Surface on Long-Span Orthotropic Steel Deck Bridges. *J. Mater. Civ. Eng.* **2016**, *28*, 1470. [CrossRef]
13. Lu, Q.; Bors, J. Alternate uses of epoxy asphalt on bridge decks and roadways. *Constr. Build. Mater.* **2015**, *78*, 18–25. [CrossRef]
14. Jamshidi, A.; White, G.; Kurumisawa, K. Rheological characteristics of epoxy asphalt binders and engineering properties of epoxy asphalt mixtures—State-of-the-art. *Road Mater. Pavement Des.* **2021**, *23*, 1–24. [CrossRef]
15. Xiang, Q.; Xiao, F. Applications of epoxy materials in pavement engineering. *Constr. Build. Mater.* **2020**, *235*, 117529. [CrossRef]
16. Chen, Y.; Hossiney, N.; Yang, X.; Wang, H.; You, Z.; Palumbo, D. Application of Epoxy-Asphalt Composite in Asphalt Paving Industry: A Review with Emphasis on Physicochemical Properties and Pavement Performances. *Adv. Mater. Sci. Eng.* **2021**, *2021*, 1–35. [CrossRef]
17. Xie, H.; Li, C.; Wang, Q. A critical review on performance and phase separation of thermosetting epoxy asphalt binders and bond coats. *Constr. Build. Mater.* **2022**, *326*, 126792. [CrossRef]
18. Huang, W.; Guo, W.; Wei, Y. Prediction of Paving Performance for Epoxy Asphalt Mixture by Its Time- and Temperature-Dependent Properties. *J. Mater. Civ. Eng.* **2020**, *32*, 3060. [CrossRef]
19. Gao, J.; Wang, H.; Liu, C.; Ge, D.; You, Z.; Yu, M. High-temperature rheological behavior and fatigue performance of lignin modified asphalt binder. *Constr. Build. Mater.* **2020**, *230*, 117063. [CrossRef]
20. Luo, S.; Qian, Z.; Yang, X.; Lu, Q. Fatigue behavior of epoxy asphalt concrete and its moisture susceptibility from flexural stiffness and phase angle. *Constr. Build. Mater.* **2017**, *145*, 506–517. [CrossRef]
21. Xu, W.; Wei, X.; Wei, J.; Chen, Z. Experimental Evaluation of the Influence of Aggregate Strength on the Flexural Cracking Behavior of Epoxy Asphalt Mixtures. *Materials* **2020**, *13*, 81876. [CrossRef] [PubMed]
22. Li, X.; Shen, J.; Ling, T.; Yuan, F. Fatigue Properties of Aged Porous Asphalt Mixtures with an Epoxy Asphalt Binder. *J. Mater. Civ. Eng.* **2022**, *34*, 4130. [CrossRef]
23. Wang, X.; Wu, R.; Zhang, L. Development and performance evaluation of epoxy asphalt concrete modified with glass fibre. *Road Mater. Pavement Des.* **2017**, *20*, 715–726. [CrossRef]
24. Liu, Y.; Su, P.; Li, M.; You, Z.; Zhao, M. Review on evolution and evaluation of asphalt pavement structures and materials. *J. Traffic Transp. Eng.* **2020**, *7*, 573–599. [CrossRef]
25. Yao, Y.; Li, J.; Ni, J.; Liang, C.; Zhang, A. Effects of gravel content and shape on shear behaviour of soil-rock mixture: Experiment and DEM modelling. *Comput. Geotech.* **2022**, *141*, 104476. [CrossRef]
26. Zhang, J.; Fan, H.; Zhang, S.; Liu, J.; Zheng, J. Time-domain elasto-dynamic model of a transversely isotropic, layered road structure system with rigid substratum under a FWD load. *Road Mater. Pavement Des.* **2021**, 1–19. [CrossRef]
27. Guo, M.; Hou, F.; Zhang, S.; Li, X.; Li, Y.; Bi, Y.; De Jesus, A. Research and Evaluation on Dynamic Response Characteristics of Various Pavement Structures. *Adv. Mater. Sci. Eng.* **2022**, *2022*, 1–15. [CrossRef]

28. Yu-Shan, A.; Shakiba, M. Flooded Pavement: Numerical Investigation of Saturation Effects on Asphalt Pavement Structures. *J. Transp. Eng. B-Pave.* **2021**, *147*, 276. [CrossRef]
29. Liu, P.; Wang, C.; Lu, W.; Moharekpour, M.; Oeser, M.; Wang, D. Development of an FEM-DEM Model to Investigate Preliminary Compaction of Asphalt Pavements. *Buildings* **2022**, *12*, 932. [CrossRef]
30. Liu, Z.; Gu, X.; Ren, H.; Wang, X.; Dong, Q. Three-dimensional finite element analysis for structural parameters of asphalt pavement: A combined laboratory and field accelerated testing approach. *Case Stud. Constr. Mater.* **2022**, *17*, e01221. [CrossRef]
31. Alae, M.; Ling, M.; Haghshenas, H.F.; Zhao, Y. Three-dimensional finite element analysis of top-down crack propagation in asphalt pavements. *Eng. Fract. Mech.* **2021**, *248*, 107736. [CrossRef]
32. Wang, H.; Wu, Y.; Yang, J.; Wang, H. Numerical Simulation on Reflective Cracking Behavior of Asphalt Pavement. *Appl. Sci.* **2021**, *11*, 7990. [CrossRef]
33. *JTG E20-2011*; Standard Test Methods of Bitumen and Bituminous Mixtures for Highway Engineering. Ministry of Transport of the People's Republic of China: Beijing, China, 2011.
34. *JTG F40-2004*; Technical Specifications for Construction of Highway Asphalt Pavements. Ministry of Transport of the People's Republic of China: Beijing, China, 2004.
35. *JTG D50-2017*; Specifications for Design of Highway Asphalt Pavement. Ministry of Transport of the People's Republic of China: Beijing, China, 2017.
36. IKE. Chinese Life Cycle Database—CLCD. 2012. Available online: http://www.ike-global.com/products-2/chinese-lca-database-clcd (accessed on 16 August 2022).
37. National Bureau of Statistics. *China Energy Statistics Yearbook*; Department of Energy Statistics: Beijing, China, 2020.
38. IPCC. *Guidelines for National Greenhouse Gas Inventories*; National Greenhouse Gas Inventories Programme: Beijing, China, 2006.
39. Cong, P.; Luo, W.; Xu, P.; Zhang, Y. Chemical and physical properties of hot mixing epoxy asphalt binders. *Constr. Build. Mater.* **2019**, *198*, 1–9. [CrossRef]
40. Luo, S.; Liu, Z.; Yang, X.; Lu, Q.; Yin, J. Construction Technology of Warm and Hot Mix Epoxy Asphalt Paving for Long-Span Steel Bridge. *J. Constr. Eng. Manag.* **2019**, *145*, 1716. [CrossRef]
41. Zhang, Y.; Guo, K.; Zhao, S.; Li, C.; Yang, Z. Determination of construction temperature for modified asphalt mixture. *J. Jiangsu Univ. Nat. Sci. Ed.* **2016**, *37*, 740–744.
42. Chen, F.; Coronado, C.F.; Balieu, R.; Kringos, N. Structural performance of electrified roads: A computational analysis. *J. Clean. Prod.* **2018**, *195*, 1338–1349. [CrossRef]
43. Jiang, X.; Zeng, C.; Gao, X.; Liu, Z.; Qiu, Y. 3D FEM analysis of flexible base asphalt pavement structure under non-uniform tyre contact pressure. *Int. J. Pavement Eng.* **2017**, *20*, 999–1011. [CrossRef]
44. Yi, X.; Chen, H.; Wang, H.; Shi, C.; Yang, J. The feasibility of using epoxy asphalt to recycle a mixture containing 100% reclaimed asphalt pavement (RAP). *Constr. Build. Mater.* **2022**, *319*, 126122. [CrossRef]
45. Chen, A.; Qiu, Y.; Wang, X.; Li, Y.; Wu, S.; Liu, Q.; Wu, F.; Feng, J.; Lin, Z. Mechanism and Performance of Bituminous Mixture Using 100% Content RAP with Bio-Rejuvenated Additive (BRA). *Materials* **2022**, *15*, 30723. [CrossRef]
46. Apostolidis, P.; Liu, X.; Erkens, S.; Scarpas, T. Oxidative aging of epoxy asphalt. *Int. J. Pavement Eng.* **2020**, *23*, 1471–1481. [CrossRef]
47. Si, J.; Wang, J.; Yu, X.; Ding, G.; Ruan, W.; Xing, M.; Xie, R. Influence of thermal-oxidative aging on the mechanical performance and structure of cold-mixed epoxy asphalt. *J. Clean. Prod.* **2022**, *337*, 130482. [CrossRef]
48. Wang, Y.; Jiao, B. Preparation and properties of UV aging resistant modified asphalt. *J. Funct. Mater.* **2021**, *52*, 1062–1067.

MDPI

Article

The Synergistic Effect of Polyphosphates Acid and Different Compounds of Waste Cooking Oil on Conventional and Rheological Properties of Modified Bitumen

Wentong Wang [1], Jin Li [2,*], Di Wang [3,4,*], Pengfei Liu [5] and Xinzhou Li [1]

1 School of Highway, Chang'an University, Xi'an 710064, China
2 School Transportation Civil Engineering, Shandong Jiaotong University, Jinan 250357, China
3 Department of Civil Engineering, Aalto University, 02150 Espoo, Finland
4 Hangzhou Telujie Transportation Technology Co., Ltd., Hangzhou 311121, China
5 Institute of Highway Engineering, RWTH Aachen University, D-52074 Aachen, Germany
* Correspondence: sdzblijin@163.com (J.L.); di.1.wang@aalto.fi (D.W.)

Abstract: In order to conserve non-renewable natural resources, waste cooking oil (WCO) in bitumen can help lower CO_2 emissions and advance the environmental economy. In this study, three different components of WCO were isolated and then, together with polyphosphoric acid (PPA), used separately as bitumen modifiers to determine the suitability of various substances in WCO with PPA. Conventional tests, including penetration, softening point temperature, and ductility, and the dynamic shear rheology (DSR) test, including temperature sweep and frequency sweep, were used to evaluate the influence of WCO/PPA on the traditional performance and rheological properties at high and low temperatures. The results indicate that WCO reduced the ductility and penetration value, when the use of PPA increased the softening point temperature and high-temperature performance. Compared to reference bitumen, the rutting factor and viscous activation energy (Ea) of bitumen modified with 4% WCO and 2% PPA has the most significant increase by 18.6% and 31.5, respectively. All components of WCO have a significant impact on improving the low-temperature performance of PPA-modified bitumen. The performance of the composite-modified bitumen at low temperatures is negatively affected by some waxy compounds in WCO, such as methyl palmitate, which tends to undergo a solid–liquid phase change as the temperature decreases. In conclusion, the inclusion of WCO/PPA in bitumen offers a fresh approach to developing sustainable pavement materials.

Keywords: modified bitumen; waste cooking oil; polyphosphate acid; rheological property; conventional property

Citation: Wang, W.; Li, J.; Wang, D.; Liu, P.; Li, X. The Synergistic Effect of Polyphosphates Acid and Different Compounds of Waste Cooking Oil on Conventional and Rheological Properties of Modified Bitumen. *Materials* **2022**, *15*, 8681. https://doi.org/10.3390/ma15238681

Academic Editor: Giovanni Polacco

Received: 4 November 2022
Accepted: 29 November 2022
Published: 5 December 2022

Publisher's Note: MDPI stays neutral with regard to jurisdictional claims in published maps and institutional affiliations.

1. Introduction

Over the past 30 years, the construction of road infrastructure has seen extraordinary progress worldwide [1]. As a result, massive amounts of non-renewable natural resources are frequently utilized to build road infrastructure, including natural aggregates and bitumen derived from fossil fuels [2], which increases the cost of building and garners significant media attention about the environment and natural resources. In the meantime, the rate of global garbage creation is increasing, with annual production reaching over 1.5 billion tons [3]. The government faces tremendous challenges when it comes to handling, storing, and transporting waste [4]. Typically, the debris was handled by piling and burying, which reduced the amount of landfill area and harmed [5]. Finding a novel method of treating garbage is, therefore, urgently needed. One promising solution is to recycle and reuse construction demolition waste, especially recycled asphalt pavement (RAP), in road construction [6]. Lowering CO_2 emissions, increasing land-use efficiency, protecting resources, and advancing the economy are all beneficial [7]. Nowadays, up to 80% of RAP can be used in bitumen road construction in Europe [8,9]. However, there are still knowledge gaps to be filled, such as whether the sorting of waste resources be efficiently applied to bitumen materials to improve engineering performance.

Household garbage is being produced in significant quantities as a result of the world population's fast rise. Waste cooking oil (WCO) is one of the most well-known types [10,11], which contains large amounts of polycyclic aromatic hydrocarbons and benzo(a)pyrene, posing a serious threat to human health and the safety of the ecological environment [12]. WCO pollutes the natural results to different degrees because there is no effective way to treat it [13,14]. Statistics reveal that in China alone, there are 7 million tons of WCO produced annually, but only 1 million tons of that is correctly reused because of technical and equipment issues [15]. The efficient utilization of WCO resources has also become a hotly debated issue among academics.

Recently, much attention has been paid to the potential application of WCO in bitumen modification and rejuvenation. Plenty of studies have shown that WCO has been used as an asphalt rejuvenation agent (ARA), modifier, additive for warm mix bitumen, and raw material for the production of bio-based bitumen [14–18]. Chen [19] added 3%, 4%, 5%, 6%, and 7% of waste cooking oil to three different types of aged bitumen and evaluated the regeneration effect of waste cooking oil by rheological performance tests. The results showed that the application of WCO in bitumen led to a decrease in viscosity, which was due to the low molecular weight of WCO, low viscosity, good flowability, and high viscosity reduction and drag reduction effect. Azahar collected WCO at different levels of use and mixed it with aged bitumen to prepare recycled bitumen. The results show that the frequency and manner of WCO use affect the performance of recycled bitumen [20]. Meanwhile, Chen studied the regeneration effect of waste oil on bitumen with the same degree of aging. The results of the study found that the type and content of fatty acids in the waste oil were important factors affecting the regeneration effect [19]. When the aforementioned literature is combined, it is simple to see that the types, sources, and qualities of WCO used in various investigations vary greatly. After varying application techniques and usage frequencies, the same edible oil's modification effect on bitumen also changes greatly. That is to say, the type and quality of WCO employed are the primary factors that affect how changed bitumen is modified. Additionally, some specific WCO components may potentially harm bitumen's performance [21]. In view of this, it is important to study the influence of the composition of WCO on bitumen properties and to pre-treat WCO to classify and manage it, thus improving the efficiency of WCO utilization [22,23].

Modifiers and additives were commonly used to improve the bitumen's rheological properties in high-performance bitumen pavements [24–26]. Among them, both physical modification and chemical modification of polyphosphoric acid (PPA) were attempted in previous studies [27]. PPA-modified bitumen is gaining interest due to the more prominent and less costly modification effect. From the analysis of the rheological test results, it was validated that the addition of PPA could benefit the high-temperature performance and rutting deformation resistance of bitumen [28,29]. Moreover, PPA can react with the alcohols in bitumen [30]. The addition of PPA was found to contribute to the increase in asphaltene content. However, as a result of the bitumen's reaction with PPA, it becomes brittle and less ductile [29,31]. Therefore, the combined use of WCO/PPA additives to modify the low-temperature properties of bitumen could be an option, this may be attributed to the high amount of light component of WCO. The interplay of WCO and PPA not only potentially balances out each other's flaws, but also offers promising opportunities for long-term use. the effect of the composition of WCO on the performance of modified bitumen is not clear. In summary, the major goal of this work was to examine the impact of various WCO constituents on the rheological characteristics of PPA-modified bitumen at high and low temperatures.

In view of the above, the WCO was firstly, fractionated into three stable fractions: light component (LC), intermediate component (IC), and heavy component (HC). Next, the three components were blended with the matrix bitumen at a 4% admixture rate, respectively. Then, the composite-modified bitumen was prepared by adding PPA (0.5%, 1%, 1.5% and 2%) using high-speed mixing. In order to determine its basic function, the conventional properties, including penetration value, softening point temperatures, and ductility, of the bitumen were evaluated. Finally, the high- and low-temperature rheological properties of WCO/PPA-modified bitumen was evaluated and discussed. Due to the many occurrences of proper nouns in this paper, all abbreviations have been listed in Table 1 for ease of reading.

Table 1. The abbreviations of proper nouns.

Abbreviations		Abbreviations	
4LC0.5PPA	4%LC + 0.5%PPA	WCO	waste cooking oil
4LC1PPA	4%LC + 1PPA	PPA	polyphosphoric acid
4LC1.5PPA	4%LC + 1.5%PPA	DSR	dynamic shear rheology
4LC2PPA	4%LC + 2%PPA	Ea	viscous activation energy
4IC0.5PPA	4%IC + 0.5%PPA	ARA	asphalt rejuvenation agent
4IC1PPA	4%IC + 1PPA	LC	light component
4IC1.5PPA	4%IC + 1.5%PPA	IC	intermediate component
4IC2PPA	4%IC + 2%PPA	HC	heavy component
4HC0.5PPA	4%HC + 0.5%PPA	TS	temperature sweep
4HC1PPA	4%HC + 1PPA	FS	frequency sweep
4HC1.5PPA	4%HC + 1.5%PPA	G*/(1-(1/TanδSinδ))	Shenoy parameter
4HC2PPA	4%HC + 2%PPA	BBR	Bending beam rheometer
G*/Sinδ	Rutting factor	GTS	grade temperature sensitivity

2. Materials and Methods

2.1. Raw Materials

2.1.1. Bitumen

SK-90# bitumen was selected as the original bitumen, and the base performance is listed in Table 2. All the parameters fit the requirement of the active Chinese national standard GB JTG E20-2011.

Table 2. Properties of SK-90#.

Technical Indexes	Unit	Results	Test Method
Ductility (5 °C, 5 cm/min)	cm	37.1	T0605
Penetration (25 °C, 100 g, 5 s)	0.1 mm	82.9	T0604
Softening Point	°C	47.3	T0604
	RTFOT		
Weight loss	%	−0.066	T0601
Residual penetration ratio	%	57.9	T0604
Residual ductility (5 °C)	cm	21.6	T0605

2.1.2. PPA

According to the literature, the commonly used PPA modifier content range from 0.5% to 2.5% by weight (wt%) of the original bitumen [27,29,32–34]. Given that PPA prepolymers have a positive impact on bitumen's high-temperature properties, this study tried to select four low PPA contents, as low as 0.5 wt%, 1 wt%, 1.5 wt%, and 2 wt%. The PPA was provided by Shandong Hui'an Chemical Co., Ltd. (Jinan, Chian), and Table 3 provides a list of PPA's fundamental characteristics.

Table 3. The technical indexes of PPA.

Index	Unit	Test Results
Density (25 °C)	g/cm^3	2.15
Viscosity (85 °C)	mPa.s	562
Iron content	%	≤0.01
Sulfate	%	0.01
P_2O_5	%	82.95

2.1.3. WCO

The quality and properties of WCO vary greatly from source to source and type to type, which is a major factor in determining the performance of modified bitumen. Therefore, in practical engineering applications, it is important to first ensure that the WCO is sourced consistently. WCO was provided by an oil refinery in Shandong. In this study, WCO was separated by vacuum distillation, and the separation process is shown in Figure 1. The collected LC, IC, and HC processes are shown in Figure 1. LC, IC, and HC were internally doped at 4 wt%. Table 4 lists the characteristics of the three components that were produced by vacuum distilling WCO. Some of the basic performance parameters are presented in Table 4.

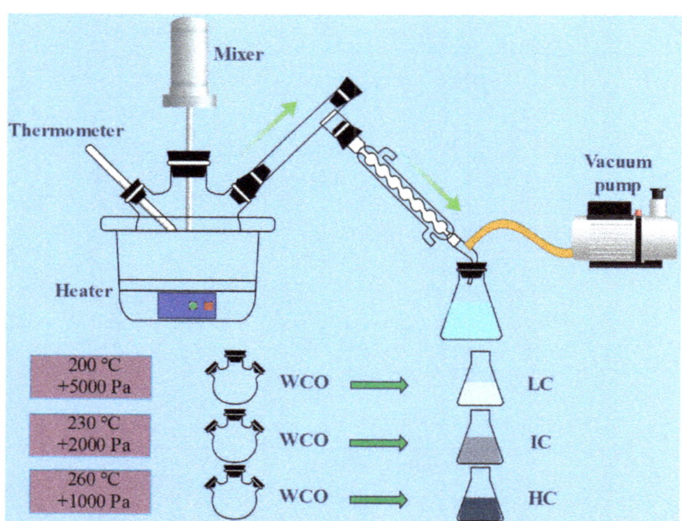

Figure 1. Schematic diagram of laboratory extraction of different compositions of WCO.

Table 4. Basic performance properties of WCO.

Index	Unit	Test Results		
		LC	IC	HC
Viscosity (50 °C)	cP	66	85	219
Flash point	°C	197	219	242
Fire point	°C	216	233	264
Density	g/cm^3	0.89	0.92	0.97
Mechanical impurity	%	0.425	0.002	0
Viscosity (50 °C)	cP	66	85	219

2.2. Preparation of Samples

A high-speed shear mixer was used to prepare the WCO/PPA-modified bitumen, and the procedure is shown in Figure 2. Firstly, the original bitumen was heated in the oven at the temperature of 135 °C. The 4% of LC, IC, and HC were added to the bitumen and mixed for 10 min at 3500 rpm to create WCO-modified bitumen. Then, 0.5%, 1%, 1.5%, and 2% of PPA were added to bitumen. After stirring for 20 min with 4000 r/min continuously, 12 kinds of bitumen modified by WCO/PPA are obtained. The WCO/PPA-modified bitumen combination and abbreviation are shown in Figure 2.

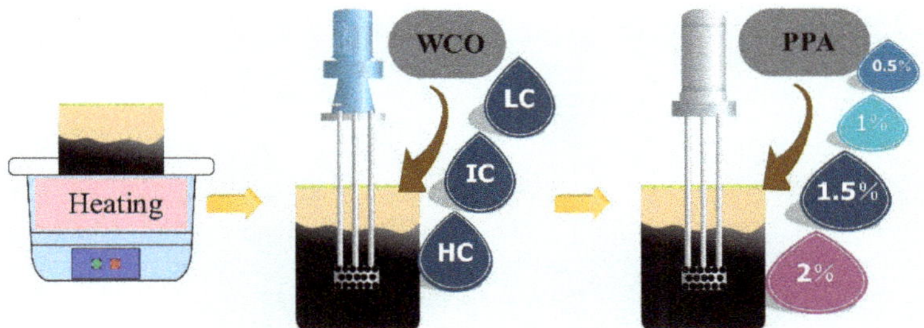

Figure 2. The preparation process of WCO\PPA-modified bitumen.

2.3. Tests Methods

2.3.1. Conventional Tests

The 25 °C penetration, softening point temperatures, and 5 °C ductility of modified bitumen were carried out by following the active Chinese standards. Three replicates were used in these experiments.

2.3.2. Brookfield Viscosity Test

The temperatures used for the Brookfield viscosity tests were 120 °C, 135 °C, 150 °C, and 165 °C. The whole collection of bitumen shown in Figure 2 was examined in three samples. The less viscous the bitumen is at a high temperature, the more accessible construction is generally.

Generally speaking, the viscous activation energy (Ea) can be used as a gauge to assess how temperature affects various types of materials. As the temperature rises, the thermal motion of molecules intensifies and the spacing between them widens. Ea is the amount of power needed to transport a deformation unit from its beginning location to a nearby "hole". As Ea increased, the temperature sensitivity decreased. The following are possible ways to determine Ea using the Arrhenius equation [35]:

$$\mu(T) = B \cdot e^{\frac{E_a}{RT}} \quad (1)$$

$$\ln(\mu(T)) = \ln(B) + \frac{E_a}{RT} \quad (2)$$

where μ is viscosity (Pa·s); B represents the regression coefficient; R represents the universal gas constant; T represents the absolute temperature.

2.4. Rheological Properties Tests

In order to produce the rheological indicator, which can be used to research the high-temperature performance and viscoelasticity of bitumen, the modified bitumen was subjected to the temperature sweep (TS) test at a temperature range of 58 °C to 82 °C with an increment of 6 °C. To determine the viscoelastic property of bitumen at 40 °C with a frequency range of 0.1 rad/s to 100 rad/s, the frequency sweep (FS) test was performed.

The maximum strain of the bitumen is a combination of recoverable and non-recoverable deformation. In order to obtain better rutting resistance, the value of non-recoverable deformation should be lower. It is known from the variational form of the Shenoy equation [36], as in Equation (3), when the Shenoy parameter ($G^*/(1 - (1/Tan\delta Sin\delta))$) should be maximized.

$$\%\gamma_{ur} = \frac{100\sigma}{G^*/\left(1 - \frac{1}{Tan\delta sin\delta}\right)} \tag{3}$$

where γ_{ur} is the unrecovered strain, σ is the applied stress level, G^* is the complex modulus, δ is the phase angle.

2.5. Bending Beam Rheometer (BBR) Test

The low-temperature characteristics of modified bitumen were assessed by using the BBR test and −12 °C and −18 °C were chosen as the trial temperatures. The evaluation indices used were the creep rate (m) and stiffness modulus (S) at 60 s of test operation.

3. Results and Discussion

3.1. Physical Properties

Figure 3 displays the bitumen's physical characteristics. All modified bitumen exhibited greater penetrations than original bitumen, as shown in Figure 3a, increasing by 35.1%, 30.4%, 13%, 2.1%, 78.4%, 70.4%, 57.9%, 35.9%, 128.9%, 119.8%, 102.2%, and 81.1%, respectively. As PPA rose, bitumen penetrations decreased. This suggested that the addition of PPA to bitumen increases its hardness, consistency, and ability to resist shear failure, but when the molecular weight of the WCO component drops, the penetration of modified bitumen increases noticeably. It is possible that the light component of WCO causes bitumen to become diluted and softer, which is the cause of this phenomenon [37]. This conclusion is further supported by the finding that 4LC0.5PPA-modified bitumen had the highest penetration.

Figure 3b illustrates a range of bitumen softening points with various WCO/PPA percentages. Compared to penetration, the degree of change was not remarkable. In comparison to the original bitumen, the modified bitumen's softening points (4LC0.5PPA and 4IC1PPA) were comparable. With decreases of 17.8% and increases of 18.4% in comparison to the original bitumen, the minimum and maximum values of softening points for additives containing 4LC0.5PPA and 4HC2PPA, respectively, were displayed. Generally speaking, with the same constituent concentration of WCO, the PPA content causes a rise in the softening point of modified bitumen. The higher the content of PPA, the greater the influence degree. According to previous studies, PPA can transform bitumen from a sol-gel to a sol-gel structure, this improvement can be attributed to this chemical reaction between bitumen and PPA [38,39].

As a crucial criterion to assess the elastic qualities of bitumen, ductility may be used. The plasticity of bitumen is better the higher the ductility. Figure 3c shows that the addition of WCO/PPA had a greater effect on ductility than on softening point and penetration. As the PPA concentration rose, the ductility decreased when the proportion of WCO components remained constant. This might be the result of the bitumen's molecular structure becoming more complicated as a result of the reaction between PPA and bitumen, which restricts bitumen molecule movement. The bitumen's ductility decreases as a result. This showed that the plastic content of WCO-modified bitumen decreases when PPA is present. Despite this, it is clear that modified bitumen has more ductility than original bitumen, increasing by 152.4%, 119.6%, 85.8%, 43.8%, 282.6%, 254.9%, 223.1%, 186.3%, −2.6%, −7.9%, −24.5%, and −42.7%, respectively. Except for LC-modified bitumen, most modified bitumen demonstrated a higher ductility than unmodified bitumen. In particular, for the same PPA content, the IC-modified bitumen has the highest ductility when compared to other samples. One explanation is that the addition of lightweight components alters how bitumen segments move and the system's free volume, improving the modified bitumen's resistance to deformation at low temperatures [25,40]. Meanwhile, the high content of light components in LC gives it excellent softening ability at high temperatures, but at low temperatures, it cures as easily as wax and may even destroy the properties of bitumen [41].

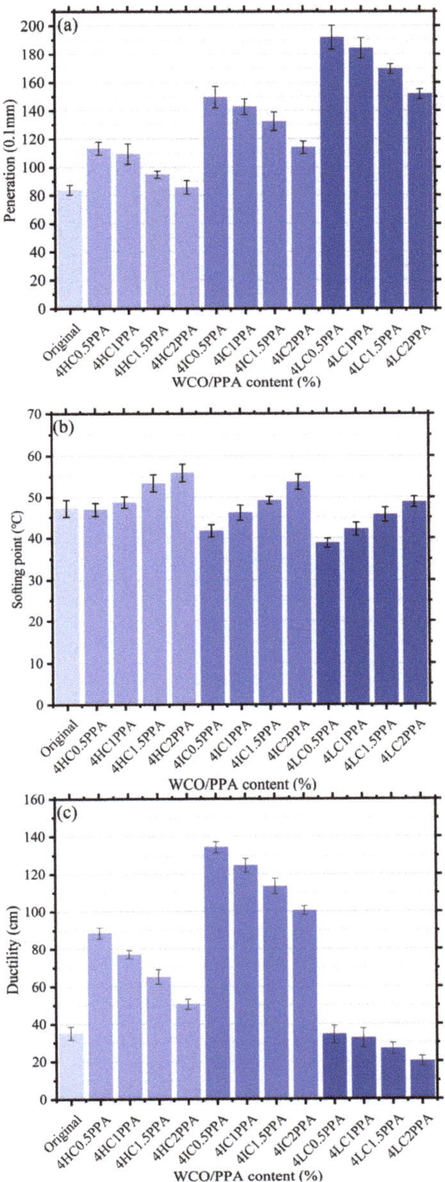

Figure 3. The results of conventional performance: (**a**) Penetration; (**b**) Softening point; (**c**) Ductility.

3.2. *Brookfield Viscosity*

The results of the Brookfield viscosity are represented in Figure 4, which is employed to assess the viscosity–temperature properties. With rising temperatures, it has been discovered that the viscosity of all bitumen kinds decreases. At various temperatures, modified bitumen containing 4HC2PPA had viscosities that were higher than that of the original bitumen by 53.8%, 87.5%, 82.6%, and 68.3%, respectively. The viscosity of 4LC0.5PPA was lower than that of the original bitumen at different temperatures by 24, 26,

6, 58, and 50.1%, respectively. The bitumen's resistance to shear deformation brought on by external pressures is strengthened by the presence of PPA, as shown by the fact that the viscosity of the bitumen increases with the addition of PPA when the WCO component is kept constant. This is because PPA causes the bi-colloidal tumen's structure to transition from a sol-gel type to a gel type.

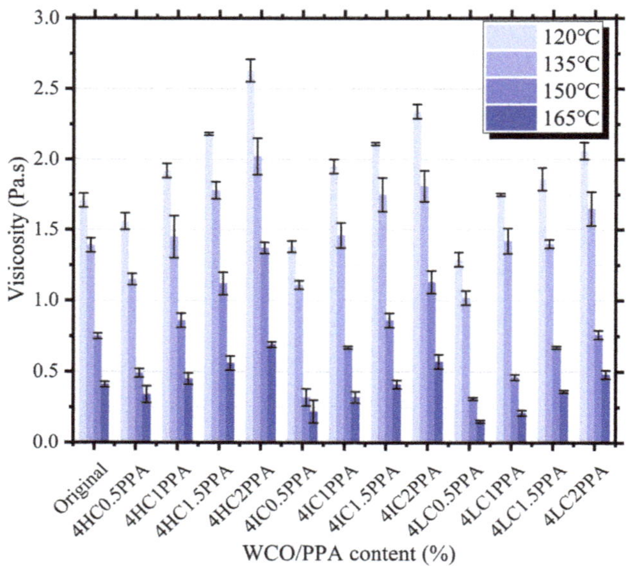

Figure 4. The Brookfield viscosity of the bitumen.

Figure 5 depicts the Ea results of different bitumen. It is clear that the Ea of the original bitumen is affected differently depending on the WCO/PPA ratio. The Ea of 4LC0.5PPA, 4LC1PPA, 4LC1.5PPA, and 4IC0.5PPA were all lower than the original bitumen by 9.7 kJ/mol, 5.6 kJ/mol, 0.8 kJ/mol, and 3.8 kJ/mol, respectively. The Ea of 4HC2PPA, on the other hand, significantly increased and increased to 31.5% in comparison to the original bitumen. As the WCO component reduced for the same proportion of PPA, the Ea of modified bitumen decreased. This could be explained by the bitumen being softened by the light component of WCO, which results in a reduction in the bitumen molecule's internal restriction.

The ASTM Ai and viscosity–temperature susceptibility (VTSi) for original and WCO/PPA-modified bitumen are summarized in Table 5. The bitumen is often more sensitive to temperature the smaller the absolute value of VTSi. The table shows that Ai and VTSi vary depending on the type and admixture of the modifier, demonstrating how significantly the modifier's type and admixture affect temperature sensitivity. The findings demonstrate that, when the WCO doping is held constant, the values of parameters Ai and |VTSi| drop as the quantity of PPA increases. However, when PPA is held constant, the WCO admixture raises the values of Ai and |VTSi|. This shows that PPA enhances bitumen's anti-sensitivity performance whereas WCO enhances bitumen's temperature sensitivity. Additionally, the LC-modified bitumen has a lower temperature sensitivity than the original bitumen, which is consistent with Ea's results. The movement of bitumen segments and the system's free volume are both impacted by the addition of WCO, which could be the cause.

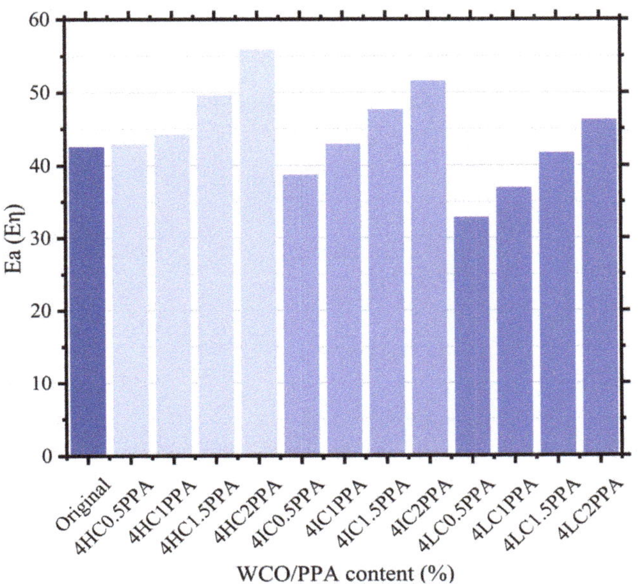

Figure 5. The Ea results of bitumen.

Table 5. ASTM Ai-VTSi values for original and modified bitumen.

	Original	4HC0.5PPA	4HC1PPA	4HC1.5PPA	4HC2PPA	4IC0.5PPA	4IC1PPA
Ai	9.251	9.279	9.263	9.011	7.935	9.418	9.284
∣VTSi∣	3.319	3.367	3.325	3.281	3.243	3.338	3.296
	4IC1.5PPA	4IC2PPA	4LC0.5PPA	4LC1PPA	4LC1.5PPA	4LC2PPA	
Ai	9.025	8.329	10.019	9.846	9.355	8.367	
∣VTSi∣	3.269	3.271	3.741	3.709	3.328	3.443	

3.3. Rheological Properties

3.3.1. Temperature Sweep

As can be seen in Figure 6a, all bitumen's phase angles were discovered to rise with temperature. That elasticity is declining is demonstrated by this. In particular, three different modifier triads (4HC2PPA, 4LC1PPA, and 4LC2PPA) were shown as an illustration. The phase angles of every modified bitumen were greater than those of the original bitumen, with the exception of 4HC2PPA. Phase angles for the modified 4HC2PPA bitumen were, respectively, 0.49%, 0.47%, 0.36%, 0.45%, 0.64%, and 0.42% smaller than those for the original bitumen within the studied temperature range. The phase angles of the 4LC and 2PPA-containing additives, however, were higher than those of the base bitumen, measuring 8.1%, 7.0%, 6.3%, 6.1%, 5.4%, and 5.0%, respectively. The phase angles for 4LC1PPA have increased at different temperatures by 6.5%, 5.8%, 5.5%, 5.1%, 4.7%, and 4.4% in comparison to those of the original bitumen. It was discovered that the elasticity of modified bitumen increased as the levels of WCO components rose at the same PPA content. When the WCO components are held constant, it is clear that the modified bitumen's phase angles decreased as the PPA concentrations increased. This phenomenon can be explained by the fact that the addition of PPA causes a change in the bitumen's sol-gel composition, which prevents WCO from softening the bitumen.

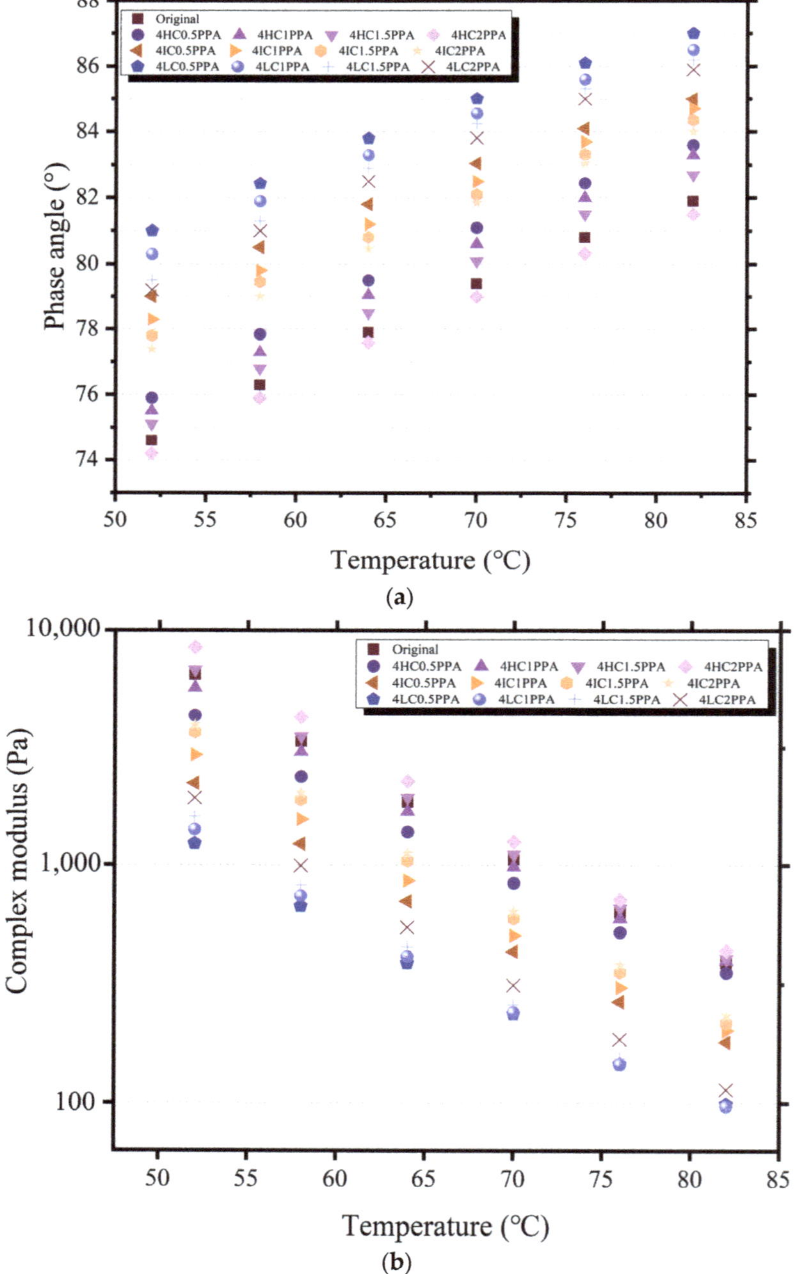

Figure 6. *Cont.*

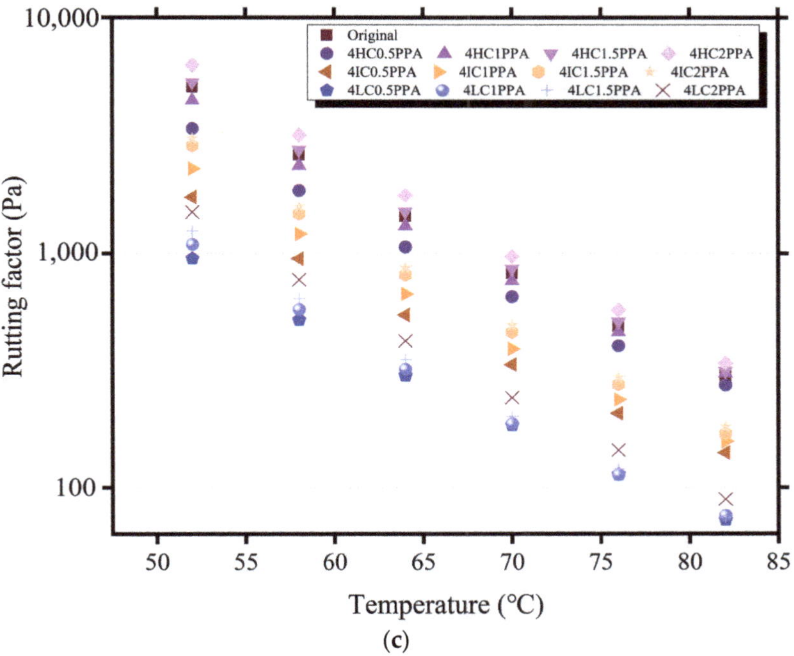

(c)

Figure 6. The temperature sweep results of modified bitumen. (**a**) Phase angle of modified bitumen. (**b**) Complex module of modified bitumen. (**c**) Rutting factor of modified bitumen.

Figure 6b shows the bitumen treated with WCO/PPA complex modulus. It is clear that the modifier's impact on the original bitumen was different. For instance, the 4LC0.5PPA's complex modulus has reductions of 58.4%, 59.7%, 58.1%, 55.9%, 53.8%, and 52.7%, respectively, compared to the original bitumen. The complex modulus of 4HC2PPA has increased by around 20.08%, 18.5%, 16.9%, 14.7%, 14.1%, and 10.7%, respectively, when compared to those of original bitumen. The complex modulus of 4LC2PPA is 37.1%, 34.8%, 32.5%, 32.3%, 31.6%, and 28.6% less than that of the original bitumen over the investigated temperature range. The complex modulus of composite-modified bitumen with a higher PPA content and higher WCO molecular weight is, hence, higher.

The rutting factors ($G^*/\text{Sin}\delta$) at different temperatures are shown in Figure 6c, and they can be used to evaluate the rutting resistance of bitumen. The findings show that a rise in temperature results in a fall in the $G^*/\text{Sin}\delta$. The $G^*/\text{Sin}\delta$ of modified bitumen increases with an increase in PPA when the WCO component is held constant, demonstrating the PPA's beneficial influence on high-temperature deformation resistance. In contrast, when the amount of WCO components decreases, the rutting resistance of composite-modified bitumen decreases. When compared to original bitumen at 70 °C, the $G^*/\text{Sin}\delta$ of WCO/PPA-modified bitumen decreased by 9.8%, 4.6%, −5.9%, −18.2%, 17.3%, 14.8%, 12.7%, 12.2%, 23.1%, and 22.4%, respectively. The fact that 4HC2PPA's $G^*/\text{Sin}\delta$ was 19.1% higher than that of the original bitumens shows that PPA's incorporation has fixed the original bitumen's low performance at high temperatures. At high temperatures, however, the difference between WCO/PPA-modified bitumen and unmodified bitumen is the smallest.

Shenoy parameters for the temperature ranges under test are also shown in Table 6. The Shenoy parameter increases dramatically when PPA content is raised to 2%. For instance, when keeping the WCO content constant, the control bitumen Shenoy parameter is 2.9137 kPa, which increased to 3.1244 kPa and 3.8137 kPa, respectively, at 64 °C with the addition of 1.5% and 2% PPA. Such a response, once more, demonstrates that the addition of PPA up to 2% may be beneficial in enhancing rutting performance.

Table 6. Average Shenoy parameter values.

	Original	4HC0.5PPA	4HC1PPA	4HC1.5PPA	4HC2PPA	4IC0.5PPA	4IC1PPA	4IC1.5PPA	4IC2PPA	4LC0.5PPA	4LC1PPA	4LC1.5PPA	4LC2PPA
52	9.1742	7.425	8.6143	9.3766	10.1087	4.5115	5.7138	6.6786	6.9724	3.1075	3.1611	3.2238	3.3047
58	5.8879	4.3825	5.4347	6.0867	6.7212	2.6934	2.7951	3.3827	3.6849	2.4372	2.4778	2.5251	2.6064
64	2.9134	2.3851	2.5038	3.1244	3.8137	2.0815	2.1751	2.2575	2.2861	1.8258	1.8592	1.8947	1.9763
70	1.7021	1.6027	1.6655	1.7203	1.7727	1.3119	1.3787	1.4503	1.4803	1.0485	1.0587	1.0902	1.1702
76	1.2156	1.1326	1.1913	1.2356	1.2846	0.8426	0.9013	0.9656	0.9956	0.5826	0.5813	0.7051	0.6856
82	0.8666	0.8264	0.8753	0.8866	0.9197	0.5362	0.5853	0.6166	0.6466	0.4738	0.4757	0.4743	0.4823

At various temperatures, the Shenoy parameter clearly has a greater value. Additionally, when the PPA content rose, the discrepancy between the Shenoy parameter and the matching G*/Sinδ increased. For example, when the main component of WCO is IC, the difference in Shenoy parameter and G*/Sinδ is 1.3742 kPa for control bitumen, and it rises to 1.6125 kPa, 1.9143 kPa, 2.29111 kPa, and 3.0182 kPa with the addition of 0.5%, 1%, 1.5%, and 2% PPA, respectively, at 58 °C. A possible explanation for this response is the bitumen's better elastic response brought on by the addition of PPA, which helped to lower the time-dependent unrecovered strain value. This means that in defining the non-recoverable strain value, the Shenoy parameter for evaluating rutting performance is relatively more sensitive than the Superpave rutting parameter to change in WCO/PPA composite-modified bitumen.

Based on the complex modulus at various temperatures, the complex modulus index (GTS) parameter is used to evaluate temperature sensitivity which can be calculated using Equation (4). In general, the greater the absolute value of GTS, the more temperature sensitivity bitumen is. As present, in Figure 7, it is obvious to see that all modified bitumen have lower GTS values than the original bitumen. This suggests that WCO/PPA positively affects the thermal stability of the original bitumen. At the same time, under the same WCO components, the GTS of modified bitumen decreases with the increase in PPA, demonstrating an improvement in bitumen's temperature stability.

$$GTS = \frac{lgG_* - C}{lgT} \tag{4}$$

where: T represents the test temperature; C represents constant.

Typically, the G*/Sinδ of 1 kPa is chosen as the temperature damage threshold for the original bitumen. As can be observed in Figure 6c, all bitumens failed at temperatures lower than 1 kPa at 70 °C, with the exception of 4HC1.5PPA and 4HC2PPA, which had failure temperatures greater than the original bitumen. The failure temperature grows as PPA rises while the WCO component stays constant. This displays how bitumen's resistance to deformation at high temperatures is improved by the addition of PPA. It is worth noting that the high-temperature rutting resistance of LC-modified bitumen is poor at the same PPA concentration, which is due to the excessive softening phenomenon of bitumen caused by the high number of lighter components in the LC. In contrast, under the conditions of this study, HC can effectively soften the bitumen without compromising its resistance to deformation, Therefore, in engineering applications, the use of this type of component in WCO should be considered first.

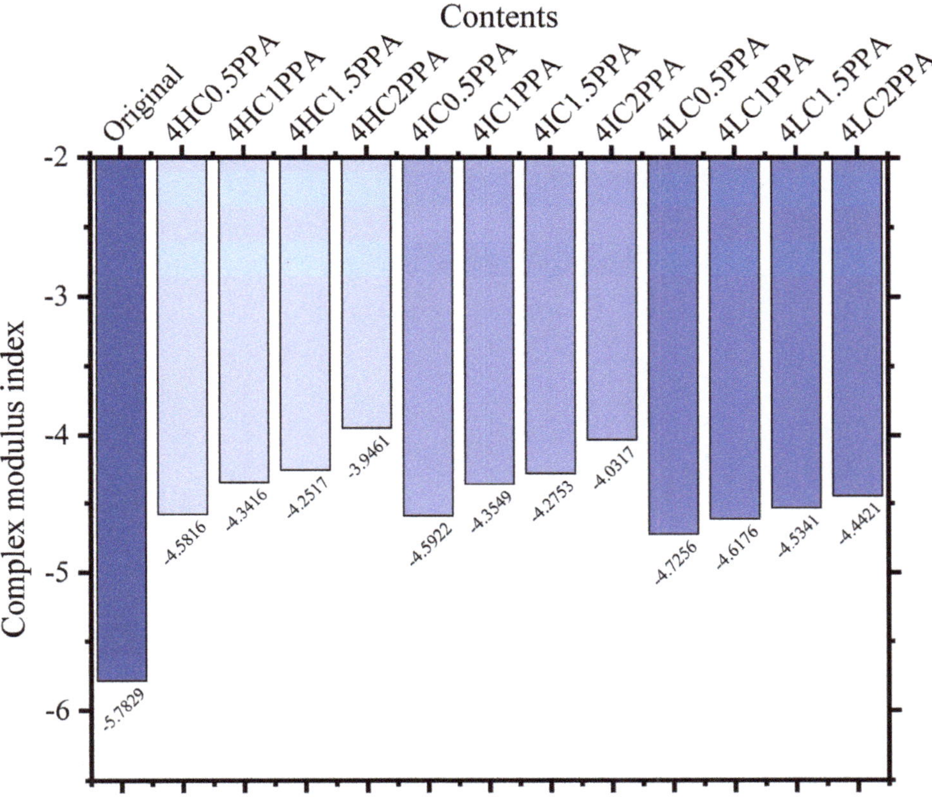

Figure 7. The complex modulus index of different types of bitumen.

3.3.2. Frequency Sweep

Regardless of the ratio of WCO/PPA, the complex modulus rose and the phase angles decreased as frequency increases, as shown in Figure 8. Meanwhile, the higher proportion of PPA can contribute to increase complex modulus and decrease phase angle, and this tendency becomes more obvious with the decrease in the WCO component. It has been established that WCO had a detrimental impact on bitumen's high-temperature performance [17]. Attribution to the WCO's softening action, the fluid characteristics of bitumen are improved. However, components with different molecular weights isolated from WCO have different modification effects on bitumen. Specifically, the molecular weight of the substance is inversely proportional to the softening effect on bitumen, due to the fact that the molecular weight determines the mobility of the substance. Therefore, under the combination of the maximum molecular weight of WCO and the maximum content of PPA, the modified bitumen has the best deformation resistance, which is caused by the synergistic effect of the two modifiers. However, the average molecular weight and dispersion coefficient are raised by the addition of PPA, which have a negative impact on the fluidity of bitumen.

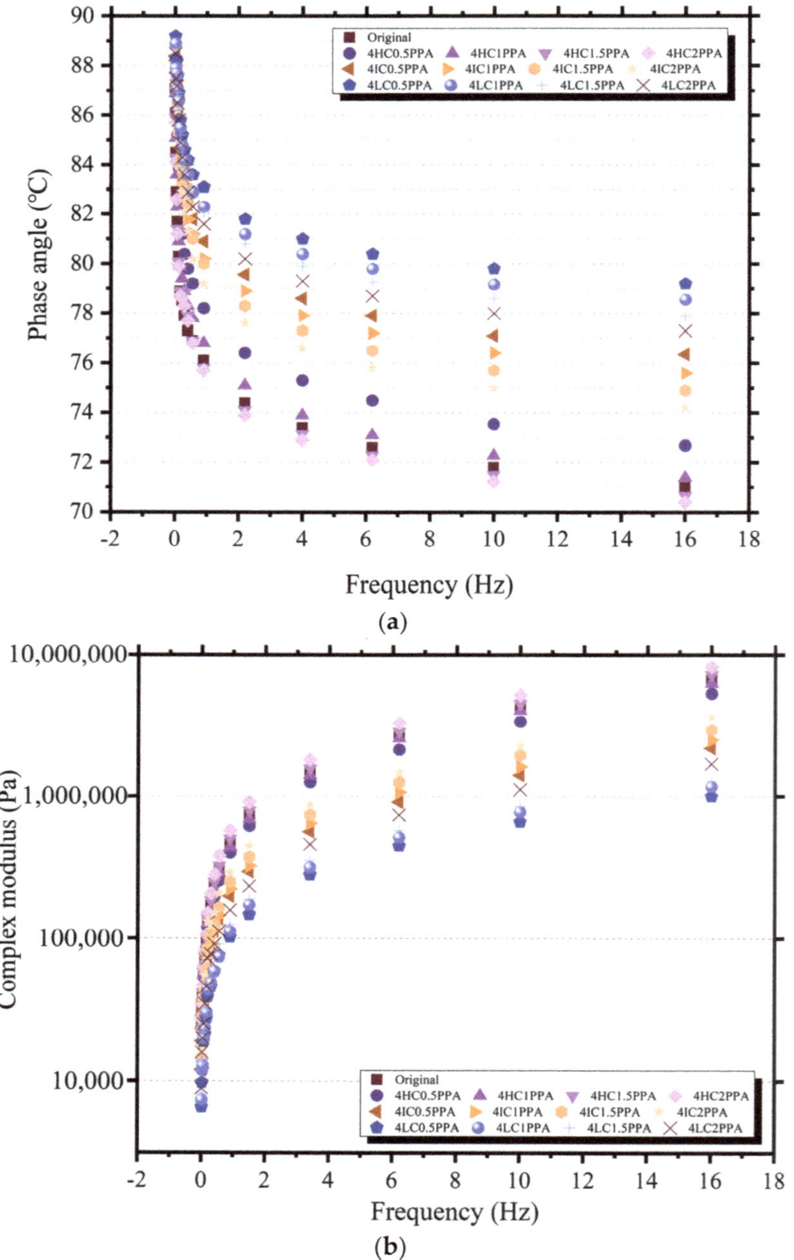

Figure 8. *Cont.*

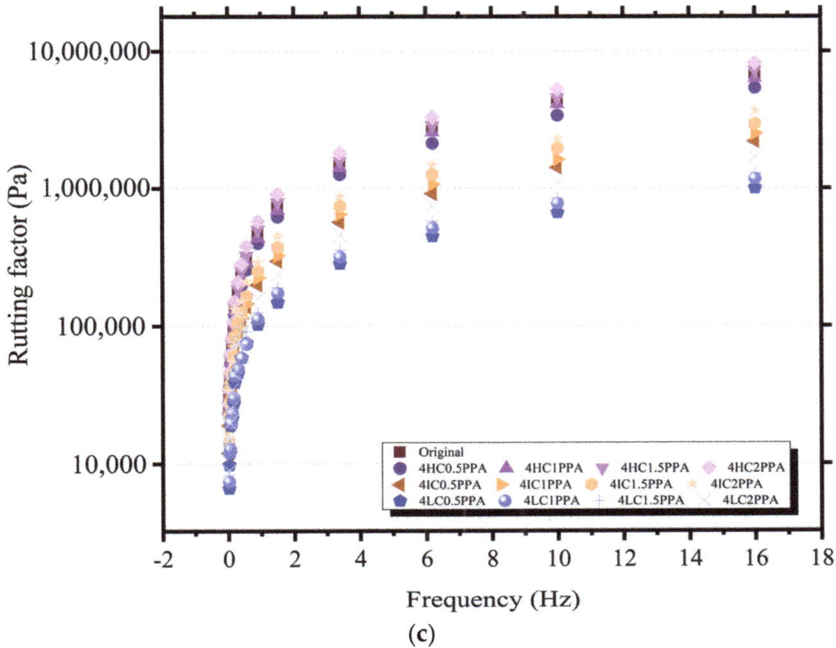

Figure 8. The frequency sweep results of modified bitumen. (**a**) Phase angle of modified bitumen. (**b**) Complex module of modified bitumen. (**c**) Rutting factor of modified bitumen.

3.4. BBR

The results of all samples at test temperatures of −12 °C and −18 °C are displayed in Figure 9. It can be shown that when the temperature drops, the S-value rises and the m-value falls. Moreover, it was found that the S of the modified bitumen increased and the m-value decreased as the PPA content increased, indicating that PPA increases the risk of bitumen cracking at low temperatures. The various WCO components, fortunately, improved the cracking resistance of PPA-modified bitumen. Impressively, the incorporation of IC and HC components isolated from WCO improved the cracking resistance of PPA-modified bitumen, except for the low molecular weight IC which weakened the low-temperature cracking performance of the modified bitumen. The crack resistance of IC-modified PPA bitumen is better than the degree of LC and HC modification. This phenomenon can be explained by the difference in molecular weight and composition of substances: (1) The capacity to increase bitumen's low-temperature qualities is limited by the fact that the heave component is a mixture of several chemicals with the longest carbon chains in WCO; (2) LC, although it has a good softening ability at high temperatures, tends to solidify like wax at low temperatures and can even destroy the high-content properties of bitumen [42]. Therefore, considering its wax-like properties, LC is not recommended to be used as a regenerating agent for RAP.

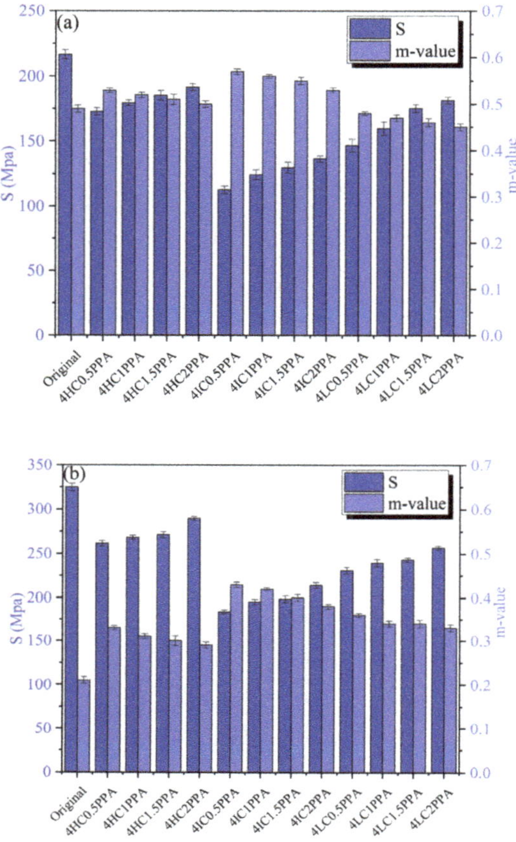

Figure 9. The S and m-value results of modified bitumen: (**a**) −12°C (**b**) −18°C.

4. Conclusions

This paper aims to investigate the synergistic effect of WCO and PPA on the properties of bitumen. The appropriate WCO/PPA proportion in bitumen was obtained according to the conventional and rheological properties. The following conclusions can be drawn:

The presence of PPA increases the WCO-modified bitumen's softening point while decreasing its penetration and ductility. PPA can mitigate the impact of WCO's detrimental effects on high-temperature performance. The penetration and softening point with 4LC0.5PPA and 4HC2PPA present extreme value, rising by 128.9%, −17.8% and 2.1%, 18.4%, the ductility with 4IC0.5PPA and 4LC2PPA present extreme value, rising by 282.6% and −42.7%.

When the WCO content remains constant, the temperature stability of bitumen evaluated by ASTM Ai and VTSi and Ea are enhanced with the raise of PPA proportion. The existence of PPA is beneficial to resist high-temperature deformation.

The results of DSR experiments indicate that PPA can enhance the high-temperature rheology of WCO-modified bitumen. The rutting factors of the 4HC2PPA were 19.1% better than those of the original bitumen, which shows the same result of $G^*/(1 - (1/\mathrm{Tan}\delta\mathrm{Sin}\delta))$.

The ability of modified bitumen to perform at low temperatures has no significant linear correlation with the molecular weight of the three WCO components. Since IC is too sensitive to temperature and prone to phase change, it will adversely affect the low-temperature performance of bitumen.

Author Contributions: Conceptualization, W.W.; methodology, D.W.; software, W.W.; validation, J.L.; formal analysis, D.W.; investigation, J.L.; resources, W.W.; data curation, W.W.; writing—original draft preparation, X.L.; writing—review and editing, P.L. All authors have read and agreed to the published version of the manuscript.

Funding: This research was funded by Fundamental Research Funds for the Central Universities-Excellent doctoral dissertation cultivation project of Chang'an University (300203211212). This work also described in this paper was supported by a grant from the German Academic Exchange Service of Germany (Grant no. 57601840).

Institutional Review Board Statement: Not applicable.

Informed Consent Statement: Not applicable.

Data Availability Statement: Not applicable.

Conflicts of Interest: The authors declare no conflict of interest.

References

1. Akhtar, A.; Sarmah, A.K. Construction and demolition waste generation and properties of recycled aggregate concrete: A global perspective. *J. Clean. Prod.* **2018**, *186*, 262–281. [CrossRef]
2. Ahmed, R.B.; Hossain, K. Waste cooking oil as an asphalt rejuvenator: A state-of-the-art review. *Constr. Build. Mater.* **2020**, *230*, 116985. [CrossRef]
3. Tunay, D.; Yildirim, O.; Ozkaya, B.; Demir, A. Effect of organic fraction of municipal solid waste addition to high rate activated sludge system for hydrogen production from carbon rich waste sludge. *Int. J. Hydrog. Energy* **2022**, *47*, 26284–26293. [CrossRef]
4. Chen, W.; Li, Y.; Chen, S.; Zheng, C. Properties and economics evaluation of utilization of oil shale waste as an alternative environmentally-friendly building materials in pavement engineering. *Constr. Build. Mater.* **2020**, *259*, 119698. [CrossRef]
5. Maturi, K.C.; Gupta, A.; Haq, I.; Kalamdhad, A.S. *Chapter 7—A glance over current status of waste management and landfills across the globe: A review, In Biodegradation and Detoxification of Micropollutants in Industrial Wastewater*; Haq, I., Kalamdhad, A.S., Shah, M.P., Eds.; Elsevier: Amsterdam, The Netherlands, 2022; pp. 131–144.
6. Banerji, A.K.; Chakraborty, D.; Mudi, A.; Chauhan, P. Characterization of waste cooking oil and waste engine oil on physical properties of aged bitumen. *Mater. Today* **2022**, *59*, 1694–1699. [CrossRef]
7. Bui, T.-D.; Tseng, J.-W.; Tseng, M.-L.; Lim, M.K. Opportunities and challenges for solid waste reuse and recycling in emerging economies: A hybrid analysis. *Resour. Conserv. Recycl.* **2022**, *177*, 105968. [CrossRef]
8. Wang, D.; Riccardi, C.; Jafari, B.; Falchetto, A.C.; Wistuba, M.P. Investigation on the effect of high amount of Re-recycled RAP with Warm mix asphalt (WMA) technology. *Constr. Build. Mater.* **2021**, *312*, 125395. [CrossRef]
9. Wang, D.; Falchetto, A.C.; Hugener, M.; Porot, L.; Kawakami, A.; Hofko, B.; Grilli, A.; Pasquini, E.; Pasetto, M.; Tabatabaee, H.; et al. *Effect of Aging on the Rheological Properties of Blends of Virgin and Rejuvenated RA Binders, RILEM International Symposium on Bituminous Materials*; Springer: Berlin/Heidelberg, Germany, 2020; pp. 3–10.
10. Rahman, M.T.; Mohajerani, A.; Giustozzi, F. Recycling of Waste Materials for Asphalt Concrete and Bitumen: A Review. *Materials* **2020**, *13*, 1495. [CrossRef]
11. Bilema, M.; Aman, M.; Hassan, N.; Al-Saffar, Z.; Mashaan, N.; Memon, Z.; Milad, A.; Yusoff, N. Effects of Waste Frying Oil and Crumb Rubber on the Characteristics of a Reclaimed Asphalt Pavement Binder. *Materials* **2021**, *14*, 3482. [CrossRef]
12. Muhbat, S.; Tufail, M.; Hashmi, S. Production of diesel-like fuel by co-pyrolysis of waste lubricating oil and waste cooking oil. *Biomass Convers. Biorefinery* **2021**, *9*, 1–11. [CrossRef]
13. Singhabhandhu, A.; Tezuka, T. The waste-to-energy framework for integrated multi-waste utilization: Waste cooking oil, waste lubricating oil, and waste plastics. *Energy* **2010**, *35*, 2544–2551. [CrossRef]
14. Landi, F.F.D.; Fabiani, C.; Castellani, B.; Cotana, F.; Pisello, A.L. Environmental assessment of four waste cooking oil valorization pathways. *Waste Manag.* **2022**, *138*, 219–233. [CrossRef] [PubMed]
15. Chen, M.; Xiao, F.; Putman, B.; Leng, B.; Wu, S. High temperature properties of rejuvenating recovered binder with rejuvenator, waste cooking and cotton seed oils. *Constr. Build. Mater.* **2014**, *59*, 10–16. [CrossRef]
16. Li, H.; Dong, B.; Wang, W.; Zhao, G.; Guo, P.; Ma, Q. Effect of Waste Engine Oil and Waste Cooking Oil on Performance Improvement of Aged Asphalt. *Appl. Sci.* **2019**, *9*, 1767. [CrossRef]
17. Taherkhani, H.; Noorian, F. Comparing the effects of waste engine and cooking oil on the properties of asphalt concrete containing reclaimed asphalt pavement (RAP). *Road Mater. Pavement Des.* **2020**, *21*, 1238–1257. [CrossRef]
18. Yan, K.; Li, Y.; Long, Z.; You, L.; Wang, M.; Zhang, M.; Diab, A. Mechanical behaviors of asphalt mixtures modified with European rock bitumen and waste cooking oil. *Constr. Build. Mater.* **2022**, *319*, 125909. [CrossRef]
19. Chen, M.; Leng, B.; Wu, S.; Sang, Y. Physical, chemical and rheological properties of waste edible vegetable oil rejuvenated asphalt binders. *Constr. Build. Mater.* **2014**, *66*, 286–298. [CrossRef]
20. Azahar, W.N.A.W.; Jaya, R.P.; Hainin, M.R.; Bujang, M.; Ngadi, N. Chemical modification of waste cooking oil to improve the physical and rheological properties of asphalt binder. *Constr. Build. Mater.* **2016**, *126*, 218–226. [CrossRef]

21. Oldham, D.; Rajib, A.; Dandamudi, K.P.R.; Liu, Y.; Deng, S.; Fini, E.H. Transesterification of Waste Cooking Oil to Produce A Sustainable Rejuvenator for Aged Asphalt. *Resour. Conserv. Recycl.* **2021**, *168*, 105297. [CrossRef]
22. Zhao, Y.; Chen, M.; Zhang, X.; Wu, S.; Zhou, X.; Jiang, Q. Effect of chemical component characteristics of waste cooking oil on physicochemical properties of aging asphalt. *Constr. Build. Mater.* **2022**, *344*, 128236. [CrossRef]
23. Yuechao, Z.; Meizhu, C.; Shaopeng, W.; Qi, J. Rheological properties and microscopic characteristics of rejuvenated asphalt using different components from waste cooking oil. *J. Clean. Prod.* **2022**, *370*, 133556. [CrossRef]
24. Fang, Y.; Zhang, Z.; Wang, S.; Li, N. High temperature rheological properties of high modulus asphalt cement (HMAC) and its definition criteria. *Constr. Build. Mater.* **2020**, *238*, 117657. [CrossRef]
25. Wang, W.; Jia, M.; Jiang, W.; Lou, B.; Jiao, W.; Yuan, D.; Li, X.; Liu, Z. High temperature property and modification mechanism of asphalt containing waste engine oil bottom. *Constr. Build. Mater.* **2020**, *261*, 119977. [CrossRef]
26. Hu, C.; Feng, J.; Zhou, N.; Zhu, J.; Zhang, S. Hydrochar from corn stalk used as bio-asphalt modifier: High-temperature performance improvement. *Environ. Res.* **2021**, *193*, 110157. [CrossRef] [PubMed]
27. Alam, S.; Hossain, Z. Changes in fractional compositions of PPA and SBS modified asphalt binders. *Constr. Build. Mater.* **2017**, *152*, 386–393. [CrossRef]
28. Baldino, N.; Gabriele, D.; Rossi, C.O.; Seta, L.; Lupi, F.R.; Caputo, P. Low temperature rheology of polyphosphoric acid (PPA) added bitumen. *Constr. Build. Mater.* **2012**, *36*, 592–596. [CrossRef]
29. Jiang, X.; Li, P.; Ding, Z.; Yang, L.; Zhao, J. Investigations on viscosity and flow behavior of polyphosphoric acid (PPA) modified asphalt at high temperatures. *Constr. Build. Mater.* **2019**, *228*, 116610. [CrossRef]
30. Huang, Y.; Bird, R.N.; Heidrich, O. A review of the use of recycled solid waste materials in asphalt pavements. *Resour. Conserv. Recycl.* **2007**, *52*, 58–73. [CrossRef]
31. Liu, H.; Chen, Z.; Wang, Y.; Zhang, Z.; Hao, P. Effect of poly phosphoric acid (PPA) on creep response of base and polymer modified asphalt binders/mixtures at intermediate-low temperature. *Constr. Build. Mater.* **2018**, *159*, 329–337. [CrossRef]
32. Han, Y.; Tian, J.; Ding, J.; Shu, L.; Ni, F. Evaluating the storage stability of SBR-modified asphalt binder containing polyphosphoric acid (PPA). *Case Stud. Constr. Mater.* **2022**, *17*, e01214. [CrossRef]
33. Yang, X.; Liu, G.; Rong, H.; Meng, Y.; Peng, C.; Pan, M.; Ning, Z.; Wang, G. Investigation on mechanism and rheological properties of Bio-asphalt/PPA/SBS modified asphalt. *Constr. Build. Mater.* **2022**, *347*, 128599. [CrossRef]
34. Qian, C.; Fan, W.; Ren, F.; Lv, X.; Xing, B. Influence of polyphosphoric acid (PPA) on properties of crumb rubber (CR) modified asphalt. *Constr. Build. Mater.* **2019**, *227*, 117094. [CrossRef]
35. Liu, Z.; Wang, Y.; Meng, Y.; Han, Z.; Jin, T. Comprehensive performance evaluation of steel fiber-reinforced asphalt mixture for induction heating. *Int. J. Pavement Eng.* **2022**, *23*, 3838–3849. [CrossRef]
36. Ashish, P.K.; Singh, D. Effect of Carbon Nano Tube on performance of asphalt binder under creep-recovery and sustained loading conditions. *Constr. Build. Mater.* **2019**, *215*, 523–543. [CrossRef]
37. Cao, Z.; Chen, M.; Liu, Z.; He, B.; Yu, J.; Xue, L. Effect of different rejuvenators on the rheological properties of aged SBS modified bitumen in long term aging. *Constr. Build. Mater.* **2019**, *215*, 709–717. [CrossRef]
38. Hossain, Z.; Alam, S.; Baumgardner, G. Evaluation of rheological performance and moisture susceptibility of polyphosphoric acid modified asphalt binders. *Road Mater. Pavement Des.* **2020**, *21*, 237–252. [CrossRef]
39. Bonemazzi, F.; Giavarini, C. Shifting the bitumen structure from sol to gel. *J. Pet. Sci. Eng.* **1999**, *22*, 17–24. [CrossRef]
40. Dong, Z.; Yang, C.; Luan, H.; Zhou, T.; Wang, P. Chemical characteristics of bio-asphalt and its rheological properties after CR/SBS composite modification. *Constr. Build. Mater.* **2019**, *200*, 46–54. [CrossRef]
41. Kovinich, J.; Kuhn, A.; Wong, A.; Ding, H.; Hesp, S.A.M. Wax in Asphalt: A comprehensive literature review. *Constr. Build. Mater.* **2022**, *342*, 128011. [CrossRef]
42. Huang, T.; He, H.; Zhang, P.; Lv, S.; Jiang, H.; Liu, H.; Peng, X. Laboratory investigation on performance and mechanism of polyphosphoric acid modified bio-asphalt. *J. Clean. Prod.* **2022**, *333*, 130104. [CrossRef]

MDPI
St. Alban-Anlage 66
4052 Basel
Switzerland
Tel. +41 61 683 77 34
Fax +41 61 302 89 18
www.mdpi.com

Materials Editorial Office
E-mail: materials@mdpi.com
www.mdpi.com/journal/materials

www.ingramcontent.com/pod-product-compliance
Lightning Source LLC
LaVergne TN
LVHW071028170726
843515LV00006B/1230

* 9 7 8 3 0 3 6 5 6 5 2 0 0 *